SCHAUM'S OUTLINE OF

THEORY AND PROBLEMS

OF

ELEMENTS OF STATISTICS

Descriptive Statistics and Probability

•

STEPHEN BERNSTEIN, Ph.D.
Research Associate
University of Colorado

RUTH BERNSTEIN, Ph.D
Associate Professor
University of Colorado

SCHAUM'S OUTLINE SERIES

McGraw-Hill

New York   Chicago   San Francisco   Lisbon
London   Madrid   Mexico City   Milan   New Delhi
San Juan   Seoul   Singapore   Sydney   Toronto

Stephen Bernstein, Ph.D., *Research Associate, Department of Environmental, Population, and Organismic Biology, University of Colorado at Boulder*

Dr. Stephen Bernstein has taught biostatistics, quantitative reasoning, and general biology in his current position. Previously as an Assistant Professor in the Department of Psychiatry at UCLA he taught animal behavior, and at the University of Wisconsin he taught statistics for psychologists. He received his B.A. from Princeton University and his Ph.D. in psychology from the University of Wisconsin. A recipient of various NIMH fellowships and awards, he attended the University of Zürich and the University of Paris for postdoctoral studies. His published research is in animal behavior, neurophysiology, and brain-body allometry. He is co-author with Ruth Bernstein of three general biology textbooks.

Ruth Bernstein, Ph.D., *Associate Professor, Department of Environmental, Population, and Organismic Biology, University of Colorado at Boulder*

Dr. Ruth Bernstein currently teaches ecology and population dynamics, and has taught general biology. Previously, she taught general biology at the University of California, Los Angeles. She received her B.S. from the University of Wisconsin and her Ph.D. in biology from UCLA. Her published research is in evolutionary ecology, with emphasis on ants and beetles. She is the co-author with Stephen Bernstein of three general biology textbooks.

Schaum's Outline of Theory and Problems of
Elements of Statistics I: Descriptive Statistics and Probability

2 3 4 5 6 7 8 9 10 11 12 13 14 15 16 17 18 19 20   PRS PRS   09 8 7 6 5 4 3

ISBN 0-07-005023-6

Sponsoring Editor: Barbara Gilson
Production Supervisor: Tina Cameron
Editing Supervisor: Maureen B. Walker

Library of Congress Cataloging-in-Publication Data
Bernstein, Stephen.
      Schaum's outline of theory and problems of elements of statistics
  I : Statistics and probability / Stephen Bernstein, Ruth Bernstein.
            p.   cm.—(Schaum's outline series)
      Includes index.
      ISBN 0-07-005023-6
      1. Statistics—Problems, exercises, etc.   2. Statistics—Outlines,
  syllabi, etc.   I. Bernstein, Ruth.   II. Title.   III. Title: Outline
  of theory and problems of elements of statistics I.   IV. Title:
  Theory and problems of elements of statistics I.   V. Title: Elements
  of statistics I.
  QA276.2.B42   1998
  519.5′076—dc21   98-49900
                                                              CIP
                                                               AC

McGraw-Hill
A Division of The McGraw·Hill Companies

Preface

Statistics is the science that deals with the collection, analysis, and interpretation of numerical information. Having a basic understanding of this science is of importance not only to every research scientist, but also to anyone in modern society who must deal with such information: the doctor evaluating conflicting medical research reports, the lawyer trying to convince a jury of the validity of quantitative evidence, the manufacturer working to improve quality-control procedures, the economist interpreting market trends, and so on.

The theoretical base of the science of statistics is a field within mathematics called *mathematical statistics*. Here, statistics is presented as an abstract, tightly integrated structure of axioms, theorems, and rigorous proofs. To make this theoretical structure available to the nonmathematician, an interpretative discipline has been developed called *general statistics* in which the presentation is greatly simplified and often nonmathematical. From this simplified version, each specialized field (e.g., agriculture, anthropology, biology, economics, engineering, psychology, sociology) takes material that is appropriate for its own numerical data. Thus, for example, there is a version of general statistics called *biostatistics* that is specifically tailored to the numerical data of biology.

All introductory courses in general statistics or one of its specialized offshoots share the same core of material: the *elements of statistics*. The authors of this book have learned these elements in courses, used them in research projects, and taught them, for many years, in general statistics and biostatistics courses. This book, developed from our experience, is a self-help guide to these elements that can be read on its own, used as a supplement to a course textbook, or, as it is sufficiently complete, actually used as the course textbook. All the mathematics required for understanding this book (aspects of high-school algebra) are reviewed in Chapter 1.

The science of statistics can be divided into two areas: *descriptive statistics* and *inferential statistics*. In descriptive statistics, techniques are provided for processing raw numerical data into usable forms. These techniques include methods for collecting, organizing, summarizing, describing, and presenting numerical information. If entire groups (*populations*) were always available for study, then descriptive statistics would be all that is required. However, typically only a small segment of the group (a *sample*) is available, and thus techniques are required for making generalizations and decisions about the entire population from limited and uncertain sample information. This is the domain of inferential statistics.

All courses in introductory general statistics present both areas of statistics in a standard sequence. This book follows this sequence, but separates these areas into two volumes. Volume I (Chapters 1–10) deals with descriptive statistics and also with the main theoretical base of inferential statistics: *probability theory*. Volume II (Chapters 11–20) deals with the concepts and techniques of inferential statistics. Each chapter of the book has the same format: first a section of *text* in outline form with fully solved problem-examples for every new concept and procedure; next a section of *solved problems* that both reviews the same material and also makes you look at the material from a different perspective; and finally a section of *supplementary problems* that tests your mastery of the material by providing answers without the step-by-step solutions. Because this is a book on general statistics, an attempt has been made throughout to have a diverse selection of problems representing many specialized fields. Also, we have tried in these problems to show how decisions are made from numerical information in actual problem-solving situations.

To master statistics you must both read the text and do the problems. We suggest that you first read the text and follow the examples, and then go back to re-read the text before going on to the solved and supplementary problems. Also, the book is cross-referenced throughout, so that you can quickly review earlier material that is required to understand later material.

If you go on to work with statistics, you will likely use a computer and one of the many available packages of statistical programs. This book does not deal with how to use such computer programs, but instead gives you the mastery required to understand which aspects of the programs to use and, as importantly, to interpret the output-results that the computer provides. A computer is not required for doing the problems in this book; all problems are solvable with an electronic calculator.

We would like to thank the following people at the McGraw-Hill Corporation who have contributed significantly to the development of this book: Barbara Gilson, Elizabeth Zayatz, John Aliano, Arthur Biderman, Mary Loebig Giles, and Meaghan McGovern. We would also like to thank the anonymous reviewers of the chapters and the individuals and organizations that gave us permission to use their published materials (specific credit is given where the material is presented).

Contents

Chapter 1

Mathematics Required for Statistics

1.1 WHAT IS STATISTICS?

Statistics is the science that deals with the collection, analysis, and interpretation of numerical information. This science can be divided into two areas: *descriptive statistics* and *inferential statistics*. In descriptive statistics, techniques are provided for processing raw numerical data into usable forms. These techniques include methods for collecting, organizing, summarizing, describing, and presenting numerical information. If entire groups (*populations*) were always available for study, then descriptive statistics would be all that is required. However, typically only a small segment of the group (a *sample*) is available, and thus techniques are required for making generalizations and decisions about the entire population from limited and uncertain sample information. This is the domain of inferential statistics.

The theoretical base of the science of statistics is a field within mathematics called *mathematical statistics*. Here, statistics is presented as an abstract, tightly integrated structure of axioms, theorems, and rigorous proofs, involving many other areas of mathematics such as calculus, probability theory, and higher algebra. To make this theoretical structure available to the nonmathematician, an interpretative discipline has been developed called *general statistics* in which the presentation is greatly simplified and often nonmathematical. From this simplified version, each specialized field (e.g., agriculture, anthropology, biology, economics, and so on) takes material that is appropriate for its own numerical data. Thus, for example, there is a limited and appropriate version of general statistics called *biostatistics* (or *biometry*) that is specifically tailored for the numerical data of biology.

This book, which consists of two volumes, is an introduction to general statistics with an attempt made to use examples and problems taken from a wide variety of specialized fields. It follows the typical outline of an introductory course in general statistics: first descriptive statistics—the collecting (Chapters 2 and 3), organizing (Chapter 4), graphing (Chapter 5), and describing (Chapters 6 and 7) of numerical data; then probability theory (Chapters 8 through 12) and sampling theory (Chapter 13); and finally, the estimation and hypothesis-testing techniques of inferential statistics (Chapters 14 through 20).

In this introduction to general statistics, all the mathematics that will be required is at the level of high-school algebra, and this first chapter will review all the elements of algebra you will need. We will also assume that you have an electronic calculator for doing the calculations.

1.2 OPERATIONS WITH FRACTIONS

If we let m and n represent two numbers, then the multiplication of m times n is indicated by these *equivalent symbols*: $m \times n$, $(m)(n)$, and $m \cdot n$. Similarly, the division of m by n is indicated by these equivalent symbols: $m \div n$, m/n, and $\dfrac{m}{n}$.

EXAMPLE 1.1 Given that $m = 4$ and $n = 2$, perform the following *operations*: $m \times n$, $m \div n$, m/n, $(m)(n)$, $m \cdot n$, $\dfrac{m}{n}$.

Solution

$$4 \times 2 = (4)(2) = 4 \cdot 2 = 8$$
$$4 \div 2 = 4/2 = \frac{4}{2} = 2$$

1

The value of a fraction remains the same (*equivalent*) if its numerator and denominator are multiplied or divided by the same number. However, adding or subtracting the same number from the numerator and denominator will typically change the value of the fraction.

EXAMPLE 1.2 Which of the following are *equivalent fractions* to $\frac{6}{8}$: $\frac{8}{9}, \frac{12}{16}, \frac{2}{4}, \frac{3}{4}, \frac{16}{18}$, or $\frac{1}{2}$?

Solution

$$\frac{6}{8} = \frac{(2)(6)}{(2)(8)} = \frac{12}{16} = \frac{6/2}{8/2} = \frac{3}{4}$$

while

$$\frac{6}{8} \neq \frac{6-4}{8-4} = \frac{2}{4} = \frac{1}{2} \quad \text{and} \quad \frac{6}{8} \neq \frac{6+10}{8+10} = \frac{16}{18} = \frac{8}{9}$$

where $\neq$ is the symbol for "not equal to."

To add or subtract fractions they must first be transformed to have a common denominator. Fractions are multiplied by separately multiplying their numerators and denominators. Fractions are divided by first inverting the divisor and then multiplying.

EXAMPLE 1.3 Perform the indicated operations: (a) $\frac{4}{5} + \frac{5}{6}$,   (b) $\frac{5}{7} \times \frac{1}{4} \times \frac{2}{3}$,   (c) $\frac{f}{g} \div \frac{m}{n}$.

Solution

(a) $\frac{4}{5} + \frac{5}{6} = \frac{4 \times 6}{5 \times 6} + \frac{5 \times 5}{6 \times 5} = \frac{24}{30} + \frac{25}{30} = \frac{49}{30} = 1\frac{19}{30}$

(b) $\frac{5}{7} \times \frac{1}{4} \times \frac{2}{3} = \frac{5 \times 1 \times 2}{7 \times 4 \times 3} = \frac{10}{84} = \frac{5}{42}$

(c) $\frac{f}{g} \div \frac{m}{n} = \frac{f}{g} \times \frac{n}{m} = \frac{fn}{gm}$

1.3 OPERATIONS WITH SIGNED NUMBERS

When adding numbers with the same signs, simply add the numbers and give the total the common sign. When adding numbers with different signs, add the $(+)$ and the $(-)$ numbers separately, subtract the smaller total from the larger, and then give the difference the sign of the larger. When subtracting signed numbers, change the signs of the numbers being subtracted.

EXAMPLE 1.4 Perform the following additions and subtractions with signed numbers: (a) $5 + 7 + 9 + 2$, (b) $5 + (-7) + 9 + (-2)$,   (c) $(-5) - (+7) + 9 - (-2)$.

Solution

(a) $5 + 7 + 9 + 2 = 23$

(b) $5 + (-7) + 9 + (-2) = 14 - 9 = 5$

(c) $(-5) - (+7) + 9 - (-2) = -5 - 7 + 9 + 2 = -1$

Multiplying numbers with the same sign gives a positive product, while multiplying numbers with different signs gives a negative product. The division of numbers with the same signs gives a positive quotient, while the division of numbers with different signs gives a negative quotient.

EXAMPLE 1.5   Perform the following multiplications and divisions: (*a*) $(-4)(-2)$,   (*b*) $(-4)(2)$,   (*c*) $(-4) \div (-2)$,   (*d*) $(-4)/(2)$.

Solution

(*a*) $(-4)(-2) = 8$

(*b*) $(-4)(2) = -8$

(*c*) $(-4) \div (-2) = 2$

(*d*) $(-4)/(2) = -2$

When division or multiplication is combined with addition or subtraction, the division or multiplication is done first. Numbers within parentheses or brackets should be treated as single numbers, so that all operations within parentheses and brackets should be done first.

EXAMPLE 1.6   Determine the order of operations and then calculate: (*a*) $2 \times 10 - 9$,   (*b*) $[(-2) + (3)] \div [(-2) + (-3)]$.

Solution

(*a*) $2 \times 10 - 9 = 20 - 9 = 11$

(*b*) $[(-2) + (3)] \div [(-2) + (-3)] = [1] \div [-5] = -0.2$

1.4   ROUNDING OFF

In *rounding off* to the nearest whole number, if the decimal fraction is less than 0.5, then it is dropped and the number to the left of the decimal point remains the same. If the decimal fraction is greater than 0.5, then the fraction is dropped and the number to the left of the decimal point is increased by one. Where the decimal fraction is exactly 0.5, it is common practice to follow this rule: If the first number to the left of the decimal place is odd, then increase it by one; if it is even, then leave it the same.

EXAMPLE 1.7   Round off the following to the nearest whole number: (*a*) 2.2,   (*b*) 1.89,   (*c*) 2.5,   (*d*) 1.50.

Solution

(*a*)   2.2 rounds off to 2

(*b*)   1.89 rounds off to 2

(*c*)   2.5 rounds off to 2

(*d*)   1.50 rounds off to 2

Essentially the same rounding-off rules apply if rounding off to a decimal place, except now they apply to the decimal fraction beyond that decimal place.

EXAMPLE 1.8   Round off: (*a*) 1.933 to two decimal places,   (*b*) 0.01791 to two decimal places,   (*c*) 1.23915 to three decimal places,   (*d*) 0.0015 to three decimal places.

Solution

(*a*)   1.933 rounds off to 1.93

(*b*)   0.01791 rounds off to 0.02

(*c*)   1.23915 rounds off to 1.239

(*d*)   0.0015 rounds off to 0.002

1.5   ABSOLUTE VALUES

The *absolute value* of the number n (indicated by the symbol $|n|$) is the numerical value of the number regardless of sign.

EXAMPLE 1.9.   Give the absolute values of: (*a*) -5,   (*b*) $10/2$,   (*c*) $\dfrac{43-52}{9}$

Solution

(*a*) $|-5| = 5$

(*b*) $|10/2| = 5$

(*c*) $\left| \dfrac{43-52}{9} \right| = 1$

1.6   FACTORIALS

The symbol $n!$ (which is read *n factorial*) represents the product of all positive integers from n to 1

$$n! = n \times (n-1) \times (n-2) \times (n-3) \times \cdots \times 1$$

where the symbol $\cdots$ indicates that not all of the multiplications are shown.

EXAMPLE 1.10   Calculate the following *factorials*:   (*a*) $2!$,   (*b*) $4!$,   (*c*) $9!$.

Solution

(*a*) $2! = 2 \times 1 = 2$

(*b*) $4! = 4 \times 3 \times (2!) = 24$

(*c*) $9! = 9 \times 8 \times 7 \times 6 \times 5 \times (4!) = 362{,}880$

1.7   RADICALS AND ROOTS

In the expression $a = \sqrt[n]{b}$, the symbol $\sqrt{}$ is called a *radical sign*, $\sqrt[n]{b}$ is called a *radical* (or *radical expression*), a is called the *nth root of b*, b is called the *radicand*, and n is called the *index*.

EXAMPLE 1.11   Solve the following radical: $\sqrt[2]{4}$.

Solution

$\sqrt[2]{4} = \sqrt{4} =$ the *second root* (or *square root*) of 4. The second root of 4 could be either $+2$ or -2, but by convention the symbol $\sqrt{4} = +2$ and the symbol $-\sqrt{4} = -2$.

The *principal nth root* of a number is its one real root, or if there is a choice between a positive and a negative root, it is the positive root.

EXAMPLE 1.12   Give the principal *nth* root of: (*a*) $\sqrt{16}$,   (*b*) $-\sqrt{16}$.

Solution

(*a*) $\sqrt{16} = 4$. This is true both by convention (see Example 1.11) and by selecting the positive root.

(*b*) $-\sqrt{16} = -4$. Here the convention indicates the minus number.

1.8   OPERATIONS WITH SQUARE ROOTS

Two square roots are multiplied by multiplying their radicands under one radical sign. In dividing one square root by another, the two radicands are divided under one radical sign.

EXAMPLE 1.13   Perform the indicated operations: (a) $\sqrt{5}\sqrt{5}$,   (b) $\dfrac{\sqrt{125}}{\sqrt{5}}$.

Solution

(a) $\sqrt{5}\sqrt{5} = \sqrt{5 \times 5} = \sqrt{25} = 5$

(b) $\dfrac{\sqrt{125}}{\sqrt{5}} = \sqrt{\dfrac{125}{5}} = \sqrt{25} = 5$

To multiply a square root by a number, multiply the radicand by the square of the number. To divide a square root by a number, divide the radicand by the square of the number.

EXAMPLE 1.14   Perform the indicated operations: (a) $2\sqrt{20.25}$,   (b) $\dfrac{\sqrt{25}}{5}$.

Solution

(a) $2\sqrt{20.25} = \sqrt{2^2}\sqrt{20.25} = \sqrt{4(20.25)} = \sqrt{81} = 9$

(b) $\dfrac{\sqrt{25}}{5} = \dfrac{\sqrt{25}}{\sqrt{25}} = \sqrt{\dfrac{25}{25}} = 1$

1.9   OPERATIONS WITH POWERS

b^n is the nth *power of b*; it is the product obtained from multiplying b times itself n times. Thus, for example, $b^2 = b \times b$ is the second power of b (*b squared*); and $b^3 = b \times b \times b$ is the third power of b (*b cubed*). In the expression b^n, n is called the *exponent*, and b is called the *base*.

EXAMPLE 1.15   Give the following powers: (a) b^0,   (b) 12^0,   (c) 12^2,   (d) 5^5.

Solution

(a)   For any nonzero number that has a zero exponent, the power is always one: $b^0 = 1$

(b) $12^0 = 1$

(c) $12^2 = 12 \times 12 = 144$

(d) $5^5 = 5 \times 5 \times 5 \times 5 \times 5 = 3{,}125$

EXAMPLE 1.16   As appropriate, express the following as a fraction, a root, or both: (a) b^{-n},   (b) $b^{1/n}$, (c) $b^{-1/n}$.

Solution

(a)   For any nonzero number that has a negative exponent, the following inverse relationship is true:

$$b^{-n} = \frac{1}{b^n}$$

(b) $b^{1/n} = \sqrt[n]{b}$

(c) $b^{-1/n} = \dfrac{1}{\sqrt[n]{b}}$

Numbers can be stated either as a power of 10 or as the product of a power of 10 with another number.

EXAMPLE 1.17   Using one digit to the left of the decimal point, state each of the following as the product of a power of ten: (a) 237,   (b) 0.000237,   (c) 116,270,000.

Solution

(a) $237 = 2.37 \times 10^2$

(b) $0.000237 = 2.37 \times 10^{-4}$

(c) $116,270,000 = 1.1627 \times 10^8$

When multiplying powers, if the bases are identical the exponents are added. When dividing powers, if the bases are identical the exponent of the denominator is subtracted from the exponent of the numerator.

EXAMPLE 1.18   Perform the following: (a) 10×10^3,   (b) $\dfrac{10^7}{10^3}$.

Solution

(a) $10 \times 10^3 = 10^{1+3} = 10^4 = 10,000$

(b) $\dfrac{10^7}{10^3} = 10^{7-3} = 10^4 = 10,000$

1.10   OPERATIONS WITH LOGARITHMS

The *logarithm* of a number n to the *base* c (where $c \neq 1$, and $c > 0$) is the exponent for c needed to obtain n. Thus, if $n = c^b$, then $\log_c n = b$.

EXAMPLE 1.19   Solve the following: (a) If $4 = 2^b$, what is $\log_2 4$?   (b) If $\log_{10} n = 2$, what is n?

Solution

(a)   If $4 = 2^b$, then $b = 2$ and $\log_2 4 = 2$

(b)   If $\log_{10} n = 2$, then $n = 10^2 = 100$

In general, if $\log_c n = b$, then the *antilogarithm* (or *antilog*) of b is n.

EXAMPLE 1.20   Give the antilogs of: (a) $\log_2 16 = 4$,   (b) $\log_{10} 10 = 1$.

Solution

(a)   antilog of $4 = 16$

(b)   antilog of $1 = 10$

When numbers are multiplied together, the logarithm of their product is the sum of their separate logarithms. The logarithm of a fraction is the logarithm of the numerator minus the logarithm of the denominator. The logarithm of a number with an exponent is the exponent times the logarithm of the number.

EXAMPLE 1.21   What is the logarithm of (a) bc,   (b) b/c,   (c) a^b?

Solution

(a) $\log(bc) = \log b + \log c$

(b) $\log (b/c) = \log b - \log c$

(a) $\log(a^b) = b(\log a)$

1.11 ALGEBRAIC EXPRESSIONS

An *algebraic expression* is any mixture of *arithmetic numbers* (having specific numerical values) and *general numbers* (symbols that represent numerical values), linked by the four fundamental processes of algebra $[(+), (-), (\times), \text{ and } (\div)]$. The *terms of an algebraic expression* are single numbers (arithmetic or general) separated from other numbers by $(+)$ or $(-)$ signs, or groups of numbers connected by multiplication or division and separated from other numbers by $(+)$ or $(-)$ signs. Algebraic expressions are called *monomial* if they have one term, *binomial* if they have two terms, and, in general, *multinomial* if they have two or more terms.

EXAMPLE 1.22   Which are the terms in the following algebraic expression: $14a^2 + 10b - 3c^4$?

Solution

The terms are $(14a^2)$, $(10b)$, $(3c^4)$

1.12 EQUATIONS AND FORMULAS

An *equation* is a statement that two algebraic expressions are equal. The following are examples of equations: $a - b = c$; $\dfrac{15}{5} = 3$; $y + 3 = 4$; $\log x + y = 2$. Each equation consists of an *equal sign* $(=)$ and expressions to its left (left member or left side of the equation) and right (right member or right side of the equation). A *formula* is an equation that states a principle or rule in algebraic terms. A familiar formula is $c = \pi d$ [the circumference of a circle $c =$ its diameter $d \times$ the constant π ($\pi = 3.14159\ldots$); here the symbol $\ldots$ indicates that the decimal fraction continues indefinitely and only the first part is shown]. An equivalent equation results when the same operations are applied to the left and right sides of an equation. This is true for all operations except divisions by zero.

EXAMPLE 1.23   Produce equivalent equations by applying the indicated operations on both sides of these equations: (*a*) $a + b = c$, add b to both sides,   (*b*) $a + b = c$, square both sides.

Solution

(*a*) $a + 2b = c + b$
(*b*) $(a + b)^2 = c^2$, or $a^2 + 2ab + b^2 = c^2$

In an *identical equation*, both sides can be made exactly the same by performing the indicated operations. Thus, $(y - 3)(y - 1) = y^2 - 4y + 3$ and $7 + 2 = 9$ are identical equations. Such equations, also called *identities*, are often indicated by using the symbol $(\equiv)$ in place of the equal sign. It can be seen that if an identity contains general numbers, it remains true for any numerical values given to the general numbers.

EXAMPLE 1.24   Show that the equation $(y - 3)(y - 1) = y^2 - 4y + 3$, is an identity by letting $y = 5$.

Solution

$$(5 - 3)(5 - 1) = (5)^2 - 4(5) + 3$$

$$8 = 8$$

While identities are true for any numerical value assigned to the general numbers, *conditional equations* are true only for certain numerical values.

EXAMPLE 1.25 Determine the value of y for which the conditional equation $2y+4=10$ is true.

Solution

$$2y + 4 = 10 \text{ is true only for } y = 3$$

To *solve an equation* means to find a value (or values) for general numbers in the equation which when substituted for them produces an identity. Such values, also called the *roots* of an equation, are said to *satisfy* the equation. For an identity such as $(y-3)(y-1)=y^2-4y+3$, any value of y solves or satisfies the equation. For the conditional equation $2y+4=10$, there is only one solution, or root: $y=3$.

EXAMPLE 1.26 Solve this equation $\dfrac{x}{10} + 18 = x$.

Solution

$$\frac{x}{10} + 18 = x$$

$$\frac{x}{10} - x = -18$$

$$x\left(\frac{1}{10} - 1\right) = -18$$

$$x(-0.9) = -18$$

$$x = 20$$

1.13 VARIABLES

A general number (see Section 1.11) in an equation or a formula that can assume different values (*vary*) in the context of a problem or discussion is called a *variable*. Thus, for the formula $c=\pi d$ (see Section 1.12), if the problem being worked on requests that the circumference should be calculated for three circles having diameters of 1, 2, and 3 inches, then both c and d vary in this problem and thus both are variables in this context.

If, on the other hand, in the context of such a problem a number in an equation or a formula can assume only one value, then it is called a *constant*. There are two types of constants: *absolute constants* and *arbitrary constants*. Absolute constants always have the same values. They are arithmetic numbers (e.g., 5, $\frac{1}{2}$, 100) or general numbers that always have the same fixed values [e.g., e (base of the natural logarithm, see Problem 1.23); π (see Section 1.12)]. Arbitrary constants are general numbers that remain fixed in the context of one problem but can vary from problem to problem.

While up to now in this book we have used lower-case letters to represent general numbers in equations and formulas, from this point on, unless otherwise indicated, *if the general number represents a variable it will be denoted by a capital letter from the end of the alphabet (Z, Y, X, etc.), and if the general number repre-sents a constant it will be denoted by a lower-case letter from the beginning of the alphabet (a, b, c, etc.).*

1.14 SINGLE-VARIABLE EQUATIONS AND THE QUADRATIC FORMULA

Single-variable linear equations (also known as *equations of the first degree*) have only constants and the first power (exponent $=1$) of a single variable. Examples of single-variable linear equations are: $2X-3=2$ and $aY-b=cY$. *Single-variable quadratic equations* (also known as *equations of the second degree*) can have constants and both the first and second powers (exponent $=2$) of a single variable. Examples of such equations are: $aX^2+bX=c$; $aX^2=c$ and $3X^2-5X=0$. If the equation has both the first and second powers (e.g., $aX^2+bX=c$) it is called a *complete quadratic equation*, and if it only has the second power (e.g., $aX^2 = c$) it is called an *incomplete quadratic equation*. If the complete quadratic equation $aX^2 + bX + c = 0$ is solved for X, the solution is

$$X = \frac{-b \pm \sqrt{b^2 - 4ac}}{2a} \tag{1.1}$$

This equation is called the *quadratic formula*, and it can be used to find the roots of any single-variable quadratic equation where $a \neq 0$ (See Problem 1.31 for the derivation of this formula.)

EXAMPLE 1.27   Use the quadratic formula to solve this equation: $2X^2 = -3X + 9$.

Solution

To solve, first put the equation in the general quadratic form: $aX^2 + bX + c = 0$. Thus, $2X^2 + 3X - 9 = 0$, and so $c = -9$, $b = 3$, $a = 2$

$$X = \frac{-b \pm \sqrt{b^2 - 4ac}}{2a}$$

$$= \frac{-3 \pm \sqrt{3^2 - [4 \times (-9) \times 2]}}{2 \times 2}$$

$$= \frac{-3 \pm \sqrt{9 - (-72)}}{4} = \frac{-3 \pm \sqrt{81}}{4} = \frac{-3 \pm 9}{4}$$

Therefore

$$X = \frac{6}{4} = \frac{3}{2} \quad \text{and} \quad X = \frac{-12}{4} = -3$$

1.15 VARIABLES IN STATISTICS

In statistics, variables are measurable characteristics of things (persons, objects, places, etc.) that vary within a group of such things. Thus, for example, if you are studying a group of children, then a variable might be the weight of each child; it is measurable (in pounds or kilograms) and it varies from child to child. Or, if you are studying a group of tomato plants, then you might consider measuring these variables for each plant: height, width, number of leaves, and number of tomatoes. Such variables are represented in the equations and formulas of statistics by the mathematical variables defined in Section 1.13. How variables are measured will be discussed in Chapter 2.

Each individual measurement of a variable (e.g., each weight of a child) is called a *variate* (or *value* or *observation*). In this book a variable and its variates will typically be represented by upper-case and lower-case versions of the same end-of-alphabet letter (e.g., X for the variable and x for its variates).

A *quantitative variable* in statistics is a characteristic whose variates can be ordered in terms of the magnitude of the characteristic (heavier, taller, richer, etc.). Thus, weight of a child is a quantitative variable, with one child weighing 45.2 lb and a heavier child weighing 50.9 lb. Similarly, number of tomatoes on a plant is a quantitative variable, with one plant having 20 and another 26.

A *qualitative variable*, on the other hand, is a characteristic with variates that are different categories which cannot be ordered by magnitude. For example, type of tree is a qualitative variable, with variates such as pine, maple, aspen, and hickory.

1.16 OBSERVABLE VARIABLES, HYPOTHETICAL VARIABLES, AND MEASUREMENT VARIABLES

Variables in statistics can be classified as being *directly measurable* or *indirectly measurable*. To understand these classifications consider the following example:

You are a geneticist studying inherited differences between male athletes who are Olympic champions in short-distance or long-distance running events. To do this you measure such variables as height, leg length, calf and thigh circumference, etc.

In this example, you are directly measuring anatomical variables in order to indirectly measure genetic variables (differences in specific genes or groups of genes). Directly measurable variables are called *observable variables*, and indirectly measurable variables are called *hypothetical variables* (or *intervening variables*).

Measurement variables are observable variables; directly measurable characteristics of the things being measured, expressed as the values on specific measurement scales. As seen in Section 1.15, measurement variables can be quantitative or qualitative. Some examples of measurement variables are: height in inches, weight in grams, number of leaves on a plant, and type of flower.

1.17 FUNCTIONS AND RELATIONS

To discuss functions and relations, we must first define two concepts: a *set* and the *real number system*.

A set is a collection of things (objects, symbols, numbers, words, etc.), and every item in the set is called an *element* (or *member*) of the set.

The real number system includes the *set of rational numbers* $\left(\text{all ratios of integers } \frac{a}{b} \text{ where } b \neq 0\right)$

and the *set of irrational numbers* (numbers that can not be written as the ratios of two integers, such as $\sqrt{2}$ and π).

If two variables X and Y are so related that for every permissible specific value x of X there is associated one and only one specific value y of Y, then it is said that Y *is a function of* X. Such a function always has: a *domain*, which is the set of all specific x values that X can assume; a *range*, which is the set of all specific y values associated with the x values; and a *rule of association*, the function itself, which associates with every permissible x value one and only one y value.

EXAMPLE 1.28 For the function $Y = X^2$, what are its domain, range, and rule of association?

Solution

For the function $Y = X^2$, it is permissible to let X be any real number, and so the domain of the function is the real number system. Then, for every x value selected there is an associated y value which must be zero or positive. Therefore, the range of this function is the set of all nonnegative real numbers. The rule of association for this function is the equation $Y = X^2$. These concepts are illustrated for this function in Fig. 1-1, for $x = -2, -1, \frac{1}{2}, 1, \sqrt{2}$, and 2. It can be seen that for every x value from the domain, the function associates one and only one y value in the range.

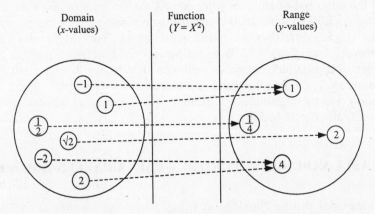

Fig. 1-1

For a function such as $Y = X^2$, once a specific x value has been selected the associated y value is automatically determined. Therefore, it can be said that the y value "depends" on the x value. This is why X is called the *independent variable* of the function and Y is called the *dependent variable*.

A *relation* (or *multiple-valued function*) differs from a function (also called a *single-valued function*) in its rule of association. Both have domains and ranges, but while the rule of association for a function always assigns one and only one value in the range to every value in the domain, the rule of association for a relation can assign more than one y value in the range to each x value in the domain.

EXAMPLE 1.29   For the relation $Y = X \pm 3$, what are its domain, range, and rule of association?

Solution

For $Y = X \pm 3$, the domain and range are all the real numbers, and the rule assigns two y values for every x value. This relation is illustrated in Fig. 1-2 for $x = 1, 2, 3$.

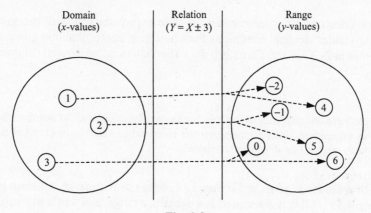

Fig. 1-2

1.18 FUNCTIONAL NOTATION

Functions can be written either as associations between variables (e.g., $Y = X^2$) or in functional notation as the associations between specific values of the variables. Thus, $Y = X^2$ would be stated in functional notation as

$$y = f(x) = x^2$$

where $y = f(x)$ is read y equals a function of x; it does not read y equals f times x. $f(x)$ is typical functional notation, but a function can be indicated by any other combination of letters: $F(z)$, $g(x)$, $h(y)$, and so on.

EXAMPLE 1.30   For $y = f(x) = -3 + 2x + x^2$, find   (a) $f(0)$,   (b) $f(1)$.

Solution

(a) $f(0) = -3 + 2(0) + 0^2 = -3 + 0 + 0 = -3$

(b) $f(1) = -3 + 2(1) + 1^2 = -3 + 2 + 1 = 0$

1.19 FUNCTIONS IN STATISTICS

A principal goal of any research effort is to study *cause and effect*; to discover the factors that *cause* something (the *effect*) to occur. For example, a botanist may want to know the soil characteristics (causes) that influence plant growth (effect); or an economist may want to determine the advertising factors (causes) that influence car sales (effect).

To study cause and effect the researcher uses statistical techniques to examine functional relationships between independent and dependent variables. Recall from Section 1.17 that in a mathematical function $y = f(x)$, Y is said to be the dependent variable and X the independent variable because the value of Y "depends" on the value of X. In the research context, the dependent variable is a measurement variable (see Section 1.16) associated with the effect that has values that to some degree "depend" on the values of an independent variable (a measurement variable associated with the cause). Thus, in the plant example above, the botanist might try to show that plant height in inches (the dependent variable) is a function of %-nitrogen in the soil (the independent variable). And, in the car-sales example, the economist might investigate whether the number of cars sold by a car company in each of the last ten years (the dependent variable) is a function of the amount of money the company spent on advertising in each of the last ten years (the independent variable).

EXAMPLE 1.31 In the following *experiment*, which is the independent variable and which is the dependent variable?

To determine the effects of water temperature on salmon growth, you raise two groups of salmon (10 in each group) under identical conditions from hatching, except that one group is kept in 20°C water and the other in 24°C water. Then, 200 days after hatching, you weigh (in grams) each of the 20 salmon.

Solution

In an experiment an independent variable is changed (*manipulated*) to see the effect on the dependent variable. In this experiment, water temperature (the independent variable) is changed between groups to see the effect on body weight (the dependent variable).

The definition of a function, given in Section 1.17, does not require an equation relating quantitative variables (see Section 1.15). All that is required is a domain, a range, and a rule that relates two variables in such a way that when a value from the domain is selected it automatically determines an associated value in the range. Thus, in a statistics class the domain of a function could be the name of each male student, the range could be their hair colors, and the rule of association could be: If a name is given, there is then one, and only one, associated hair color.

1.20 THE REAL NUMBER LINE AND RECTANGULAR CARTESIAN COORDINATE SYSTEMS

The *real number line* (or *number line* or *real axis*) is a straight line that graphically represents all of the real number system (see Section 1.17).

EXAMPLE 1.32 Plot the following numbers on the real number line: -4, $-\sqrt{2}$, $-\frac{1}{3}$, 0, $\sqrt{2}$, $2\frac{3}{4}$, 4.

Solution

Every number in the real number system can be represented by a point on the real number line. The numbers for this problem are plotted with exaggeratedly large dots in Fig. 1-3. This line was constructed by first arbitrarily selecting a location for the *origin* (the zero value) and then marking off equal-length spaces to

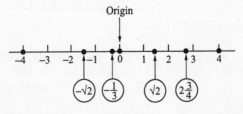

Fig. 1-3

the right and left of the origin. The positive integers increase outward to the right of the origin, and the negative integers increase outward to the left of the origin. All fractions and irrational numbers can be represented by points at appropriate distances between the integers.

A *rectangular Cartesian coordinate system* (also known as a *rectangular coordinate system*) is constructed by making two real number lines perpendicular to each other, such that their point of intersection (the *origin*) is the zero point of both lines. Such a coordinate system is called Cartesian in honor of the French mathematician and philosopher René Descartes (1596–1650), who was the first to propose its use (in 1637). An example of a rectangular Cartesian coordinate system is shown in Fig. 1-4. On the horizontal line (called the *X axis*) the numbers are positive and increasing to the right of the origin and negative and increasing to the left of the origin. On the vertical line (called the *Y axis*) the numbers are positive and increasing above the *X* axis and negative and increasing below the *X* axis. The two lines determine a plane (the *XY plane*) which they divide into four equal areas called the *first* (I), *second* (II), *third* (III), and *fourth* (IV) quadrants. For each point on the *XY* plane defined by a rectangular coordinate system there is a pair of real numbers (x, y) that are the coordinates of that point. The first number (or *x coordinate*, or *abscissa*) is the horizontal distance of the point from the *Y* axis and the second number (or *y coordinate*, or *ordinate*) is the vertical distance of the point from the *X* axis.

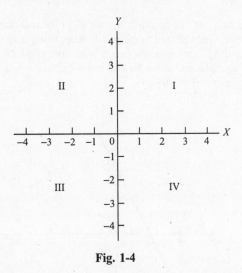

Fig. 1-4

EXAMPLE 1.33   From their *coordinates*, plot the following points on a rectangular coordinate system: A $(0, 0)$; B $(-1, 3)$; C $(1, -3)$; D $(2, 1)$; E $(-4, -2)$.

Solution

The locations of these points in a rectangular system are shown in Fig. 1-5, where the dotted lines show the distances from the points to the *X* and *Y* axes.

1.21   GRAPHING FUNCTIONS

A *graph* is a pictorial representation of the relationship between the variables of a function. By convention, on a rectangular coordinate system the dependent variable of a function is the *Y* axis topped by the symbol $f(x)$ and the independent variable is the *X* axis with the variate symbol x to its right. Functions are classified like equations (see Section 1.14), so the function $y = f(x) = c + bx$ is a *linear* or *first-degree function*, and its graph is a straight line.

EXAMPLE 1.34   Graph the function $y = f(x) = 4 + 2x$ on a rectangular coordinate system.

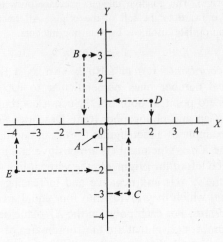

Fig. 1-5

Solution

Both the domain and range of this function include all of the real numbers, but a linear function like this one is typically graphed over a very limited range and domain, using only two points. The graph shown in Fig. 1-6 was drawn using the x and y *intercepts*, the points where the graph crosses the X and Y axes. To find the x intercept, set $y = 0$ in the equation and solve for x: $0 = 4 + 2x$, so $x = -2$. To find the y intercept, set $x = 0$ in the equation and solve for y: $y = 4 + 2(0) = 4$.

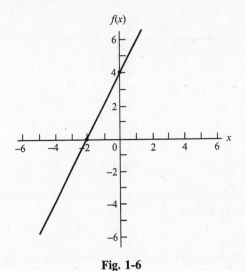

Fig. 1-6

A two-variable function in the form $y = f(x) = c + bx + ax^2$ (where a, b, and c are real numbers and $a \neq 0$) is called a *quadratic function*. The graph of a quadratic function is a *parabola* that is symmetric about a line parallel to the Y axis; this line is called the *axis of symmetry*. The parabola has: x and y intercepts, and either a *maximum value* (if it opens downward) or a *minimum value* (if it opens upward). (See Problem 1.40 for how to graph quadratic functions.)

1.22 SEQUENCES, SERIES, AND SUMMATION NOTATION

In mathematics, a *sequence* is a function with a domain that consists of all, or some part of, the consecutive positive integers. If the domain is all positive integers, the sequence is called an *infinite sequence*. If the domain is only a part of the consecutive positive integers, the sequence is called a *finite sequence*.

An example of an infinite sequence is the function $f(i) = i + 1$, for $i = 1, 2, 3, \ldots, \infty$, where ∞ is the symbol for infinity. The first number in the sequence is $f(1) = 1 + 1 = 2$; the second is $f(2) = 2 + 1 = 3$; the third is $f(3) = 3 + 1 = 4$; and so on. The sequence extends from $f(1)$ to $f(\infty)$: 2, 3, 4, $\ldots, \infty$. Each number in the sequence is called a *term of the sequence*.

An example of a finite sequence is the function $f(i) = x_i$, for $i = 1, 2, 3$, where the i in the x_i is called a *subscript* or an *index*, and x_i is read "x sub i." There are three terms in this sequence: x_1, x_2, x_3.

EXAMPLE 1.35   What are the terms of this sequence: $f(i) = i^2 - 3$, for $i = 2, 3, 4$?

Solution

The three terms are: $f(2) = (2)^2 - 3 = 1$; $f(3) = (3)^2 - 3 = 6$; $f(4) = (4)^2 - 3 = 13$.

In mathematics, a *series* is the sum of the terms of a sequence. Thus, for the infinite sequence $f(i) = i + 1$, for $i = 1, 2, 3, \ldots, \infty$, the series is the sum

$$2 + 3 + 4 + \cdots + \infty$$

For the finite sequence $f(i) = x_i$, for $i = 1, 2, 3$, the series is

$$x_1 + x_2 + x_3$$

The notation $\sum\limits_{i=1}^{n} x_i$ is called *summation notation*, and it is a symbolic representation of the series: $x_1 + x_2 + x_3 + \cdots + x_n$. The symbol $\sum$ is the capital letter *sigma* in Greek, and it indicates that the sequence function to its right should be summed. The letter below the $\sum$ is called the *index of summation* or the *variable of summation* and its numerical value indicates the lower limit of the domain of the sequence function. The numerical value above the $\sum$ gives the upper limit. For $\sum\limits_{i=1}^{n}$ the limits indicate that the sequence should be summed from $i = 1$ to $i = n$ (from x_1 to x_n). Typically the lower-case letters i, j, or k are used as indices of summation, but any letter can be used.

EXAMPLE 1.36   Find the following sum: $\sum\limits_{i=1}^{4} i^2$.

Solution

$$\sum_{i=1}^{4} i^2 = (1)^2 + (2)^2 + (3)^2 + (4)^2 = 1 + 4 + 9 + 16 = 30$$

If x_i represents a measurement of variable X, then in statistics an entire set of n such measurements is typically summed from x_1 to x_n. This series is indicated by $\sum\limits_{i=1}^{n} x_i$. However, where it is clear that it is the entire set being summed, the lower and upper limits of the summation are often omitted. When this is clear, then $\sum\limits_{i=1}^{n} x_i = \sum x_i = \sum x$.

EXAMPLE 1.37   The heights of five boys in a third-grade class form the following sequence: $x_1 = 2.1$ ft, $x_2 = 2.0$ ft, $x_3 = 1.9$ ft, $x_4 = 2.0$ ft, $x_5 = 1.8$ ft. For this set of measurements, find $\sum x_i$.

Solution

$$\sum x_i = \sum_{i=1}^{n} x_i = \sum_{i=1}^{5} x_i = x_1 + x_2 + x_3 + x_4 + x_5$$
$$= 2.1 \text{ ft} + 2.0 \text{ ft} + 1.9 \text{ ft} + 2.0 \text{ ft} + 1.8 \text{ ft}$$
$$= 9.8 \text{ ft}$$

1.23   INEQUALITIES

The symbols $(<, >, \leq, \geq)$ are called *inequality symbols*. Any mathematical statement that utilizes these symbols to show an *inequality relationship* (less than, greater than, less than or equal to, greater than or equal to) between two algebraic expressions is called an *inequality*. In an inequality, the pointed end of the inequality symbol always points to the smaller expression.

EXAMPLE 1.38   Interpret in words the following mathematical statements: (a) $3 < 4$,   (b) $5 > 2$,   (c) $b > a$, (d) $b \geq a$,   (e) $b \leq a$.

Solution

(a) $3 < 4$ means 3 is less than 4

(b) $5 > 2$ means 5 is greater than 2

(c) $b > a$ means b is greater than a

(d) $b \geq a$ means b is greater than or equal to a

(e) $b \leq a$ means b is less than or equal to a

EXAMPLE 1.39   Place the appropriate inequality symbol ($<$ or $>$) between each of the following number pairs: (a) 1   2,   (b) -1 -2,   (c) -1 2,   (d) 1 -2.

Solution

To solve this problem we must refer back to the real number line (see Section 1.20). For any two numbers on this line, the number to the right is always the larger number. Therefore: (a) $1 < 2$, (b) $-1 > -2$,   (c) $-1 < 2$,   (d) $1 > -2$.

If the same number is added to or subtracted from both sides of an inequality, the inequality remains valid. If both sides of an inequality are multiplied or divided by the same positive number, the inequality remains valid. If both sides of an inequality are multiplied or divided by the same negative number, the inequality relationship is reversed.

EXAMPLE 1.40   For the inequality $8 > 6$:   (a) add 5 to both sides,   (b) multiply both sides by 3,   (c) multiply both sides by -3.

Solution

(a) $8 + 5 > 6 + 5$, or $13 > 11$

(b) $8 \times 3 > 6 \times 3$, or $24 > 18$

(c) $8 \times (-3) < 6 \times (-3)$, or $-24 < -18$

For a single-variable inequality, a number is a *solution to the inequality* if, when the variable is replaced by that number, the inequality remains valid. All numbers that are solutions to such an inequality form a *solution set* for that inequality. Such an inequality is solved by using the above algebraic rules for inequalities, to isolate the variable on one side of the inequality symbol.

EXAMPLE 1.41   Solve the inequality: $X + 7 > -3$.

Solution

Inequality	Steps in solution
$X + 7 > -3$	
$X > -10$	subtract 7 from both sides

Therefore, the solution set contains all real numbers greater than -10.

Solved Problems

OPERATIONS WITH FRACTIONS

1.1   If we let a and b represent two numbers, then which of the following are equivalent fractions to $\dfrac{a}{2b}$: $\dfrac{a^2}{2ab}$, $\dfrac{2a}{2b}$, $\dfrac{1}{(2b)/(a)}$, or $\dfrac{a-1}{2b-1}$?

Solution

$$\frac{a}{2b} = \frac{(a)(a)}{(a)(2b)} = \frac{a^2}{2ab} = \frac{(a)/(a)}{(2b)/(a)} = \frac{1}{(2b)/(a)}$$

while

$$\frac{a}{2b} \neq \frac{2a}{2b} \quad \text{and} \quad \frac{a}{2b} \neq \frac{a-1}{2b-1}$$

1.2   Perform the indicated operations: (a) $\dfrac{g}{h} - \dfrac{a}{b}$,   (b) $\dfrac{1}{2} \div 1\dfrac{1}{2}$.

Solution

(a) $\dfrac{g}{h} - \dfrac{a}{b} = \dfrac{b \times g}{b \times h} - \dfrac{a \times h}{b \times h} = \dfrac{bg - ah}{bh}$

(b) $\dfrac{1}{2} \div 1\dfrac{1}{2} = \dfrac{1}{2} \times \dfrac{2}{3} = \dfrac{2}{6} = \dfrac{1}{3}$

OPERATIONS WITH SIGNED NUMBERS

1.3   Perform the following additions and subtractions: (a) $(-12) + (-3) + (-5) + (-6)$, (b) $14 - (-5) + (+2) - (+3)$.

Solution

(a) $(-12) + (-3) + (-5) + (-6) = -26$

(b) $14 - (-5) + (+2) - (+3) = 21 - 3 = 18$

1.4   Perform the following multiplications and divisions: (*a*) $(0.4)(-0.002)$, (*b*) $(-0.29) \cdot (-0.36)$, (*c*) $(-0.009) \div (-0.03)$,   (*d*) $(4.2) \div (-1.2)$.

Solution

(*a*) $(0.4)(-0.002) = -0.0008$

(*b*) $(-0.29) \cdot (-0.36) = 0.1044$

(*c*) $(-0.009) \div (-0.03) = 0.300$

(*d*) $(4.2) \div (-1.2) = -3.5$

1.5   Determine the order of operations and then calculate: (*a*) $2 \div 10 + 9$,   (*b*) $(0.004/0.002) + (0.9/0.003)$.

Solution

(*a*) $2 \div 10 + 9 = 0.2 + 9 = 9.2$

(*b*) $(0.004/0.002) + (0.9/0.003) = (2) + (300) = 302$

1.6   Perform the following operations with zero: (*a*) $4 + 0 + 3$,   (*b*) $4 \times 0 \times 3$,   (*c*) $12/0$.

Solution

(*a*) $4 + 0 + 3 = 7$

(*b*) $4 \times 0 \times 3 = 0$

(*c*)   It is not permitted to divide zero into a number.

ROUNDING OFF

1.7   Round off the following to the nearest whole number: (*a*) 13.499990,   (*b*) 13.50000.

Solution

(*a*)   13.499990 rounds off to 13

(*b*)   13.50000 rounds off to 14

1.8   Round off: (*a*) 40.195 to two decimal places,   (*b*) 0.020936 to three decimal places.

Solution

(*a*)   40.195 rounds off to 40.20

(*b*)   0.020936 rounds off to 0.021

ABSOLUTE VALUES

1.9   Show that $|c| + |d| \neq |c + d|$.

Solution

This can be done by using positive and negative values in the equation, say $c = 5$ and $d = -4$.

$$|5| + |-4| \neq |5 + (-4)|$$
$$9 \neq 1$$

FACTORIALS

1.10 Calculate the following factorials: (*a*) 0!, (*b*) 1!.

Solution

(*a*) By convention, this is defined: $0! = 1$

(*b*) Again by convention: $1! = 1$

1.11 Perform the indicated operations: (*a*) $\dfrac{6!}{(4-2)!}$, (*b*) $\dfrac{3!}{(3!)(2!)}$.

Solution

(*a*) $\dfrac{6!}{(4-2)!} = \dfrac{6 \times 5 \times 4 \times 3 \times 2 \times 1}{2!} = \dfrac{6 \times 5 \times 4 \times 3}{1} = 360$

(*b*) $\dfrac{3!}{(3!)(2!)} = \dfrac{1}{2!} = \dfrac{1}{2}$

RADICALS AND ROOTS

1.12 Solve the following radical: $\sqrt[3]{125}$.

Solution

$\sqrt[3]{125} =$ the *third root* (or *cube root*) of 125. Here $+5$ is the only possible answer.

1.13 Give the principal *n*th root of $\sqrt[3]{-8}$.

Solution

$\sqrt[3]{-8} = -2$. Here there is only one real root.

OPERATIONS WITH SQUARE ROOTS

1.14 Perform the indicated operations: (*a*) $\sqrt{28} + \sqrt{63}$, (*b*) $5\sqrt{(2 - 1/25)}$.

Solution

(*a*) $\sqrt{28} + \sqrt{63} = (\sqrt{4}\sqrt{7}) + (\sqrt{9}\sqrt{7}) = 2\sqrt{7} + 3\sqrt{7} = 5\sqrt{7}$

(*b*) $5\sqrt{(2 - 1/25)} = \sqrt{25}\sqrt{(2 - 1/25)} = \sqrt{25(2 - 1/25)} = \sqrt{50 - 1} = \sqrt{49} = 7$

OPERATIONS WITH POWERS

1.15 Express the following as roots: (*a*) $4^{1/2}$, (*b*) $8^{1/3}$.

Solution

(*a*) $4^{1/2} = \sqrt{4} = 2$

(*b*) $8^{1/3} = \sqrt[3]{8} = 2$

1.16 Express the following as fractions: (*a*) 7^{-2}, (*b*) $4^{-1/2}$.

Solution

(a) $7^{-2} = \dfrac{1}{7^2} = \dfrac{1}{49}$

(b) $4^{-1/2} = \dfrac{1}{\sqrt{4}} = \dfrac{1}{2}$

1.17 Give values for the following: 10^6, 10^4, 10^2, 10^0, 10^{-2}, 10^{-4}, 10^{-6}.

Solution

$10^6 = 1,000,000$; $10^4 = 10,000$; $10^2 = 100$; $10^0 = 1$; $10^{-2} = 0.01$; $10^{-4} = 0.0001$; $10^{-6} = 0.000001$

1.18 Perform the following: (a) $10^3 \times 10^{-4}$, (b) $\dfrac{7^8}{7^{-3}}$.

Solution

(a) $10^3 \times 10^{-4} = 10^{3-4} = 10^{-1} = 0.1$

(b) $\dfrac{7^8}{7^{-3}} = 7^{8-(-3)} = 7^{11}$

1.19 Perform the following: (a) $(10^2)^2$, (b) $(10^{-2})^{-2}$, (c) $(ab)^n$.

Solution

(a) $(10^2)^2 = 10^{2 \times 2} = 10^4 = 10,000$

(b) $(10^{-2})^{-2} = 10^{(-2)(-2)} = 10^4 = 10,000$

(c) $(ab)^n = (a^n)(b^n)$

1.20 Convert the following either from exponent to radical or from radical to exponent: (a) $10^{4/5}$, (b) $\sqrt[3]{5^6}$.

Solution

(a) $10^{4/5} = \sqrt[5]{10^4}$

(b) $\sqrt[3]{5^6} = 5^{6/3} = 5^2 = 25$

OPERATIONS WITH LOGARITHMS

1.21 If $\log_c 1,000 = 3$, what is c?

Solution

If $\log_c 1,000 = 3$, then $c^3 = 1,000$ and $c = \sqrt[3]{1,000} = 10$.

1.22 Give the antilog of: (a) $\log_a n = d$, (b) $\log_5 25 = 2$.

Solution

(a) antilog of $d = n$

(b) antilog of $2 = 25$

1.23 Use an electronic calculator to find: (*a*) the *common logarithm* of 100,   (*b*) the *natural logarithm* of 100.

Solution

(*a*) Most electronic calculators have two logarithm keys with the symbols [LOG] and [LN]. Pressing [LOG] will give the logarithm of a number to the base 10, which is also called the *common logarithm* of the number. Thus, if [LOG] is pressed when 100 is in the display, it will produce 2 in the display, which is $\log_{10} 100$. The common logarithm is often written without the base 10, as in log 100 = 2.

(*b*) Pressing [LN] will give the logarithm of a number to the base e, where $e = 2.71828\ldots$. Logarithms to the base e are called *natural* or *Naperian logarithms*. Thus, to five decimal places, (the natural logarithm of 100) = $\log_e 100 = 4.60517$.

1.24 Use an electronic calculator to solve this problem: If 1.69897 is the common logarithm of a number a (to five decimal places), what is its antilogarithm?

Solution

The relationship for the common logarithm is: If $\log_{10} n = x$, then $n = 10^x$. Therefore, the calculator antilog key for common logarithms will be [10^x]. Using this key here

$$\text{(antilog of 1.69897)} = 10^{1.69897} = 49.9999995, \text{ or } 50$$

1.25 Use an electronic calculator to solve this problem: If 3.91202 is the natural logarithm of a number b (to five decimal places), what is its antilogarithm?

Solution

The relationship for the natural logarithm is: If $\log_e n = x$, then $n = e^x$. Therefore, the antilog key for natural logarithms will be [e^x]. Using this key here

$$\text{(antilog of 3.91202)} = e^{3.91202} = 49.9998497, \text{ or } 50$$

1.26 Find the common logarithms of the following: (*a*) $4bc$,   (*b*) $7/5$,   (*c*) $1\frac{5}{27}$.

Solution

(*a*) $\log(4bc) = \log 4 + \log b + \log c = 0.60206 + \log b + \log c$

(*b*) $\log(7/5) = \log 7 - \log 5 = 0.84510 - 0.69897 = 0.14613$

(*c*) $\log\left(1\frac{5}{27}\right) = \log(32/27) = \log 32 - \log 27 = 1.50515 - 1.43136 = 0.07379$

1.27 Find the common logarithm of the following: $\sqrt[4]{\dfrac{(49)(27)}{(3.1)^3}}$

Solution

$$\log\left(\sqrt[4]{\frac{(49)(27)}{(3.1)^3}}\right) = \log\left[\left(\frac{(49)(27)}{(3.1)^3}\right)^{\frac{1}{4}}\right] = \frac{1}{4}[(\log 49 + \log 27) - 3(\log 3.1)]$$

$$= \frac{1}{4}[(1.69020 + 1.43136) - 3(0.49136)]$$

$$= \frac{1}{4}(3.12156 - 1.47408)$$

$$= 0.41187$$

ALGEBRAIC EXPRESSIONS

1.28 Which are the terms in the following algebraic expression: $\dfrac{2\sigma}{a} - \mu$?

Solution

The terms are: $\left(\dfrac{2\sigma}{a}\right), (\mu)$.

EQUATIONS AND FORMULAS

1.29 Produce equivalent equations by applying the indicated operations on both sides of these equations: (a) $a+b=c$, subtract b from both sides; (b) $ab=c$, divide both sides by a; (c) $\dfrac{b}{a}=c$, multiply both sides by a.

Solution

(a) $a = c - b$

(b) $b = \dfrac{c}{a}$

(c) $b = ca$

1.30 Solve these equations: (a) $x+5=10$, (b) $(x-2)^2=4$.

Solution

(a) $x + 5 = 10$

$x = 5$

(b) $(x-2)^2 = 4$

$(x-2) = 2$

$x = 4$

SINGLE-VARIABLE EQUATIONS AND THE QUADRATIC FORMULA

1.31 Derive the quadratic formula [see equation (1.1)] by solving this complete quadratic equation: $aX^2 + bX + c = 0$.

Solution

To solve the equation and thus derive the quadratic formula, the following steps are required.

(1) Multiply both sides of the equation by $\dfrac{1}{a}$:

$$\frac{1}{a}(aX^2 + bX + c) = \frac{1}{a}(0)$$

$$X^2 + \frac{b}{a}X + \frac{c}{a} = 0$$

(2)   Add $\left[-\left(\dfrac{c}{a}\right)+\left(\dfrac{b}{2a}\right)^2\right]$ to both sides:

$$X^2+\frac{b}{a}X+\frac{c}{a}-\frac{c}{a}+\left(\frac{b}{2a}\right)^2=-\frac{c}{a}+\left(\frac{b}{2a}\right)^2$$

$$X^2+\frac{b}{a}X+\left(\frac{b}{2a}\right)^2=-\frac{c}{a}+\frac{b^2}{4a^2}$$

(3)   Rearrange the left side:

$$\left(X+\frac{b}{2a}\right)^2=-\frac{c}{a}+\frac{b^2}{4a^2}$$

(4)   Sum the fractions on the right side:

$$\left(X+\frac{b}{2a}\right)^2=-\frac{4a\times c}{4a\times a}+\frac{b^2}{4a^2}$$

$$=-\frac{4ac}{4a^2}+\frac{b^2}{4a^2}$$

$$=\frac{b^2-4ac}{4a^2}$$

(5)   Take the square root of both sides:

$$X+\frac{b}{2a}=\pm\sqrt{\frac{b^2-4ac}{4a^2}}$$

$$=\pm\frac{\sqrt{b^2-4ac}}{2a}$$

(6)   Solve for X:

$$X=-\frac{b}{2a}\pm\frac{\sqrt{b^2-4ac}}{2a}$$

$$=\frac{-b}{2a}\pm\sqrt{\frac{b^2-4ac}{2a}}$$

1.32   Use the quadratic formula [equation (1.1)] to solve: $X^2=12X-36$.

Solution

$$X^2-12X+36=0,\qquad\text{and so}\qquad a=1,b=-12,c=36$$

$$X=\frac{12\pm\sqrt{12^2-(4\times36\times1)}}{2\times1}$$

$$=\frac{12\pm\sqrt{144-144}}{2}=\frac{12}{2}=6$$

VARIABLES IN STATISTICS

1.33   Is color a quantitative or qualitative variable?

Solution

The characteristic of color in an object can be measured to produce either quantitative or qualitative variates. The physical basis of color is the wavelength of light, which can be expressed as an arithmetic number [e.g., deep red is roughly $8,000 \times 10^{-10}$ meters (or 8,000 angstroms)]. If the measurements taken are in such wavelengths, then color is a quantitative variable. If the color variates are unordered categories, however, then the color variable is qualitative. Thus, for example, a group of people might be classified by hair color as having black, red, blond, brown, gray, or white hair.

FUNCTIONS AND RELATIONS

1.34 What are the domain and range of this function: $Y = \dfrac{1}{X}$?

Solution

For $Y = \dfrac{1}{X}$, both the domain and the range are all real numbers other than 0.

1.35 For $y = f(x) = 2^x$, find: (a) $f(0)$, (b) $f(3)$, (c) $f(6)$.

Solution

(a) $f(0) = 2^0 = 1$
(b) $f(3) = 2^3 = 8$
(c) $f(6) = 2^6 = 64$

1.36 From the following table giving three x values from the domain of a function and the y values associated with each x value, find $f(x)$:

x	2	3	4
y	2	4	6

Solution

$$y = f(x) = -2 + 2x$$

1.37 For the function $y = f(x) = -5 + 5x^3$, fill in the range in the following table:

x	1	2	3
y			

Solution

x	1	2	3
y	0	35	130

THE REAL NUMBER LINE AND RECTANGULAR CARTESIAN COORDINATE SYSTEMS

1.38 Find the coordinates of the points shown in Fig. 1-7.

Solution

To find the coordinates of a point on a rectangular coordinate system, draw lines from the point perpendicular to both the X axis and the Y axis, and where these lines meet the axes are the coordinates of the point. Using this technique here, the coordinates are: A $(1, 2\frac{1}{2})$; B $(1, -1)$; C $(-2, -3)$; D $(-2, 2)$.

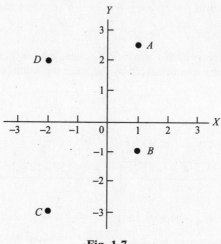

Fig. 1-7

GRAPHING FUNCTIONS

1.39 Graph the function $y = f(x) = 4 + 2x$ on a rectangular coordinate system using its *slope* and *y* intercept.

Solution

For a linear function written in the form $y = f(x) = c + bx$ where c and b are real numbers and c and b are not both zero, c is the y intercept of the straight line and b is its slope (the rise or drop over the run; the distance the line moves vertically for a given distance that it moves horizontally). For the function $y = f(x) = 4 + 2x$, the y intercept is (0, 4) (see Example 1.34) and the slope is $+2$. A slope of $+2$ means that the line rises two units for each unit it moves horizontally in the positive direction. The graph of this function, using the y intercept and a second point determined by the slope [out one x unit from the y intercept and up two y units (1, 6)], is shown in Fig. 1-8.

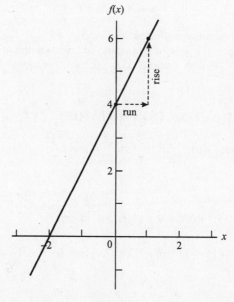

Fig. 1-8

1.40 Graph the quadratic function $y=f(x)=-4+3x+x^2$ on a rectangular coordinate system.

Solution

To find the two x intercepts of this quadratic function, put $y=0$ in the function: $0=-4+3x+x^2$. The equation is now in a form that can be solved by the quadratic formula [equation (1.1)].

$$x=\frac{-b\pm\sqrt{b^2-4ac}}{2a}$$

Here, $a=1$, $b=3$, and $c=-4$

$$x=\frac{-3\pm\sqrt{3^2-[4\times(-4)\times1]}}{2\times1}=\frac{-3\pm\sqrt{9-[-16]}}{2}=\frac{-3\pm\sqrt{25}}{2}=\frac{-3\pm5}{2}$$

So the coordinates of the two x intercepts are $\left(x=\frac{-3+5}{2}=1, y=0\right)$ and $\left(x=\frac{-3-5}{2}=-4, y=0\right)$.

The x intercept of the axis of symmetry (where it crosses the X axis) can also be found using the quadratic formula written in the form

$$x=\frac{-b}{2a}\pm\frac{\sqrt{b^2-4ac}}{2a}$$

The term $\frac{-b}{2a}$ is the x intercept for the axis of symmetry, and here

$$\frac{-b}{2a}=\frac{-3}{2\times1}=-1.5$$

Thus, the coordinates of this x intercept are $(-1.5, 0)$.

The y intercept for this parabola is found by setting $x=0$ in the function: $y=-4+3(0)+(0)^2=-4$. Thus, the coordinates of the y intercept are $(0, -4)$.

Whether a parabola opens downward or upward is determined by whether the constant a in the function is positive or negative: If a is positive the parabola opens upward, and if a is negative it opens downward. Here $a=1$, so the parabola opens upward and has a minimum value. This minimum value is the only point on the parabola that is also on the axis of symmetry, and therefore it has an abscissa of -1.5. Then, to find the y coordinate for the minimum point, put -1.5 in the function and solve for y.

$$y=-4+3(-1.5)+(-1.5)^2=-6.25$$

Thus, the coordinates for the minimum value are $(-1.5, -6.25)$.

The x and y intercepts, the axis of symmetry, and the minimum value are plotted on a rectangular coordinate system in Fig. 1-9, with a smooth curve drawn to represent all points of the parabola.

SEQUENCES, SERIES, AND SUMMATION NOTATION

1.41 Find the following sums: (a) $\sum_{i=3}^{6} i^2$, (b) $\sum_{i=4}^{7}(i-1)$.

Solution

(a) $\sum_{i=3}^{6} i^2=(3)^2+(4)^2+(5)^2+(6)^2=9+16+25+36=86$

(b) $\sum_{i=4}^{7}(i-1)=(4-1)+(5-1)+(6-1)+(7-1)=3+4+5+6=18$

1.42 Find: (a) $\sum_{i=1}^{5} 6$, (b) $\sum_{i=2}^{5} 6$.

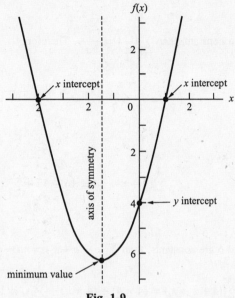

Fig. 1-9

Solution

(a)   In general, if a is a constant, then $\sum_{i=1}^{n} a = na$. Therefore

$$\sum_{i=1}^{5} 6 = 6 + 6 + 6 + 6 + 6 = 5 \times 6 = 30$$

(b)   Here, the lower limit is 2 rather than 1, so $\sum_{i=2}^{n} a = (n-1)a$. Therefore

$$\sum_{i=2}^{6} 6 = (5 - 1)6 = 24$$

1.43   Show that $\sum_{i=1}^{3} 3x_i = 3 \sum_{i=1}^{3} x_i$.

Solution

In general, if a is a constant, then $\sum_{i=1}^{n} ax_i = a \sum_{i=1}^{n} x_i$. Therefore

$$\sum_{i=1}^{3} 3x_i = 3x_1 + 3x_2 + 3x_3 = 3(x_1 + x_2 + x_3) = 3 \sum_{i=1}^{3} x_i$$

1.44   Show that

$$\sum_{i=1}^{3} \left(\frac{x_i}{2}\right) = \frac{\sum_{i=1}^{3} x_i}{2}$$

Solution

In general, if a is a constant, then $\sum_{i=1}^{n}\left(\frac{x_i}{a}\right) = \frac{\sum_{i=1}^{n}x_i}{a}$. Therefore

$$\sum_{i=1}^{3}\left(\frac{x_i}{2}\right) = \frac{x_1}{2} + \frac{x_2}{2} + \frac{x_3}{2} = \frac{x_1 + x_2 + x_3}{2} = \frac{\sum_{i=1}^{3}x_i}{2}$$

1.45 Show that

$$\sum_{i=1}^{4}(4 + \pi + x_i) = 16 + 4\pi + \sum_{i=1}^{4}x_i$$

Solution

In general, if a and b are constants, then $\sum_{i=1}^{n}(a + b + x_i) = na + nb + \sum_{i=1}^{n}x_i$. Therefore

$$\sum_{i=1}^{4}(4 + \pi + x_i) = (4 \times 4) + (4 \times \pi) + \sum_{i=1}^{4}x_i = 16 + 4\pi + \sum_{i=1}^{4}x_i$$

1.46 Show that

$$\sum_{i=1}^{3}(3x_i + 2y_i) = 3\sum_{i=1}^{3}x_i + 2\sum_{i=1}^{3}y_i$$

Solution

In general, if a and b are constants, then $\sum_{i=1}^{n}(ax_i + by_i) = a\sum_{i=1}^{n}x_i + b\sum_{i=1}^{n}y_i$. Therefore

$$\sum_{i=1}^{3}(3x_i + 2y_i) = 3\sum_{i=1}^{3}x_i + 2\sum_{i=1}^{3}y_i$$

INEQUALITIES

1.47 For the inequality $6 < 7$:   (a) add 8 to both sides,   (b) subtract 8 from both sides.

Solution

(a) $6 + 8 < 7 + 8$, or $14 < 15$

(b) $6 - 8 < 7 - 8$, or $-2 < -1$

1.48 For the inequality $12 > 8$:   (a) multiply both sides by 2,   (b) divide both sides by 2.

Solution

(a) $12 \times 2 > 8 \times 2$, or $24 > 16$

(b) $12/2 > 8/2$, or $6 > 4$

1.49 For the inequality $2 > 1$:   (a) multiply both sides by -1,   (b) divide both sides by -1.

Solution

(a) $2 \times (-1) < 1 \times (-1)$, or $-2 < -1$

(b) $2/-1 < 1/-1$, or $-2 < -1$

1.50 Solve this inequality: $-3X - 2 < 2X + 5$.

Solution

Inequality	Steps in solution
$-3X - 2 < 2X + 5$	
$-2 < 5X + 5$	add $3X$ to both sides
$-7 < 5X$	subtract 5 from both sides
$\dfrac{-7}{5} < X$	divide both sides by 5

Therefore, the solution set contains all real numbers greater than $\dfrac{-7}{5}$

1.51 Solve this inequalitiy: $\dfrac{X}{2} + 1 \leq X - 1$.

Solution

Inequality	Steps in solution
$\dfrac{X}{2} + 1 \leq X - 1$	
$\dfrac{X}{2} + 2 \leq X$	add 1 to both sides
$2 \leq X - \dfrac{X}{2}$	subtract $\dfrac{X}{2}$ from both sides
$2 \leq \dfrac{X}{2}$	simplify
$4 \leq X$	multiply both sides by 2

Therefore, the solution set contains all real numbers that are greater than or equal to 4.

1.52 Interpret these single-variable inequalities: (a) $-9 \leq X < 2$,   (b) $-2 < X \leq -1$.

Solution

(a) $-9 \leq X < 2$ indicates that the variable X can assume values that are less than 2 and greater than or equal to -9. This means that the solution set for this inequality contains all real numbers between -9 and 2, including -9 but not 2.

(b) $-2 < X \leq -1$ indicates that the variable X can assume values that are less than or equal to -1 and greater than -2. This means that the solution set for this inequality contains all real numbers between -2 and -1, including -1 but not -2.

1.53 What is wrong with this single-variable inequality: $-7 > X > 7$?

Solution

This inequality is invalid because there is no solution; there is no number that is simultaneously greater than 7 and less than -7.

Supplementary Problems

OPERATIONS WITH FRACTIONS

1.54 Which of the following are equivalent fractions to $\dfrac{4 \div 3}{2 \times 3} : \dfrac{2}{9}, \dfrac{4/2}{27 \div 3}, \dfrac{(2)(2)}{36/2},$ or $\dfrac{3 \cdot 6}{6\frac{1}{2}}$?

   Ans. $\dfrac{4 \div 3}{2 \times 3} = \dfrac{2}{9} = \dfrac{4/2}{27 \div 3} = \dfrac{(2)(2)}{36/2}$, while $\dfrac{4 \div 3}{2 \times 3} \neq \dfrac{3 \cdot 6}{6\frac{1}{2}}$

1.55 Perform the indicated operations: (a) $\dfrac{8 \div 3}{(2)(3)} - \dfrac{7/2}{6 \times 4}$,   (b) $\dfrac{(14)(3)}{19/2} \div \dfrac{18 \times 4}{16 \div 3}$.

   Ans. (a) $\dfrac{43}{144}$,   (b) $\dfrac{56}{171}$

OPERATIONS WITH SIGNED NUMBERS

1.56 Perform the following additions and subtractions: (a) $1.3 + (-1.7) - (-2.3) - (+4.2) - (-3.1)$, (b) $(-0.93) + (-0.26) - (-3.91) + (-2.1)$.

   Ans. (a) 0.8, (b) 0.62

1.57 Perform the following multiplications and divisions: (a) $(1,800) \div (-0.2)$, (b) $(-3.63)(-0.0001)$, (c) $(-0.0004)/(-0.002)$.

   Ans. (a) $-9,000.0$, (b) 0.000363, (c) 0.2000

1.58 Determine the order of operations and then calculate: $[(-3) \times (-4) + (2)] \div [(2) - (-0.5)]$.

   Ans. 5.6

1.59 Perform the following operations with zero: (a) $4 \times 0 + 3$, (b) $0/12 + 3$.

   Ans. (a) 3, (b) 3

ROUNDING OFF

1.60 Round off the following to the nearest whole number: (a) 24.501, (b) 24.50.

   Ans. (a) 25, (b) 24

1.61 Round off: (a) 2.125 to two decimal places, (b) 4.93250 to three decimal places.

   Ans. (a) 2.12, (b) 4.932

ABSOLUTE VALUES

1.62 Give the absolute values of the following numbers: (a) $10/-2$, (b) 0, (c) $7 - 10$.

   Ans. (a) 5, (b) 0, (c) 3

FACTORIALS

1.63 Perform the indicated operations: $\dfrac{6!}{5(3!)}$.

   Ans. 24

RADICALS AND ROOTS

1.64   Solve the following radical: $\sqrt[4]{1}$.

   Ans.   There are two possible answers, $+1$ and -1, and again by convention $\sqrt[4]{1} = +1$ and $-\sqrt[4]{1} = -1$.

1.65   Give the principal nth root of $\sqrt[4]{625}$.

   Ans.   By convention, the principal 4th root is: $\sqrt[4]{625} = 5$.

OPERATIONS WITH SQUARE ROOTS

1.66   Perform the indicated operations: (a) $\sqrt{27} - \sqrt{12}$,   (b) $\dfrac{3\sqrt{45} - \sqrt{20}}{\sqrt{125/5}}$.

   Ans.   (a) $\sqrt{3}$,   (b) 7

OPERATIONS WITH POWERS

1.67   Express the following as fractions: (a) 8^{-3},   (b) $8^{-1/3}$.

   Ans.   (a) $\dfrac{1}{512}$,   (b) $\dfrac{1}{2}$

1.68   State the following as the product of a power of 10:   (a) 0.0237,   (b) 0.11627,   (c) 11,627.
   Ans.   (a) 2.37×10^{-2},   (b) 1.1627×10^{-1},   (c) 1.1627×10^{4}

1.69   Perform the following: (a) $(3 \times 7^3)(2 \times 7^{-2})$,   (b) $\dfrac{6 \times 10^{-3}}{3 \times 10^{-1}}$.

   Ans.   (a) 42,   (b) 0.02

1.70   Perform the following: (a) $(3 \times 5)^3$,   (b) $(3 \times 5)^{-2}$.
   Ans.   (a) 3,375,   (b) 0.00444

1.71   Convert the following from exponent to radical or from radical to exponent: (a) $9^{4/3}$,   (b) $\sqrt[5]{9^2}$.
   Ans.   (a) $\sqrt[3]{9^4}$,   (b) $9^{2/5}$

OPERATIONS WITH LOGARITHMS

1.72   If $\log_c 0.001 = -3$, what is c?
   Ans.   10

1.73   If 3.17609 is the common logarithm of a number a (to five decimal places), what is its antilogarithm?
   Ans.   1,500

1.74   If -1.89712 is the natural logarithm of a number b (to five decimal places), what is its antilogarithm?
   Ans.   0.15

1.75   What are the common logarithms of: (a) $\dfrac{5}{7}$,   (b) 111^{12}?

   Ans.   (a) -0.14613,   (b) 24.54388

1.76 Find the common logarithm of $\left(\dfrac{(15)^3\sqrt{22}}{(15)^{1/3}}\right)^{-1/3}$.

Ans. -1.26915

ALGEBRAIC EXPRESSIONS

1.77 Which are the terms in the following algebraic expression: $14 + 2(a + b)$?

Ans. (14), $(2a)$, $(2b)$

EQUATIONS AND FORMULAS

1.78 Determine the value (values) for y for which the conditional equation $2y^2 = 8$ is true.

Ans. $y = 2$ or $y = -2$

1.79 Solve these equations for c: (a) $a = b\,(1 + c)$,   (b) $c^2a - b = 2b$.

Ans.   (a) $c = \dfrac{a}{b} - 1$,   (b) $c = \sqrt{\dfrac{3b}{a}}$

SINGLE-VARIABLE EQUATIONS AND THE QUADRATIC FORMULA

1.80 Use the quadratic formula [equation (1.1)] to solve $4X^2 = 1$.

Ans. $X = \dfrac{0 \pm \sqrt{16}}{8} = \pm\dfrac{4}{8} = \pm\dfrac{1}{2}$

FUNCTIONS AND RELATIONS

1.81 What are the domain and range of this function: $Y = \sqrt{X - 2}$?

Ans.   The domain is all real numbers greater than or equal to 2, and the range is the set of all nonnegative real numbers.

1.82 For $y = f(x) = 7$, find: (a) $f(0)$,   (b) $f(3)$,   (c) $f(6)$.

Ans.   No matter what the x value, $y = f(x) = 7$.

THE REAL NUMBER LINE AND RECTANGULAR CARTESIAN COORDINATE SYSTEMS

1.83 What are the abscissas and ordinates of points A and B in Fig. 1-7?

Ans. A (1 is abscissa, $2\frac{1}{2}$ is ordinate); B (1 is abscissa, -1 is ordinate)

GRAPHING FUNCTIONS

1.84 Graph the linear function $y = f(x) = 3 - 0.5x$ on a rectangular coordinate system using its slope and y intercept.

Ans.   The graph in Fig. 1-10 was drawn using the y intercept $(0,3)$ and, as the slope is negative (-0.5), a point that is out four x units and down two y units from the y intercept.

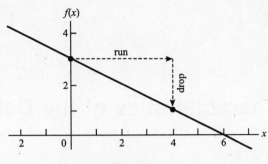

Fig. 1-10

SEQUENCES, SERIES, AND SUMMATION NOTATION

1.85 Find the following sums: (a) $\sum_{i=3}^{6} i - 1$, (b) $\sum_{i=1}^{5} x_i^2$.

Ans. (a) 17, (b) $x_1^2 + x_2^2 + x_3^2 + x_4^2 + x_5^2$

1.86 For the heights in Example 1.37, find: (a) $\sum x_i^2$, (b) $\left(\sum x_i\right)^2$.

Ans. (a) 19.26 ft^2, (b) 96.04 ft^2

1.87 Find $\sum_{i=2}^{4}(5 + 3i + i^2)$.

Ans. 71

1.88 Find $\sum_{i=1}^{3}\left(\dfrac{i^2}{4} + 3i^2\right)$.

Ans. 45.5

INEQUALITIES

1.89 For the inequality $b \geq a$: (a) add 5 to both sides, (b) subtract 5 from both sides.

Ans. (a) $b + 5 \geq a + 5$, (b) $b - 5 \geq a - 5$

1.90 For the inequality $3 < 4$: (a) multiply both sides by 3, (b) divide both sides by 2.

Ans. (a) $3 \times 3 < 4 \times 3$, or $9 < 12$, (b) $3/2 < 4/2$, or $1\frac{1}{2} < 2$

1.91 For the inequality $6 < 7$: (a) multiply both sides by -2, (b) divide both sides by -7.

Ans. (a) $6 \times (-2) > 7 \times (-2)$, or $-12 > -14$, (b) $6/-7 > 7/-7$, or $-\dfrac{6}{7} > -1$

Chapter 2

Characteristics of the Data

2.1 MEASUREMENT SCALES

A *measurement scale* is a tool that is applied to an observable variable to produce a measurement variable (see Section 1.16). Thus, a ruler marked off in inches is a measurement scale which, when placed against an object, can produce the measurement variable of length in inches. Using such a measurement scale, a specific value on the scale can be assigned to each thing being measured.

Measurement scales that produce qualitative measurement variables (see Section 1.15) simply divide the observable variable into a set of unique, unordered categories. Thus, for the observable variable of hair color, the measurement scale might include these categories: black, red, blond, brown, gray, and white. Such scales should have sufficient categories to allow each thing being measured to be classified into one, and only one, category.

Measurement scales that produce quantitative measurement variables (see Section 1.15) also consist of a set of unique categories, but now the categories can be ordered from small to large. Typically, the categories are serially increasing numerical values. Thus, for example, to measure the observable variable of height in a group of people, one could use an increasing scale graduated in 0.1 centimeter steps which place each person in one, and only one, category on the scale (e.g., 170.1 cm).

2.2 OPERATIONAL DEFINITION OF A MEASUREMENT

Measurement is the logic and procedures involved in applying a measurement scale to an observable variable to produce a measurement variable. It includes the rules used to assign each thing being measured to one category on a measurement scale. An *operational definition of a measurement* indicates the exact sequence of steps (or *operations*) that are followed in taking a measurement: applying a measurement scale to an observable variable. The definition should be sufficiently precise and detailed so that everyone who uses the procedure will achieve essentially the same measurement.

EXAMPLE 2.1   You are asked to measure the height of each person in your statistics class to the nearest millimeter, using two meter sticks, masking tape, and a stepladder. Give an operational definition for the measurement.

> **Solution**
>
> The sequence of steps (operations) could be as follows:
>
> (1)   Select a doorjamb and create a vertical two-meter scale on it by taping the meter sticks against the jamb, one on top of the other, such that the bottom scale line of the upper stick coincides with the top scale line of the lower stick.
>
> (2)   Have each member of the class, in turn, take off their shoes and stand straight with their backs against the doorjamb.
>
> (3)   Standing on the stepladder with your eyes at the level of the top of each head, read the heights from the two-meter scale to the nearest millimeter.

2.3 LEVELS AND UNITS OF MEASUREMENT

The *four levels of measurement* are four types of measurement scales: *nominal* (see Section 2.4), *ordinal* (see Section 2.5), *interval* (see Section 2.6), and *ratio* (see Section 2.7). Nominal scales produce qualitative measurement variables, while ordinal, interval, and ratio scales produce quantitative measurement variables.

With the exception of ordinal-level measurement, all measurement scales used to produce quantitative measurement variables have uniform and standard *units of measurement*. These units both identify the type of observable variable being measured (e.g., length, mass, time, temperature) and give a distance on the measurement scale as a standard reference for comparisons between measurements. The two basic systems of units used in this book are the *English system* (e.g., inch, pound, second) and the *metric system* (or *International System of Units;* e.g., meter, gram, second).

2.4 NOMINAL-LEVEL MEASUREMENT

Nominal-level measurement is the most basic level of measurement, in which the things being measured are simply classified into unique categories. These categories are *mutually exclusive* (no thing can be placed in more than one category) and *totally inclusive* (every thing can be placed in at least one category). Mathematically, the property of being classifiable into one and only one category can be symbolized by the equal-to and not-equal-to symbols ($=$, $\neq$). Categories on nominal scales are not ordered in any way (e.g., from small to large), and numbers are used only as labels for categories. Thus, car license numbers are an example of a nominal scale. The minimum number of categories on a nominal scale is two (e.g., whether a coin lands heads or tails) and there can be as many categories as needed. Other examples of nominal scales are: type of fish (e.g., shark, flounder, trout); presence or absence of disease; and type of industrial injury.

2.5 ORDINAL-LEVEL MEASUREMENT

Ordinal-level measurement is the next level above nominal. Its scales retain the nominal level property of classifying things into one and only one category ($=$, $\neq$), but now the categories are ordered: ranked according to the magnitude of the characteristic being measured. Each category can now be said to be greater than ($>$) or less than ($<$) its neighbor, depending on the amount of the characteristic it represents. Some examples of ordinal scales are: ranking the size of a set of objects on a three-number scale ($1 =$ small, $2 =$ medium, $3 =$ large); ranking the quality of movies on a five-number scale (from $1 =$ very bad, to $5 =$ excellent); and ranking the aggressiveness of children at play on a ten-number scale (from $1 =$ unaggressive, to $10 =$ very aggressive).

While ordinal scales produce quantitative measurement variables, these variables are not *isomorphic* (identical) to the underlying observable variable because they do not have standard and uniform measurement units. Instead, intervals on ordinal scales are determined subjectively and thus may differ for all users of the scale. Ordinal measurement, therefore, can indicate only the relative amount of a characteristic in each thing being measured, but not exactly how much more of the characteristic one thing has versus another.

2.6 INTERVAL-LEVEL MEASUREMENT

Interval level is the next higher level of measurement above ordinal level. Its scales include the properties of nominal ($=$, $\neq$) and ordinal ($<$, $>$) scales, and in addition have uniform and standard reference units. Such units eliminate subjectivity in quantitative measurements, producing scales with constant and equal intervals. With such interval scales it is possible to determine exact distances between two things with regard to the characteristic being measured, by addition or subtraction between the scale values ($+$, $-$). Interval scales always produce quantitative measurement variables that are isomorphic with the observable variable being measured, with the exception that interval scales have *arbitrary* and not *absolute zero points*.

One example of an interval scale is the *Celsius* (or *centigrade*) *scale* for temperature. On this scale, zero temperature ($0°C$) is arbitrarily defined as the freezing point of water, and the unit ($°C$) is then defined as 1/100th of the distance on the scale to the boiling point of water ($100°C$). Exact distances can be determined on the Celsius scale through addition or subtraction; thus it can be said that an object that is 40°C is 10°C hotter than an object that is 30°C (40°C − 30°C = 10°C).

Another example of an interval scale is the *Fahrenheit scale* for temperature, which again uses the freezing and boiling points of water to determine the zero values and the scale units (°F). On the Fahrenheit scale, however, the distance from the freezing point (32°F) to the boiling point (212°F) is 180°F, and zero (0°F) is 32°F below the freezing point. Other examples of interval scales are various time scales, such as the calendar year with its culturally determined zero point, and time of day on the 24-hour clock which arbitrarily has 12 midnight as its zero point.

2.7   RATIO-LEVEL MEASUREMENT

Ratio level is the highest level of measurement. Its scales include the properties of nominal ($=$, $\neq$), ordinal ($<$, $>$), and interval ($+$, $-$) scales, and now in addition also have *absolute zeros*. This means that at the zero value on a ratio scale, the characteristic being measured has decreased to the point where it is not present or at least it is not observable. Because numbers on such scales now represent distances from an absolute zero, it is legitimate to calculate ratios between measurements on the scale: to express one measurement as a multiple of another.

An example of a ratio scale is the *Kelvin scale* for temperature. Zero on the Kelvin scale is an absolute zero, defined as the temperature at which no pressure can be detected in an ideal gas: when the average kinetic energy per gas molecule is zero. The unit on the Kelvin scale [degree Kelvin (K); by convention a degree symbol (°) is not used with Kelvin temperature] is the same distance as the unit on the Celsius scale (1/100 of the distance between the freezing and boiling points of water), but zero on the Kelvin scale is equivalent to −273.15°C. While ratios are not legitimate between either Celsius or Fahrenheit measurements, it is legitimate to calculate ratios between Kelvin measurements. Thus, for example, it can be said that 300 K is twice as hot as 150 K.

Other examples of ratio scales are: weight in grams, length in centimeters, time in seconds, miles per hour, and many other scales in common use.

EXAMPLE 2.2   Demonstrate, by converting the Kelvin measurements 300 K and 150 K into equivalent °C values, why ratios are not legitimate on the ordinal-level Celsius scale.

Solution

The absolute zero on the Kelvin scale allows legitimate ratios whereas the arbitrary zero on the Celsius does not. To demonstrate, we convert 300 K and 150 K to equivalent °C values, using the following relationship between the two scales: °C = K − 273.15. Therefore for 300 K

$$°C = 300 − 273.15 = 26.85$$

and for 150 K

$$°C = 150 − 273.15 = −123.15$$

The resulting ratio, 26.85°C/−123.15°C = −0.218, leads to the meaningless statement that 26.85°C is −0.218 times hotter than −123.15°C.

2.8   CONTINUOUS AND DISCRETE MEASUREMENT VARIABLES

To understand how *continuous measurement variables* differ from *discrete measurement variables*, consider the following measurement variables you could acquire from your statistics class: the height of each student in centimeters, which is a continuous measurement variable; and the number of students attending each lecture, which is a discrete measurement variable.

Height in centimeters is a continuous measurement variable because the characteristic of height (the observable variable) is a *continuum* without gaps. This means that, theoretically, there is an infinite number of possible intermediate measurements between any two height measurements. To see why this is true, consider the height of 165.2 cm, which indicates that your measuring technique is sensitive to tenths of a centimeter. Say, however, you want a better measurement, and with considerable effort you increase the sensitivity to thousandths of a centimeter, and get the value 165.195 cm. Theoretically, because height is a continuum without gaps, you could continue this process of increasing the sensitivity until you achieved a measurement with an infinite number of decimal places: 165.19534216239867...cm. For continuous measurement variables, there is theoretically (but not practically) an infinite number of possible values between any two values on the scale.

The most common discrete measurement variables (also called *discontinuous* or *meristic measurement variables*) are obtained by counting the number of things in some set of things. Thus, the number of students attending each lecture is a discrete measurement variable because the characteristic of number of students (the observable variable) is not a continuum, there are gaps in this characteristic. If there are 52 students enrolled in the course, then the number attending a lecture could be 52, or 37, or 25, but not 52.1 or 48.639. The measurements on a discrete measurement variable must be one of a fixed set of values, without the possibility of intermediate values. Counting is ratio-level measurement, because it has all the properties of nominal ($=$, $\neq$), ordinal ($<$, $>$), and interval ($+$, $-$) measurement, a scale unit (the number one), and an absolute zero. It is called *discrete ratio-level measurement* to distinguish it from *continuous ratio-level measurement* (e.g., height of students in centimeters).

EXAMPLE 2.3 The objects in Fig. 2-1 can be measured on each of the four levels of measurement. Give a measurement scale that could be used on these objects for each of the following types of measurement: (*a*) nominal, (*b*) ordinal, (*c*) interval, (*d*) continuous ratio, and (*e*) discrete ratio.

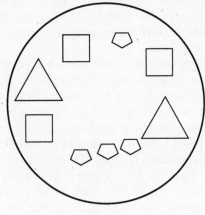

Fig. 2-1

Solution

(*a*) A possible nominal scale would be the three-category, object-shape scale: triangle, square, pentagon. This scale simply classifies the objects into unordered categories.

(*b*) A possible ordinal scale would be ranking the size of the objects on this three-number scale: 1 = small, 2 = medium, 3 = large. This scale classifies the objects into ordered categories that are separated by subjective intervals.

(*c*) A possible interval scale would be the position of each object on a 360° circular scale that shows compass direction. This scale has units, the 360 equal intervals called degrees, but an arbitrary zero.

(*d*) A possible continuous ratio scale would be the length of the longest dimension of each object with a scale that has inch or centimeter units.

(*e*) A possible discrete ratio scale would be a count of the number of objects in the figure.

2.9 TYPES OF DATA

In general, *data* is the term used to describe sets of factual information collected as part of some study. In statistics, the term refers to sets of measurements. Using the categories from Sections 2.3 through 2.8, such measurement sets can be described as *quantitative data* or *qualitative data, discrete data* or *continuous data,* or by their level of measurement (e.g., *interval-level data*).

Data can also be described as *classificatory, categorical*, or *enumeration*. These terms are different names for the same type of data: counts, for a group of things being measured, of how many of them can be placed in each category of a nominal measurement scale. An example of such data would be measuring 10 flowers on a three-category color scale (red, blue, yellow), and finding that there were 3 red, 4 blue, and 3 yellow. They are called classificatory or categorical data because on nominal scales things are simply classified into categories. They are called enumeration data because the number of things classified into each category is counted (enumerated). This use of counting is not the same as discrete ratio-level measurement (see Section 2.8), where the possible count-numbers themselves are the categories on the measurement scale.

2.10 THE APPROXIMATE NATURE OF MEASUREMENT

All real-life continuous data (see Sections 2.8 and 2.9) only approximate the *true measurements* and are therefore called *approximate measurements*. If, for example, you followed the operational definition of height measurement in Example 2.1 and measured the heights of members of your statistics class, one height might be 172.7 cm. This result states that the height is 172 cm plus roughly 7/10 of the distance between 172 cm and 173 cm. While the true measurement is an exact distance somewhere within that interval, with an infinite number of decimal places, using a relatively crude instrument (meter stick) you can only estimate it to be 7/10 of the distance. By increasing the sensitivity of your measuring instrument you could add on more and more decimal places, getting closer and closer to the true measurement, but in reality there is a limit to even the most powerful measuring instruments. Therefore, all continuous data must be considered approximate measurements.

For every approximate measurement, the last digit of the measurement (sometimes called the *doubtful digit*) is typically at the limit of sensitivity of the measuring instrument. Thus, for the height measurement 172.7 cm, the first three digits were easily acquired from the meter stick and are fairly certain, but the fourth and last digit is an uncertain estimate: 7/10 of the distance between 172 cm and 173 cm. By convention, this estimated value, while stated as a single digit, is actually considered to be an interval called the *implied range*. This range has the estimated value as its midpoint and extends above and below the estimated value by one-half of the smallest scale unit. Here, the smallest scale unit is 0.1 cm so the implied range is 172.65 cm to 172.75 cm. This form is typical of how the implied range is stated, and the form that will be used in this book. It should be noted, however, that some statisticians, to avoid overlapping boundaries between adjacent intervals, would write this range as 172.65000...cm to 172.74999...cm.

EXAMPLE 2.4   What is the implied range of this approximate measurement: 1.2965 mg?

> **Solution**
>
> A method for determining the implied range for any approximate measurement is:
>
> (1) Using the smallest measurement unit, set up a three-value scale that has the measurement as its midpoint and values one unit above and below this midpoint.
>
> (2) Determine the halfway points between the three values, which are the lower and upper boundaries of the implied range.
>
> Therefore here

3-value scale	*halfway points*	*implied range*
1.2964		
	← 1.29645	
1.2965		1.29645 mg to 1.29655 mg
	← 1.29655	
1.2966		

Discrete ratio-level measurements (see Section 2.8) can produce *exact measurements*. Thus, if one counts the number of eggs in a bird's nest, there might be exactly 7 but not $6\frac{1}{2}$ or $7\frac{1}{4}$. It is often the case, however, that counts of very large numbers of things are estimated, and these estimates are approximate measurements. Thus, for example, a soil biologist might estimate the number of organisms in a cubic meter of soil to be 36,000 to the nearest 1,000, or an economist might estimate the number of new houses started in the United States in the month of June to be 230,000 to the nearest 10,000.

2.11 SIGNIFICANT DIGITS

The *significant digits* (also called *significant figures*) in a measurement are all the digits actually obtained from the measurement scale. The nonsignificant digits are those zeros in a measurement that are there only to indicate the position of the decimal point. For measurements that are larger than or equal to one: All nonzero digits are significant; zeros between significant digits are significant; and if there is a decimal point, then zeros to the right of the last nonzero digit are significant. For measurements that are between zero and one, the same rules apply with the exception that: All zeros between the decimal and the first nonzero digit to its right are nonsignificant.

EXAMPLE 2.5   How many significant digits are there in the following approximate measurements: (*a*) 1.12 mg, (*b*) 1.02 g,   (*c*) 920.02080 mi,   (*d*) 0.0900 mg?

Solution

(*a*) 3,   (*b*) 3,   (*c*) 8,   (*d*) 3

EXAMPLE 2.6   Indicate the implied range and number of significant digits in the following measurements: (*a*) 102 horses,   (*b*) 10,100 insects to the nearest 100.

Solution

(*a*)   102 horses is an exact measurement, and therefore does not have an implied range. It has three significant digits.

(*b*)   10,100 insects to the nearest 100 is an estimated discrete ratio measurement (see Section 2.10) that has an implied range of 10,050 insects to 10,150 insects, and has three significant digits.

2.12 SCIENTIFIC NOTATION AND ORDER OF MAGNITUDE

Scientific notation expresses any number as the product of two factors. The first factor is a number with only one nonzero digit to the left of the decimal point, and the second factor is the appropriate power of 10 (see Section 1.9). Thus, for example, the number 100.0 in scientific notation would be 1.000×10^2, and the number 0.36 would be 3.6×10^{-1}. When a number is written in scientific notation, its power of 10 indicates its size, and is called the *order of magnitude* of the number. Numbers with the same power of 10 are of the same order of magnitude. Therefore, for example, 1.0×10^3, 4.213×10^3, and 9.237456×10^3 are all of the same order of magnitude.

EXAMPLE 2.7   For the following approximate measurement written in scientific notation, which are the significant digits and what is the implied range: 4.2×10^3 m?

Solution

By convention, all of the significant digits in a measurement written in scientific notation are placed in the first factor, and there are no nonsignificant digits in the first factor. Therefore, for this measurement the significant digits are: 4 and 2.

To determine implied ranges for measurements written in scientific notation, use the technique given in Example 2.4 on the first factors, and then multiply each of the resulting values by the second factors.

$$\begin{array}{c} 4.1 \\ \end{array}$$

Therefore, for 4.2×10^3 m, the three-value scale would be 4.2 the halfway points would be 4.15 to
$$\begin{array}{c} 4.3 \end{array}$$
4.25, and the implied range would be 4.15×10^3 m to 4.25×10^3 m.

2.13 SYSTEMATIC AND RANDOM ERRORS OF MEASUREMENT

In Section 2.10 we indicated that all real-life continuous data only approximate the true measurements and are therefore called approximate measurements. One reason given for this was the sensitivity (or *detection*) limit for even the most powerful measuring instrument. Thus, no matter how many decimal places the instrument provides, it can only crudely approximate the infinite number of decimal places in the true measurement. Besides this sensitivity limit, another factor that separates any real-life measurement from its true value is error of measurement present in the real-life data. Such errors may be *systematic* or *random*.

Systematic errors of measurement, also called *bias*, result from flaws in the measurement procedure that consistently produce distortions in one direction, making the measurements either always too large or always too small. If, for example, in following the operational definition in Example 2.1 the upper meter stick was accidentally taped one centimeter below the top of the lower stick, then all of the height measurements would consistently be one centimeter too large. Such bias is typically hard to detect, but it can be done if the measurement can be compared to some empirical or theoretical standard.

The other type of error that can occur in any approximate measurement is random (or *chance*) error. Such errors are produced by random variations in the measurement procedures that create inconsistent, nonsystematic distortions from measurement to measurement; the measurements are sometimes too large and sometimes too small. If, for example, you measured the height of the same student ten times, you could get random errors produced by inconsistent changes in his posture against the scale or your viewing angle from the ladder.

Because of the limits of sensitivity and these possibilities for error, each approximate measurement should be thought of as this equation.

$$\text{(Approximate measurement)} = \text{(True measurement to limit of sensitivity)} + \text{(Systematic error)} + \text{(Random error)}$$

If a discrete ratio-level measurement (see Section 2.8) is a direct count of a limited set of things (e.g., 6 eggs in a bird's nest, or 36 students at a lecture), then it is unlikely this count would include either systematic or random errors of measurement. If, however, the count is an estimate (see Section 2.10), then it may include such errors.

2.14 ACCURACY AND PRECISION IN STATISTICS

In statistics, the phrase *accuracy of a measurement* refers to how close the measurement is to the true measurement. From Section 2.13 it can be seen that accuracy is determined by both the sensitivity of the measuring instrument and the presence of errors of measurement, particularly systematic errors.

The *precision of a measurement* has quite a different meaning in statistics. It refers to the similarity of repeated "identical" measurements of the same things. Systematic errors typically remain the same from measurement to measurement, so precision, or the variability of repeated measurements, is determined primarily by the presence and amount of random errors.

From these definitions you can see that a measurement can be accurate and precise (close to the true measurement and repeatable with minimum variation); inaccurate and imprecise (far from the true measurement with great variation in repeated measurements); accurate and imprecise; or inaccurate and precise.

The significant digits reported in an approximate measurement are a statement about the accuracy (closeness to the true value) of the measurement. If, for example, you report the weight of an object to be

4.32 mg, then what you are saying is that you are reasonably certain that the object weighs 4.3 mg, and that the true value lies somewhere in an *implied interval* from 4.315 mg to 4.325 mg, a range of 0.01 mg. Because of what it says about accuracy, the implied interval is often called the *implied range of accuracy*. Assuming little or no bias, the accuracy available for the measurement is determined by the sensitivity limit of the measuring instrument as well as the presence and amount of precision-limiting random error.

2.15 ACCURACY AND PRECISION IN THE PHYSICAL SCIENCES

The statistical definitions of these concepts (see Section 2.14) come from statistical estimation theory which is introduced in Chapter 14. The definitions for these measurement properties that are used in the physical sciences (chemistry, physics, etc.) are related but somewhat different.

In the physical sciences the level of accuracy of a measurement is the number of significant digits in the measurement, while the level of precision of a measurement is the size of its smallest measuring unit. Thus, 4.32 mg, 43.2 mg, and 432 mg have all been measured at the same level of accuracy, which can be expressed in words as an accuracy of three significant digits. And 423.3 mg, 43.2 mg, and 4.3 mg have been measured at the same level of precision, which can be expressed in words as precise to the nearest 0.1 mg.

EXAMPLE 2.8 The following is a set of measurements written in scientific notation: 2.531×10^2 cm, 2.531×10 cm, 2.5316×10^2 cm, and 2.53167×10^3 cm. Using the physical-sciences definition, which of these measurements are at the same levels of: (a) accuracy, (b) precision, and (c) order of magnitude?

Solution

(a) 2.531×10^2 cm (or 253.1 cm) and 2.531×10 cm (or 25.31 cm) both have an accuracy of four significant digits.

(b) 2.531×10 cm (or 25.31 cm), 2.5316×10^2 cm (or 253.16 cm), and 2.53167×10^3 cm (or 2,531.67 cm) are all precise to the nearest 0.01 cm.

(c) 2.531×10^2 cm and 2.5316×10^2 cm both have 2 as their power of ten, so they are at the same order of magnitude.

When adding or subtracting approximate measurements, the answer should be rounded off (see Section 1.4) to the same level of precision (physical-sciences definition) as the least precise measurement in the problem. When multiplying or dividing with approximate measurements, the answer should be rounded off to the same level of accuracy (physical-sciences definition) as the least accurate measurement in the problem.

EXAMPLE 2.9 Perform the indicated algebraic operations and then round off the answers to the correct number of digits: (a) 7.123 kg + 8.9 kg, (b) 72 kg × 0.01 kg.

Solution

(a) 7.123 kg + 8.9 kg = 16.023 kg, which should be rounded off to 16.0 kg

(b) 72 kg × 0.01 kg = 0.72 kg^2, which should be rounded off to 0.7 kg^2

2.16 UNIT CONVERSIONS

The *metric system*, also called the *International System of Units*, or *SI units*, has the reference units meter (m), gram (g), and second (sec). Standard multiples and submultiples of these units have specific names and abbreviations.

EXAMPLE 2.10 Give the names and abbreviations for the following quantities: (a) 10^{-2} m, (b) 10^{-3} sec, (c) 10^{-6} g, (d) 10^{-9} m, (e) 10^3 g, (f) 10^6 sec (g) 10^9 m.

Solution

(a) 10^{-2} m is centimeter, abbreviated cm

(b) 10^{-3} sec is millisecond, abbreviated msec

(c) 10^{-6} g is micrograms, abbreviated μg

(d) 10^{-9} m is nanometer, abbreviated nm

(e) 10^{3} g is kilogram, abbreviated kg

(f) 10^{6} sec is megasecond, abbreviated Msec

(g) 10^{9} m is gigameter, abbreviated Gm

The standard method for converting *metric system (SI) units* into *English units*, or vice versa, is to multiply the measurement by the appropriate conversion factor (see Table 2.1).

Table 2.1

Conversion Factors	
English to SI	SI to English
1 inch = 2.540 cm	1 cm = 0.3937 in
1 inch = 2.540 × 10^{-5} km	1 km = 3.937 × 10^{4} in
1 ft = 0.3048 m	1 m = 3.281 ft
1 mi = 1.609 km	1 = 0.6215 mi
1 lb = 0.4536 kg	1 kg = 2.205 lb
1 lb = 453.6 g	1 g = 2.205 × 10^{-3} lb
1 gal = 3.785 liters (l)	1 l = 0.2642 gal

EXAMPLE 2.11 Using the appropriate conversion factor, determine: How far is 100 yards in centimeters?

Solution

$$100 \text{ yd} \times \left(3\frac{\text{ft}}{\text{yd}}\right) \times \left(12\frac{\text{in}}{\text{ft}}\right) \times \left(2.540\frac{\text{cm}}{\text{in}}\right) = 300 \text{ ft} \times \left(12\frac{\text{in}}{\text{ft}}\right) \times \left(2.540\frac{\text{cm}}{\text{in}}\right)$$

$$= 3,600 \text{ in} \times \left(2.540\frac{\text{cm}}{\text{in}}\right)$$

$$= 9,144 \text{ cm, or } 9.144 \times 10^{3} \text{ cm}$$

This number is not rounded off because the only approximate number in the calculation, the conversion factor, is given to four significant digits.

Solved Problems

LEVELS AND UNITS OF MEASUREMENT

2.1 Why do the physical sciences restrict the term "measurement" to interval- and ratio-level measurement?

Solution

In chemistry, physics, and other physical sciences, all measurements must include both a numerical value and a unit (e.g., 20 feet, or 15 pounds). In most of the social sciences (psychology, sociology, economics, etc.), however, measurements are often taken at the two levels without units, below interval and ratio. Therefore, in this book we use the more general definition of measurement that includes nominal and ordinal levels.

2.2 Determine the actual set of enumeration data from the objects in Fig. 2-1 for the nominal scale proposed in Example 2.3(*a*).

Solution

For the suggested three-category, object-shape scale, the enumeration data would be: 2 triangles, 3 squares, and 4 pentagons.

2.3 For each of the following, first indicate its level of measurement (nominal, ordinal, interval, continuous ratio, or discrete ratio) and then explain your choice: (*a*) the attitude of Americans toward immigrants as measured on a five-number scale from 1 (= unfavorable) to 5 (= highly favorable), (*b*) the gender of 40 clerks in a department store, (*c*) the types of birds that arrive at a feeder each day, (*d*) the day in the year when each of 50 students was born, (*e*) the time it takes for a woman runner to complete a 100-meter dash, (*f*) the body temperature in °C.

Solution

(*a*) Ordinal; this five-number attitude scale has ordered categories but the intervals between categories are subjective.

(*b*) Nominal; gender is an unordered, two-category scale: male, female.

(*c*) Nominal; types of birds form an unordered scale with as many categories as needed.

(*d*) Interval; calendar year has a unit of measurement (days) but an arbitrary zero.

(*e*) Continuous ratio; this time dimension has a standard unit (seconds) and an absolute zero (the start of the race).

(*f*) Interval; temperature measured on the Celsius scale has a standard unit (°C) but an arbitrary zero.

2.4 For each of the following, indicate the level of measurement (nominal, ordinal, interval, continuous ratio, or discrete ratio): (*a*) the price per gallon of gas in selected areas, (*b*) the type of vitamin (e.g., vitamin E), (*c*) the jersey number assigned to each member of a football team, (*d*) the sweetness of apples, judged on a four-number scale from 1 (= not sweet) to 4 (= very sweet), (*e*) the amount of sugar (in milligrams) in each of a set of apples.

Solution

(*a*) Discrete ratio

(*b*) Nominal

(*c*) Nominal

(*d*) Ordinal

(*e*) Continuous ratio

THE APPROXIMATE NATURE OF MEASUREMENT

2.5 What is the implied range of the approximate measurement 0.19032 cm?

Solution

Using the technique from Example 2.4:

3-value scale	halfway points	implied range
0.19031		
	← 0.190315	
0.19032		0.190315 cm to 0.190325 cm
	← 0.190325	
0.19033		

2.6 What is the implied range of the approximate measurement 700.3 kg?

Solution

Using the technique from Example 2.4:

3-value scale	halfway points	implied range
700.2		
	← 700.25	
700.3		700.25 kg to 700.35 kg
	← 700.35	
700.4		

2.7 What are the implied ranges for these approximate measurements: (a) 1,000,000.0 g, (b) 0.0001 m?

Solution

(a) 999,999.95 g to 1,000,000.05 g

(b) 0.00005 m to 0.00015 m

SIGNIFICANT DIGITS

2.8 How many significant digits are there in the following approximate measurements: (a) 1.20 kg, (b) 1.00000 cm,   (c) 0.0056 mg,   (d) 0.04003 mg?

Solution

(a) 3,   (b) 6,   (c) 2,   (d) 4

2.9 Indicate the implied range and the number of significant digits in the following measurements: (a) 100,000 trees,   (b) 100,001 trees,   (c) 100,000 trees.

Solution

(a) For a number that does not have a decimal point, the significant digits can be indicated by placing a dot over the last significant digit. Therefore, for 100,000 trees, the implied range is 99,500 trees to 100,500 trees, and there are three significant digits.

(b) 100,001 trees is an exact measurement, with no implied range and six significant digits.

(c) Another way to indicate significant zeros in a number that does not have a decimal point is to make the nonsignificant zeros smaller than the significant zeros. Therefore, for 100,000 trees the implied range is again 99,500 trees to 100,500 trees, and again there are three significant digits.

2.10 Indicate the implied range and number of significant digits in the following measurements: (a) 10,000 workers,   (b) 103,000.0 mi.

Solution

(a) As given, this measurement does not provide sufficient information to determine implied range or number of significant digits.

(b) The implied range is 102,999.95 mi to 103,000.05 mi, and there are seven significant digits.

SCIENTIFIC NOTATION AND ORDER OF MAGNITUDE

2.11 For the following measurement written in scientific notation, which are the significant digits and what is the implied range: 8.7961×10^{-2} m?

Solution

The significant digits are: 87961.

Using the technique from Example 2.4 for determining the implied range: For 8.7961×10^{-2} m, the

8.7960

three-value scale would be 8.7961, the halfway points would be 8.79605 and 8.79615, and therefore the

8.7962

implied range would be 8.79605×10^{-2} m to 8.79615×10^{-2} m.

2.12 For the following measurements written in scientific notation, which are the significant digits and what are the implied ranges: (a) 9.99×10^{-6} kg,   (b) 2.0×10^{6} kg?

Solution

(a) The significant digits are 999, and the implied range is 9.985×10^{-6} kg to 9.995×10^{-6} kg.

(b) The significant digits are 20, and the implied range is 1.95×10^{6} kg to 2.05×10^{6} kg.

2.13 Express in scientific notation the following number written in *decimal notation:* 0.000000060 mm.

Solution

To convert a decimal number that is less than one to scientific notation, move the decimal point to the right until there is one nonzero digit to its left. For 0.000000060 mm, the decimal point has to be moved eight places 000000006.0. The first factor then, which includes all significant digits in the measurement, is 6.0. The second factor is the necessary power of 10, which here would have a negative exponent equal to the number of places that the decimal point was moved: 10^{-8}. Therefore, 0.000000060 mm expressed in scientific notation is 6.0×10^{-8} mm.

2.14 Express in scientific notation the following number written in decimal notation: 4,000,000,000.0 kg.

Solution

To convert a decimal number that is 10 or larger to scientific notation, move the decimal to the left until only one nonzero digit is to its left. For 4,000,000,000.0 kg, the decimal has to be moved nine places 4.0000000000. The first factor then, including all significant digits, is 4.0000000000. The second factor is a positive power of ten with an exponent equal to the number of places the decimal was moved: 10^{9}. Therefore, 4,000,000,000.0 kg is written in scientific notation as $4.0000000000 \times 10^{9}$ kg.

2.15 Express in decimal notation the following measurement written in scientific notation: 4.92×10^{-7} m.

Solution

When the scientific notation includes a negative power of 10, it is converted to decimal notation by moving the decimal in the first factor to the left the number of places equal to the negative exponent. Therefore, 4.92×10^{-7} m is 0.000000492 m.

2.16 Express in decimal notation the following measurement written in scientific notation: 6.0×10^7 kg.

Solution

When the scientific notation includes a positive power of 10, it is converted by moving the first-factor decimal point to the right the number of places equal to the exponent. Therefore, 6.0×10^7 kg is 60,000,000 kg.

ACCURACY AND PRECISION IN THE PHYSICAL SCIENCES

2.17 Perform the indicated algebraic operations and then round off the answers to the correct number of digits: (*a*) 9.99623 kg $-$ 8.12 kg, (*b*) 9.99 kg $\div$ 8 kg, (*c*) 4.23 kg $\times$ 100.0039 kg.

Solution

(*a*) 9.99623 kg $-$ 8.12 kg $=$ 1.87623 kg, which should be rounded off to 1.88 kg

(*b*) 9.99 kg $\div$ 8 kg $=$ 1.24875, which should be rounded off to 1

(*c*) 4.23 kg $\times$ 100.0039 kg $=$ 423.016497 kg^2, which should be rounded off to 423 kg^2

2.18 Perform the indicated algebraic operations and then round off the answers to the correct number of digits: (*a*) 91.26 g $\times$ 1.1 g, (*b*) 452.1 g $-$ 21.239 g.

Solution

(*a*) 91.26 g $\times$ 1.1 g $=$ 100.386 g^2, which should be rounded off to 100 g^2

(*b*) 452.1 g $-$ 21.239 g $=$ 430.861 g, which should be rounded off to 430.9 g

2.19 Perform the indicated algebraic operations and then round off the answers to the correct number of digits: (4 mm $+$ 2.92 mm) $\times$ 8.397 mm.

Solution

Using the order of operation rules (see Section 1.3), the first step would be the addition within the parenthesis: 4 mm $+$ 2.92 mm $=$ 6.92 mm; which when rounded off to the level of the least precise measurement is 7 mm. This number is then multiplied by 8.397 mm: 7 mm $\times$ 8.397 mm $=$ 58.779 mm^2; which should then be rounded off to one significant digit: 60 mm^2.

2.20 Perform the indicated algebraic operations and then round off the answer to the correct number of digits: 72.916 mm $\times$ 4.21 mm $-$ 6 mm^2.

Solution

The order of operation rules indicate that in a sequence involving multiplication and subtraction, the multiplication should be done first. So: 72.916 mm $\times$ 4.21 mm $=$ 306.97636 mm^2; which should then be rounded off to 307 mm^2. Now, subtracting 6 mm^2 from this gives 301 mm^2, which should be left as is because it is at the proper level of precision.

2.21 Perform the indicated algebraic operations and then round off the answer to the correct number of digits: $(3.926 \times 10^2 \text{ kg}) \times (4.29 \times 10^3 \text{ kg})$.

Solution

Using a hand calculator, two ways to perform multiplications and divisions with numbers written in scientific notation are: (1) directly on the calculator, using the scientific notation key; or (2) by doing the algebra on the component parts of the scientific notation. Solving it directly on the calculator:

$$(3.926 \times 10^2 \text{ kg}) \times (4.29 \times 10^3 \text{ kg}) = 1,684,254.00 \text{ kg}^2$$

which should be rounded off to 1,680,000 kg^2, or 1.68×10^6 kg^2.

Doing the algebra on the component parts:

$$(3.926 \times 10^2 \text{ kg}) \times (4.29 \times 10^3 \text{ kg}) = (3.926 \times 4.29)(10^2 \times 10^3)(\text{kg} \times \text{kg})$$

$$= 16.84254 \times 10^5 \text{ kg}^2$$

$$= 1.684254 \times 10^6 \text{ kg}^2$$

which should be rounded off to 1.68×10^6 kg^2.

2.22 Perform the indicated algebraic operations and then round off the answers to the correct numbers of digits: $(2.9 \times 10^{-3} \text{ ft}) + (4.26 \times 10^{-4} \text{ ft})$.

Solution

Using a hand calculator, two ways to perform additions and subtractions on numbers written in scientific notation are: (1) directly on the calculator using the scientific notation key, or (2) by converting the numbers to decimal notation. Solving it directly on the calculator:

$$(2.9 \times 10^{-3} \text{ ft}) + (4.26 \times 10^{-4} \text{ ft}) = 0.003326 \text{ ft}$$

which should be rounded off to 0.0033 ft, or 3.3×10^{-3} ft.

Solving it by converting the numbers to decimal notation:

$$(2.9 \times 10^{-3} \text{ ft}) + (4.26 \times 10^{-4} \text{ ft}) = 0.0029 \text{ ft} + 0.000426 \text{ ft}$$

$$= 0.003326 \text{ ft}$$

which should be rounded off to 0.0033 ft, or 3.3×10^{-3} ft.

UNIT CONVERSIONS

2.23 One side of a house is 9 meters long. Express this distance in: (a) centimeters,   (b) millimeters.

Solution

(a) $9 \text{ m} \times \left(\dfrac{100 \text{ cm}}{1 \text{ m}}\right) = 900 \text{ cm}$

(b) $9 \text{ m} \times \left(\dfrac{1,000 \text{ mm}}{1 \text{ m}}\right) = 9.000 \text{ mm}$

2.24 Using the appropriate conversion factors, determine: How many feet are there in 350 kilometers?

Solution

$$350 \text{ km} \times \left(1{,}000 \ \frac{\text{m}}{\text{km}}\right) \times \left(3.281 \ \frac{\text{ft}}{\text{m}}\right) = (3.50000 \times 10^5 \text{ m})\left(3.281 \ \frac{\text{ft}}{\text{m}}\right)$$

$$= 1.148350 \times 10^6 \text{ ft}$$

which should be rounded off to 1.148×10^6 ft

2.25 Using the appropriate conversion factor, determine: How many pounds are there in 12.9 kilograms?

Solution

$$12.9 \text{ kg} \times \left(2.205 \ \frac{\text{lb}}{\text{kg}}\right) = 28.4445 \text{ lb}$$

which should be rounded off to 28.44 lb

2.26 Using the appropriate conversion factor, determine: How many kilograms are there in 9,920 pounds?

Solution

$$\left(9.920 \times 10^3 \text{ lb}\right)\left(4.536 \times 10^{-1} \ \frac{\text{kg}}{\text{lb}}\right) = 4.49971 \times 10^3 \text{ kg}$$

which should be rounded off to 4.500×10^3 kg

2.27 Using the appropriate conversion factor, determine: How many liters are there in 20 gallons?

Solution

$$20 \text{ gal} \times \left(3.785 \ \frac{\text{liter}}{\text{gal}}\right) = 75.70 \text{ liters}$$

Supplementary Problems

LEVELS AND UNITS OF MEASUREMENT

2.28 For each of the following, indicate its level of measurement: (*a*) the diameters, in millimeters, of a set of snail shells, (*b*) the grades of student essays on a six-number scale from 1 (= very good) to 6 (= bad), (*c*) the yearly sales of passenger cars from one manufacturer.

Ans. (*a*) Continuous ratio, (*b*) ordinal, (*c*) discrete ratio

2.29 For each of the following, indicate its level of measurement: (*a*) the number of words per minute in typing a sample section, (*b*) −273.15°C, (*c*) the handedness of a child (right- or left-handed).

Ans. (*a*) Discrete ratio, (*b*) interval, (*c*) nominal

2.30   For each of the following, indicate its level of measurement: (*a*) the relative temperatures of pieces of iron, using a four-color scale (gray = cold = 1,   yellow = warm = 2,   red = hot = 3,   and   white = very hot = 4),   (*b*) the sales volume of a product in dollars per month,   (*c*) the number of workers in a labor force who are unemployed.

   Ans.   (*a*) Ordinal,   (*b*) discrete ratio,   (*c*) discrete ratio

THE APPROXIMATE NATURE OF MEASUREMENT

2.31   What are the implied ranges for these approximate measurements: (*a*) 4,926.22 cm,   (*b*) 0.1920 sec, (*c*) 41.00001 in?

   Ans.   (*a*) 4,926.215 cm to 4,926.225 cm,   (*b*) 0.19195 sec to 0.19205 sec, (*c*) 41.000005 in to 41.000015 in

SIGNIFICANT DIGITS

2.32   How many significant digits are there in the following approximate measurements: (*a*) 4,930,200.0 km, (*b*) 0.8001 mg?

   Ans   (*a*) 8,   (*b*) 4

2.33   How many significant digits are there in the following approximate measurements: (*a*) 0.0006000 mg, (*b*) 0.000047 mg?

   Ans   (*a*) 4,   (*b*) 2

2.34   Indicate the implied range and the number of significant digits in the measurement 52,000 moths.

   Ans.   The implied range is 51,500 moths to 52,500 moths, and there are two significant digits.

2.35   Indicate the implied range and the number of significant digits in the measurement 94,000 mi to the nearest 100.

   Ans.   The implied range is 93,950 mi to 94,050 mi, and there are three significant digits.

SCIENTIFIC NOTATION AND ORDER OF MAGNITUDE

2.36   For the following measurement written in scientific notation, which are the significant digits and what is the implied range: 1.0000×10^{-1} kg?

   Ans.   The significant digits are 10000, and the implied range is 9.9995×10^{-2} kg to 1.00005×10^{-1} kg.

2.37   Express this number in scientific notation: 80,888 m.

   Ans. 8.0888×10^4 m

2.38   Express this number in scientific notation: 0.90009 lb.

   Ans. 9.0009×10^{-1} lb

2.39   For the following measurements, express those written in scientific notation in decimal notation, and those written in decimal notation in scientific notation: (*a*) 2.000×10^{-2} mm,   (*b*) 1,001.00 kg.

   Ans.   (*a*) 0.02000 mm,   (*b*) 1.00100×10^3 kg

2.40   Express the following measurements in decimal notation: (*a*) 8.11×10^5 sec,   (*b*) 5.1×10^{-9} in.

   Ans.   (*a*) 811,000 sec,   (*b*) 0.0000000051 in

ACCURACY AND PRECISION IN THE PHYSICAL SCIENCES

2.41   Perform the indicated algebraic operations and then round off the answers to the correct number of digits: (*a*) 0.39247 kg + 0.0000007 kg + 0.21 kg,   (*b*) 2.1 kg ÷ 0.000056.

   Ans.   (*a*) 0.6024707 kg, which should be rounded off to 0.60 kg,
   (*b*) 37,500.0 kg, which should be rounded off to 38,000 kg

2.42   Perform the indicated algebraic operations and then round off the answers to the correct number of digits: (*a*) 1.26 g ÷ 312.92,   (*b*) 892 g + 2.263 g.

   Ans.   (*a*) 0.00402659 g, which should be rounded off to 0.00403 g,
   (*b*) 894.263 g, which should be rounded off to 894 g

2.43   Perform the indicated algebraic operations and then round off the answer to the correct number of digits: (6.1 in × 2.936 in) ÷ 18.23914 in^2.

   Ans.   18 in^2 ÷ 18.23914 in^2 = 0.99

2.44   Perform the indicated algebraic operations and then round off the answer to the correct number of digits: 3.2937 in^2 − 22.3 in^2 + 8.421 in × 39.213 in.

   Ans.   − 19.0 in^2 + 330.2 in^2 = 311.2 in^2

2.45   Perform the indicated algebraic operations and then round off the answer to the correct number of digits: $(1.926 \times 10^5 \text{ kg}) \div (9.1 \times 10^4)$.

   Ans.   2.11648 kg, which should be rounded off to 2.1 kg

2.46   Perform the indicated algebraic operations and then round off the answer to the correct number of digits: $(5.27 \times 10^3 \text{ ft}) - (8.838 \times 10^2 \text{ ft})$.

   Ans.   4,386.2 ft,   which should be rounded off to 4,390 ft, or 4.39×10^3 ft

UNIT CONVERSIONS

2.47   For the house in Problem 2.23, express the 9 meter distance in: (*a*) micrometers,   (*b*) kilometers.

   Ans.   (*a*) 9,000,000 μm,   (*b*) 0.009 km

2.48   Using the appropriate conversion factor, determine: How many kilometers are there in 111 miles?

   Ans.   178.599 km, which should be rounded off to 178.6 km

2.49   Using the appropriate conversion factor, determine: How many pounds are there in 103 grams?

   Ans. 2.27115×10^{-1} lb, which should be rounded off to 2.271×10^{-1} lb

2.50   Using the appropriate conversion factor, determine: How many gallons are there in 1,350 liters?

   Ans.   356.670 gal, which should be rounded off to 356.7 gal

Chapter 3

Populations, Samples, and Statistics

3.1 PHYSICAL AND MEASUREMENT POPULATIONS

The term *population* has many nonstatistical meanings. Thus, in biology the term denotes a group of individuals of the same species that live in the same geographic area and can or do interbreed. In the social sciences, it means all the people living in a country, region, or community. And in physics, it means all the particles at a particular energy level.

In statistics, the term population (or *universe*) has a different and very specific meaning related to the fundamental task of statistics: the analysis of measurement data. You will recall that every specific measurement is defined by an operational definition (see Section 2.2). This definition must include an exact description of the items being measured: age, time, temperature, place, or whatever feature characterizes the items. The set (see Section 1.17) of all such items that satisfy the description, in the past, present, or future, is called the *physical population* for this measurement. The set of these measurements taken from every conceivable member of the physical population is called the *measurement population*. Statistical techniques of data analysis focus on the measurement population, which is why *after this chapter, when the term population is used without modifiers it will mean measurement population.*

To understand these concepts let us examine a specific measurement. Say you are a plant geneticist who has developed a new type of corn plant, and one measurement you want to take is plant height at maturity in late summer. In your operational definition of this measurement, besides stating the specific measurement steps, you also exactly describe the plant to be measured: how the seed should be selected and prepared, growth conditions (soil preparation, planting technique, fertilizing and irrigating schedules, etc.), and that the height should be measured 4 months after planting. The set of all such plants grown under these conditions, in the past, present, and future, is the physical population for this measurement. The set of height measurements for all conceivable members of this physical population is the measurement population.

3.2 FINITE, INFINITE, AND HYPOTHETICAL POPULATIONS

Every item in a population is called an *element* of the population. If there is an upper limit to the number of elements in either a physical population or a measurement population, then the population is said to be *finite* (or *fixed*). If, on the other hand, there is no upper limit to the number of elements in a population, then the population is said to be *infinite*.

An example of a finite physical population would be the eight men who make the finals of a 100-meter running race in an Olympic Games. A finite measurement population linked to this finite physical population would be the time it takes for each man to run the 100 meters. An example of an infinite physical population would be the 4 month-old corn plants (past, present, and future) that satisfy the description in Section 3.1, and the heights of all these plants would be an infinite measurement population linked to this infinite physical population. It is generally the case that infinite populations do not actually exist but are instead *hypothetical* (or *imaginary*, or *conceptual*).

EXAMPLE 3.1 You roll a six-faced die and count the number of dots that appear on the upward face when the die stops rolling. From this information you can determine: (*a*) the item being measured, (*b*) the measurement taken, (*c*) the physical population for this measurement, (*d*) the linked measurement population, (*e*) whether these populations are finite or infinite.

Solution

(*a*)   The item being measured is the upward face of the die at the end of the roll.

(*b*)   The measurement is the discrete ratio measurement (see Section 2.8) of the number of dots.

(*c*)   The physical population for this measurement is the hypothetical population that includes all such final upward faces conceivably possible in the past, present, or future.

(*d*)   The measurement population is all dot counts from this physical population.

(*e*)   Both the physical population and its linked measurement population are infinite.

3.3   SAMPLES

As with the concept of a population, the term *sample* in statistics also refers to two interlinked sets of items: a *physical sample* and a *measurement sample*. The physical sample is any subset of a physical population, and any measurement taken on all members of a physical sample will produce a measurement sample. As the term population in future chapters will refer to a measurement population, so also *the term sample will refer to a measurement sample*.

Often whether a set of items is considered to be a physical population or a physical sample will depend on the context in which the items are examined. Thus, for example, the unemployed workers in the labor force of a large city could be a physical population if the interest is exclusively in the economic state of the specific city, or it could be a physical sample if there is a broader interest in the economic state of a region or country.

EXAMPLE 3.2   For the following samples, indicate whether they are physical samples or measurement samples: (*a*) yearly family income for 15 of the families in a 500-family housing unit,   (*b*) % fat in a batch of 10 sausages from the weekly output of a meat-processing plant,   (*c*) 5 trucks from a factory's daily truck production,   (*d*) 10 of the prisoners who have been executed in the United States since 1985.

Solution

(*a*)   Measurement sample

(*b*)   Measurement sample

(*c*)   Physical sample

(*d*)   Physical sample

3.4   PARAMETERS AND STATISTICS

Sets of measurements have characteristics that can themselves be measured and described, such as: the distance between the smallest and largest values, whether the data are evenly spread across the distance or densely clustered at one or more locations, the most typical or representative value in the data set, and so on. Any numerical value calculated from an entire measurement population that describes some characteristic of this population is called a *parameter* (or *population parameter*). Similarly, any such numerical descriptive measure calculated from a measurement sample is called a *statistic* (or *sample statistic*, or *statistical measure*).

The most familiar example of a numerical descriptive measure is the *arithmetic mean*, which measures the most representative or average value in a data set. This measure, which will be discussed in detail with other measures of *central tendency, average value*, and *location* in Chapter 6, is calculated for a set of measurements by taking the sum of these measurements and then dividing this sum by the number of measurements in the set. If the set is a measurement sample, then the arithmetic mean is a statistic calculated with the formula

$$\bar{x} = \frac{\sum_{i=1}^{n} x_i}{n} \tag{3.1}$$

where $\bar{x}$ (x with a bar over it, read "x bar") is the symbol for the sample mean, n is the symbol for the *sample size* (number of elements in the measurement sample), and $\sum\limits_{i=1}^{n} x_i$ is the familiar summation notation (see Section 1.22).

If the set of measurements is the entire measurement population, then the arithmetic mean is a parameter calculated with this formula

$$\mu = \frac{\sum\limits_{i=1}^{N} x_i}{N} \tag{3.2}$$

where μ (the Greek lowercase letter *mu*) is the symbol for the population mean, and N is the symbol for *population size* (number of elements in the measurement population).

Many such paired parameters and statistics are given throughout this book, and generally, as here with the arithmetic mean, the parameter will be symbolized by a Greek letter and the statistic will be symbolized by a Roman letter. For each pairing, the parameter of the measurement population will always have just one fixed value while the value of the statistic will vary from measurement sample to measurement sample.

EXAMPLE 3.3 Calculate arithmetic means for the following: (a) measurement sample $x_1 = 7$, $x_2 = 5$, $x_3 = 6$, $x_4 = 6$, (b) measurement population $x_1 = 0.2$, $x_2 = 0.6$, $x_3 = 0.4$.

Solution

$$(a) \quad \bar{x} = \frac{\sum\limits_{i=1}^{n} x_i}{n} = \frac{\sum\limits_{i=1}^{4} x_i}{4} = \frac{7+5+6+6}{4} = 6$$

$$(b) \quad \mu = \frac{\sum\limits_{i=1}^{N} x_i}{N} = \frac{\sum\limits_{i=1}^{3} x_i}{3} = \frac{0.2+0.6+0.4}{3} = 0.4$$

3.5 THE SCIENCE OF STATISTICS

It should now be clear that the term *statistics* has several meanings. It can refer to more than one numerical descriptive measure from sample data (see Section 3.4); it can refer to collections of facts; or it can refer to the science devoted to the analysis and interpretation of measurement data. It is this last meaning, the science of statistics, that is the subject of this book.

Facts have been collected since early in recorded history. The term statistics is derived from the Latin word for state (*status*), and it originally referred to government-acquired facts from different regions in a country, such as the taxes collected or the crops grown. This meaning of statistics has now been broadened to include any collection of facts, such as labor statistics (e.g., the number of unemployed automobile workers) or sports statistics (e.g., the number of hits made by a baseball player in one season).

The science of statistics deals with more than the accumulation of facts. In their pure and unprocessed form, masses of facts are overwhelming and essentially unusable. It is the science of statistics that provides the theories and techniques necessary to make factual information usable. As we indicated in Section 1.1, the science has two divisions: *descriptive statistics* and *inferential statistics*.

We deal with the component elements of descriptive statistics in Volume I: measuring things (see Chapter 2), organizing and presenting the data in tables (see Chapter 4) and graphs (see Chapter 5), and calculating descriptive statistics and parameters (see Chapters 6 and 7). We then go on in the remainder of Volume I and in Volume II to give an introduction to the component elements of inferential statistics. This area of the science of statistics provides the techniques and logical framework for making *inferences* (generalizations) about the characteristics of entire populations from the characteristics of available samples. Beyond this inferential leap into the unknown, from sample to population, inferential statistics also provides procedures involving *probability theory* (see Chapters 8 to 12) and *sampling theory* (see Chapter 13) for determining the quality of the inferences—how close they probably are to the truth.

3.6　ESTIMATION PROBLEMS AND HYPOTHESIS-TESTING PROBLEMS

Inferential statistics is based on a theory called *statistical decision theory* (or *decision theory*). This theory can be divided into two major areas: *estimation theory* and *hypothesis-testing theory*.

Estimation theory provides techniques for solving estimation problems. In these problems, unknown parameters of the measurement population are estimated by using sample statistics from linked measurement samples (e.g., using $\bar{x}$ to estimate μ). Beyond this, a complete solution to an estimation problem also shows how good the estimate is—how certain one can be of it—by bracketing the estimate within an interval called a *confidence interval*. Estimation theory is introduced and discussed in Chapters 14 and 15.

Hypothesis-testing theory provides techniques for determining whether *statistical hypotheses* should be accepted or rejected. A statistical hypothesis is an assumption (or guess) about unknown properties of one or more measurement populations, typically either about their parameters or about how they are spread (*distributed*) from smallest to largest values. Considering parameters, it could be hypothesized, for example, that the arithmetic mean of a measurement population is equal to some constant value a ($\mu = a$). The test of such a hypothesis uses measurement-sample information to determine the probable truth of the hypothesis. Hypothesis-testing theory as applied to single measurement samples is discussed in Chapter 16.

The sequence, for one sample from one population, by which estimation and hypothesis-testing problems are solved is diagrammed schematically in Fig. 3-1. The arithmetic mean μ of a measurement population is unknown. In order to estimate it and to test hypotheses about it: a physical sample is taken from the linked physical population; the measurement of interest is taken from all elements in the physical sample, producing a measurement sample; $\bar{x}$ and other descriptive statistics are calculated from the measurement sample; and finally, using inferential statistics and the $\bar{x}$ estimate, μ is bracketed within a confidence interval (confidence: $\underline{\ \mu\ }$) and the probable truth of a statistical hypothesis ($\mu = a$) is determined.

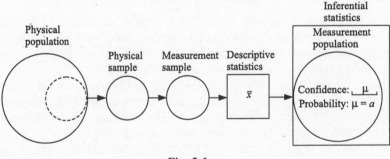

Fig. 3-1

3.7　STATISTICAL HYPOTHESES AND RESEARCH HYPOTHESES

Statistical hypotheses (see Section 3.6) and *research hypotheses* (also called *scientific hypotheses* or *working hypotheses*) are alike in certain ways and very different in others. They are alike in that both are assumptions about populations, but while statistical hypotheses are assumptions about measurement populations (e.g., $\mu = a$), research hypotheses are assumptions about the physical population under study. And while statistical hypotheses are abstract assumptions about the characteristics of a set of measurements, research hypotheses deal with real-world, cause-and-effect relations (see Section 1.19)—the factors that cause some thing to occur. Thus, for example, from many studies of lung-cancer patients, medical researchers were able to state this cause-and-effect research hypothesis: Cigarette smoking can cause lung cancer. Similarly, sociologists who studied children in many contexts developed the hypothesis: Extensive television watching in young children can cause inappropriate aggression.

Both statistical and research hypotheses are tested by using *deductive reasoning* (see Problem 3.6) to derive their consequences: what must also be true if the hypothesis is true. The mathematical, deductive reasoning for testing statistical hypotheses is part of hypothesis-testing theory, which is introduced in Chapter 16. The deductive reasoning for testing a research hypothesis is unfortunately not provided by a convenient theory; each investigator must develop the logical consequences of his or her own hypotheses. Thus, an obvious example of a deductive prediction is: If cigarette smoking can cause lung cancer, then it is logical to expect more cases of lung cancer in a population that smokes cigarettes than in a population that does not.

Statistical hypotheses are abstract, mathematical tools used for making decisions about the characteristics of measurement populations. Once established, these characteristics are then used for making decisions about research hypotheses.

3.8   EXPLORATORY RESEARCH AND HYPOTHESIS-TESTING RESEARCH

The development of understanding in any field is a process of forming and testing research hypotheses. *Exploratory research* (or *descriptive research*) is the hypothesis-formation stage in which research hypotheses are developed, through inductive reasoning (see Problem 3.6), from observations, measurements, and analyses of the data. By contrast, *hypothesis-testing research*, as the name implies, is the stage in which developed research hypotheses are tested by testing their predictions.

3.9   EXPLORATORY EXPERIMENTS

In Example 1.31 we indicated that in an experiment an independent variable is changed (manipulated) to see the effect on the dependent variable. We used a salmon-growth experiment as an example, in which two groups of salmon were raised under identical conditions except one was kept in 20°C water while the other was in 24°C water. The water temperature was the independent variable, and growth (as measured by weight at 200 days) was the dependent variable.

The values of an independent variable, here 20°C and 24°C, are called the *levels of the variable*. They are also called *treatments*. Thus, in the salmon experiment it could be said either that two levels of the independent variable (water temperature) were applied to the fish to see the effect on the dependent variable (body weight) or that two water-temperature treatments were applied.

An *exploratory experiment* is one form of exploratory research. The object of such an experiment is to develop research hypotheses from the results. Thus, for example, from the salmon-growth experiment, a biologist could develop hypotheses about the effects of water temperature on fish growth.

The sequence by which descriptive and inferential statistics can be used to analyze data from the salmon-growth experiment is diagrammed schematically in Figure 3-2. Here, the researcher is interested in the mean population weight at 200 days for both levels of the independent variable (μ_{20} and μ_{24}), and whether these means differ between the conditions.

The two fish groups are the physical samples from their infinite physical populations. Weighing every fish in the two groups produces two measurement samples, from which descriptive statistics produces two sample means, $\bar{x}_{20}$ and $\bar{x}_{24}$, as well as other sample information. Using inferential statistics, it is then possible to solve mean-estimation problems for each measurement population (confidence: μ_{20} and confidence: μ_{24}), to estimate the difference between the means (confidence: $\mu_{20} - \mu_{24}$), and to test such statistical hypotheses as $\mu_{20} = \mu_{24}$. The theory and techniques for solving such two-sample estimation and hypothesis-testing problems will be discussed in Chapter 17.

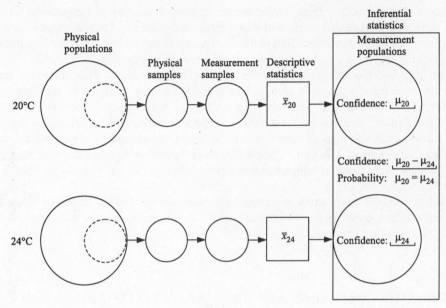

Fig. 3-2

3.10 CONTROLLED EXPERIMENTS

The term *controlled experiment* has three meanings in statistics. It can refer to experiments that have: (1) all *extraneous variables* controlled, (2) one or more *control groups*, or (3) treatment applications controlled by the experimenter.

Extraneous variables are all *potential independent variables*, other than the one under study, that could influence the experiment. Thus, for example, the salmon-growth experiment (see Section 3.9) could be affected by differing levels of food, human contact, living space, light, water chemistry, and so on. An extraneous variable becomes a *confounding variable* when it has different levels that correspond to different levels of the *true independent variable* (the one being manipulated by the experimenter). To avoid confounding variables, an attempt is made to exert control over extraneous variables, which means keeping them as near as possible to constant levels throughout the experiment.

Complete control of all extraneous variables is an ideal that is rarely achieved even in the most precisely regulated laboratories. Recognizing this, the effects of extraneous variables are offset by the use of *experimental* and *control groups* that are in some way *randomly selected* (see Section 3.17). The objects assigned to the various groups of an experiment are called *subjects* if they are human or animal and *experimental units* if they are neither.

As entire populations are rarely available, most experiments are done on physical samples, and each sample that receives one of the treatment conditions of the experiment is called an *experimental group*. In the exploratory experiment in Section 3.9 there were two such groups, the 20°C group and the 24°C group. By contrast, a control group typically receives no treatment; instead it provides *normative* or *baseline data* against which one or more experimental groups (or *treatment groups*) can be compared. Say, for example, an investigator wants to determine whether a new vaccine against the common cold actually works. She might then, from adult volunteers, randomly select two groups of subjects, treating both groups identically except that one of the groups (the experimental group) is injected with the vaccine in a saline (salt water) solution while the other group (the control group) is injected with the saline solution without the vaccine. Then, by comparing numbers of colds in the two groups over a set time, she can determine the effectiveness of the vaccine. While exploratory experiments are used in the hypothesis-formation stages of research, experiments comparing a control group (or groups) with an experimental group (or groups) are typically used in the hypothesis-testing stage (see Section 3.8).

The third meaning of a controlled experiment is that treatment applications are controlled by the experimenter. The experimenter controls which subjects or experimental units receive which treatments. This is the principal difference between controlled experiments and *observational studies* (see Section 3.11).

3.11 OBSERVATIONAL STUDIES

As we have indicated, in an experiment the investigator manipulates an independent variable to see the effect on a dependent variable. This differs from an observational study where the investigator has no control over the independent variable but instead simply observes the existing phenomenon and tries to isolate relationships between independent and dependent variables. Observational studies can be done either in the exploratory (hypothesis-formation) stage of research or in the hypothesis-testing stage.

An example of an exploratory observational study would be a study of the characteristics of successful students in an introductory course in statistics. If success is measured by the dependent variable, total points in the course, then to isolate possible performance-influencing independent variables the investigator might measure for each student: time spent each week on homework, number of previous math courses, grade point for previous math courses, and so on. The statistical techniques most commonly used to analyze such data for possible relationships between variables are called *regression* and *correlation* (see Chapter 19).

A hypothesis-testing observational study is done when, for ethical or practical reasons, it is not possible to test an existing research hypothesis with a true controlled experiment. What is sought for the test, then, is an existing *natural experiment* (or *quasi-experiment*) in which there are different levels of the hypothesized independent variable and measurable levels of the hypothesized dependent variable. Thus, for example, if the hypothesis is that a type of female birth-control pill leads to elevated blood-pressure levels, then the investigator might study blood-pressure levels (the dependent variable) in two existing populations that differ in the use of this pill (the independent variable), say users and nonusers of the pill.

3.12 SURVEYS AND CENSUSES

While in common usage a *survey* is any attempt to get information about the characteristics of a group of things, in statistics the term refers to getting information from a sample or population by asking questions: market surveys, public-opinion surveys, telephone surveys, and so on. If all members of a population are included in the survey, then it is called either a *100% survey* or a *census*. While the survey and the census are methods used typically in observational studies, they can also be used in experiments (e.g., surveying an audience on attitudes before and after a movie).

3.13 PARAMETRIC AND NONPARAMETRIC STATISTICAL TECHNIQUES

The inferential techniques appropriate for analyzing a given set of data are determined by a complex set of factors. Some of these factors have been discussed: for experiments, the number of groups that receive treatments and the use of control groups (see Sections 3.9 and 3.10); for observational studies, whether the data are exploratory or hypothesis-testing (see Section 3.11). Now we consider two more factors: the required characteristics of the underlying measurement populations, and the data's level of measurement.

Parametric statistical techniques (or *parametric statistics*) are based on very precise and restrictive assumptions about the characteristics of the measurement populations and the measurement samples being investigated. These assumptions, called *parametric assumptions*, state required features of the populations under study, such as the nature of their parameters and the shapes of their distributions, and they indicate the type of sample that must be taken. If these assumptions can be satisfied, and if the sample data are interval or ratio level (see Sections 2.6 and 2.7), then parametric techniques should be used in inferential analyses. Such techniques are always preferred because: (1) they provide the most sensitive tests of statistical hypotheses, (2) they extract the most information from the data, and (3) they can analyze the most sophisticated and complex research designs. Parametric techniques are so-named because both the statistical hypotheses being tested (e.g., $\mu = a$) and many of the assumptions deal with parameters.

Parametric statistics were the first inferential techniques developed. It was soon realized, however, that there was a need for equivalent, more broadly applicable inferential techniques with weaker and less restrictive assumptions, and this led to the development of *nonparametric statistical techniques* (or *nonparametric statistics*). They are called nonparametric because their assumptions and statistical hypotheses do not deal with parameters; some nonparametric statistics are also called *distribution-free statistics*, because their assumptions do not require that the underlying population distribution have any specific shape.

Broadly speaking, nonparametric techniques can be divided into techniques for nominal-level data and techniques for ordinal-level data (see Chapter 20). These techniques can also be used for interval-level and ratio-level data if the analysis is restricted to either the nominal ($=$, $\neq$) or ordinal ($<$, $>$) properties of the data (see Sections 2.4 and 2.5). Because parametric techniques are always preferable for interval-level and ratio-level data, they should be used for that level of data unless there are severe deviations from the parametric assumptions. Even then, it is often possible to transform the data to allow parametric analysis.

The flow chart in Fig. 3-3 summarizes how the appropriate technique should be chosen. Given interval-level or ratio-level data, if the parametric assumptions are satisfied, then the data should be analyzed with parametric techniques. If these assumptions are not satisfied, then a transformation should be attempted. If this works, do a parametric analysis; if not, then analyze the data with either ordinal-level or nominal-level nonparametric analyses. If the data are ordinal level or nominal level to begin with, then use the appropriate level of nonparametric analysis. Other decision factors (number of groups, presence of control groups, whether observational studies are exploratory or hypothesis testing) become important after these parametric/nonparametric decisions have been made.

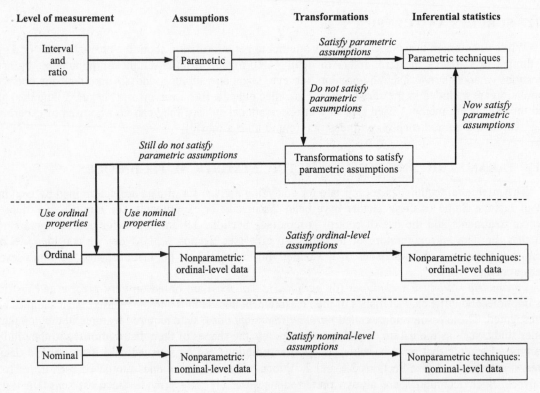

Fig. 3-3

3.14 MATHEMATICAL STATISTICS AND GENERAL STATISTICS

All forms of mathematics begin with sets of assumptions (*axioms*) from which *theorems* and whole *integrated systems* are derived by means of deductive reasoning (see Problem 3.6). *Mathematical statistics* (see Section 1.1), which makes use of elements from many areas of mathematics (e.g., probability theory, calculus, advanced algebra), provides such an integrated system for statistics. While mathematical statistics is a mathematically sophisticated, tightly integrated system of axioms and theorems, *general statistics* (see Section 1.1) presents most of the same materials at a much simpler level. In general statistics, descriptive and inferential techniques are taken from mathematical statistics but little attempt is made to show how each technique is derived from the integrated mathematical system. Because general statistics deals primarily with nonmathematical discussions of statistical concepts and techniques, it is said to be at the *intuitive level* of presentation rather than at the *mathematical level*. This book is a general statistics book presented at the intuitive level, and all the mathematics required to understand the book are reviewed in Chapter 1.

While this book, or any presentation based on general statistics, has the advantage of not requiring mathematical derivations for every technique, it also has a disadvantage. Concepts and techniques are continually brought up from the mathematical level, and students are asked to accept each of them as true as they appear, without proof. Throughout this book, therefore, and throughout any other intuitive-level book or course, such phrases as "it must be accepted as true" or "this can be proven mathematically" are repeated. Only in a book or course in mathematical statistics is the entire integrated system presented.

3.15 SAMPLING DESIGNS

So far in this chapter we have introduced three theoretical components of inferential statistics: probability theory (discussed in Chapters 8–12); estimation theory (discussed in Chapters 14 and 15), and hypothesis-testing theory (discussed in Chapter 16). Now we introduce a fourth theoretical component: *statistical sampling theory* or *sampling theory*. This theory, which deals with theoretical relationships between measurement populations and the measurement samples taken from them, is discussed in Chapter 13.

One aspect of sampling theory deals with required sampling procedures for inferential statistics—the sample-taking methods that yield data that can legitimately be analyzed for sample-to-population inferences (see Section 3.5). These abstract, mathematical sampling procedures, called *sampling designs*, assume the existence of measurement populations (called simply populations) and specify for a given inferential technique how measurement samples (called simply samples) must be taken from these populations. This theoretical sequence from population-to-sample-to-population is illustrated schematically in Fig. 3-4.

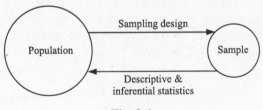

Fig. 3-4

While sampling theory describes a pure, mathematical world, actual sampling problems are solved under the messy conditions of the real world. To solve such a problem, the physical population of interest is first defined. A physical sample is then taken from it, following an appropriate theoretical sampling design as closely as possible. Then measurements are taken from the physical sample to produce a measurement sample. Finally, the data are analyzed with descriptive and inferential statistics to make inferences about the measurement population. This real-world sequence is illustrated schematically in Figs. 3-1 and 3-2.

We now go on to an overview of important theoretical sampling designs and how they are used in practical situations.

3.16 PROBABILITIES FOR SAMPLING: WITH AND WITHOUT REPLACEMENT

All aspects of inferential statistics, including sampling designs, are based on the theory of probability. This theory will be presented in detail in Chapters 8 to 12, but here we will briefly introduce some of its concepts in order to discuss sampling designs.

In essence, the *probability of an event* is the likelihood or chance that the event will occur. It is stated as a number from (0) to (1), where a probability of (0) means that the event cannot possibly occur and a probability of (1) means that the event is certain to occur. Thus, there is a probability of (0) that a man can give birth to a baby, and a probability of (1) that if the sun rises tomorrow it will rise in the east. A probability of (0.5) means there is a fifty-fifty chance the event will occur.

As we will discuss in Chapter 8, there are four different methods for interpreting and calculating such probabilities: the *classical interpretation*, the *relative frequency interpretation*, the *set theory interpretation*, and the *subjective interpretation*. For this brief introduction we will only consider the classical interpretation, which is the oldest and most familiar.

The classical interpretation was developed in the nineteenth century from studies of the games of chance used in gambling: throwing dice, flipping coins, picking cards, and so on. It applies to any "game" where on any *trial* of the game all possible *outcomes* are known, equally likely, and mutually exclusive (see Section 2.4). These outcomes can be classified into different categories called *events* and the probability on a given trial of any event A, denoted symbolically by $P(A)$, can be calculated with the equation

$$P(A) = \frac{\text{number of outcomes that produce } A}{\text{total number of possible outcomes}} \tag{3.3}$$

To understand these concepts, let us consider the game of picking a card from a well-shuffled standard 52-card deck of playing cards. Say we want to determine the probability on such a pick (trial) of the event of picking an ace, $P(\text{ace})$. We know the total number of possible outcomes for the pick is 52—any one of the 52 cards is a possible outcome—and we also know that the number of outcomes that can produce an ace is 4. Thus, using equation (3.3)

$$P(\text{ace}) = \frac{4}{52} = \frac{1}{13}$$

Let us say we picked an ace on this trial, and that we now want to determine the probability if we pick again from the deck of picking an ace. There are two possible ways in which this second trial can proceed: (1) replace the ace in the deck, reshuffle, and pick again; (2) do not replace the card, but do reshuffle, and pick again. The first way is an example of what is called *sampling with replacement*, and the second is an example of what is called *sampling without replacement*. If the sampling is done with replacement, then the probability of an ace on the new pick is again

$$P(\text{ace}) = \frac{1}{13}$$

If, however, the sampling is without replacement, then the probability of an ace on the new pick is now

$$P(\text{ace}) = \frac{3}{51} = \frac{1}{17}$$

If the population being sampled is infinitely large ($N = \infty$), then the composition of the population is considered to remain constant during sampling, whether the sampling is with or without replacement. For practical purposes, this is also considered to be true when n (sample size) is very small compared to N (roughly when n is no more than 5% of N). Only for finite populations such as a deck of cards in which N is not much larger than n does the change in population over sampling need to be considered.

3.17 RANDOM SAMPLING

The objective of most research efforts is to discover general truths about entire populations. As whole populations are rarely available, however, it is typically necessary to take a sample from the population and then use statistical techniques to make inferences from the sample back to the population. For these inferences to be legitimate, the samples must be taken under the precise conditions of theoretical sampling designs called *random sampling designs* or simply *random sampling*.

A population can be sampled by taking elements from it one at a time or in groups. Either way, the basic unit taken is called the *sampling unit*. Random sampling, then, is any sampling method in which every sampling unit in the population has a known nonzero probability of being included in the sample. Because the probabilities of inclusion are specified before sampling, random sampling is also called *probability sampling*. It is called random sampling because in order to achieve these specified probabilities, it is necessary to introduce *randomness* or *chance* into the selection process. This is typically done by using randomly determined sequences of numbers from *tables of random numbers* (see Section 3.23) or by using *random-number generating functions* available with computers and many calculators.

In the theoretically ideal case, all of the sampling units in the population of interest, called the *target population*, are known and recorded on a list called the *sampling frame* or *frame*. Then a random sampling design is selected that specifies the probabilities of inclusion of all sampling units in the sampling frame. Finally, some random selection technique is used that selects sampling units according to the specified probabilities.

We will now deal with four important random sampling designs: simple random sampling, stratified random sampling, systematic random sampling, and cluster random sampling.

3.18 SIMPLE RANDOM SAMPLING

Simple random sampling is a mathematical concept that we discuss in detail in Chapter 13 following our discussion of probability theory. At this point, as part of an overview of random sampling, we simply give this nonmathematical, intuitive-level (see Section 3.14) definition that can be found in many statistics books:

Simple random sampling is a method of sampling in which at every selection from the population all remaining sampling units in the population have the same probability of being included in the sample. A sample taken with this method is called a *simple random sample*.

To understand this definition, consider taking a simple random sample of three people from a known target population of 35 that are listed in a sampling frame. If we define the sampling unit as individual members of the population, then we could write each of the 35 names on separate identical pieces of paper, place the papers in a bowl, and blindly select one-at-a-time and without replacement, three papers from the bowl. This would be a simple random sample because every name would have the same probability of inclusion on the first pick, 1/35, and on the second, 1/34, and on the last, 1/33.

Most methods in elementary inferential statistics are based on the assumption that samples are taken by simple random sampling, and that they are therefore simple random samples. Because this assumption is so basic and common, *in the rest of the book when the terms sample or random sample are used without modifiers it means simple random sample*.

3.19 STRATIFIED RANDOM SAMPLING

For sample-to-population inferences to be valid, the sample must be representative of the population. Simple random sampling usually provides representative samples, but there are instances when a random sampling method called *stratified random sampling* provides a more representative sample. This is the case when the population contains several nonoverlapping, mutually exclusive groups called *strata* (plural of *stratum*), caused by such factors as age, gender, race, and geographic location. To use stratified random sampling, the strata should be relatively homogeneous, with greater differences between strata than within each stratum. If such strata are present, then a stratified random sampling design takes a random sample [simple, systematic (see Section 3.20), cluster (see Section 3.21), or some other] from each stratum. In *proportional stratified random sampling*, the size of the random sample taken from each stratum is made proportional to the relative size of the stratum in the population (stratum size/population size). In *disproportional stratified random sampling*, there is no attempt to make the random sample from each stratum proportional to the relative size of the stratum.

To understand stratified random sampling, consider the problem of determining the average income in a male social group where 5% of the men are high income, 65% are middle income, and 30% are low income. If we consider these income levels as three strata in the population, then a proportional stratified random sample would have 5% high, 65% middle, and 30% low income. A disproportional stratified random sample could have any distribution of percentages from the strata, say equal percentages ($33\frac{1}{3}$%).

3.20   SYSTEMATIC RANDOM SAMPLING

Systematic random sampling is often used when a random sample is to be taken from a very long sampling frame. In this type of sampling, every kth unit in the frame is taken, starting with a randomly selected list-position within the first k units, called the *starting unit*. If, for example, we decide to take every 50th unit, then the first step is to take a simple random sample of one from the numbers 1–50, to determine the starting unit. If, say, this number is 20, then the sample includes: the 20th number on the list, the 70th, the 120th, and so on through the list. This sample is a random sample because the nonzero probability of inclusion for each unit in the population was known in advance $\left(\frac{1}{k} = \frac{1}{50}\right)$.

Systematic random sampling should be avoided if there are clear periodic or cyclic patterns in the sampling frame. Thus, for example, in taking a survey of houses it could be true that every 25th house on a block is the corner house that is the most expensive house on the block.

3.21   CLUSTER RANDOM SAMPLING

If a population is very large and widely dispersed, *cluster random sampling* can provide a relatively inexpensive random sample. In cluster random sampling the population is first divided into mutually exclusive groups (called *clusters*) that are each as heterogeneous as possible (unlike the homogeneous strata of stratified random sampling). Then, in *single-stage cluster random sampling*, some form of random sample of the clusters is taken and all the elements within the selected clusters are included in the sample. In *two-stage cluster random sampling*, after the random sample of clusters is taken, a second random sample is taken from each cluster. *Multistage cluster random sampling* is the term for such a sampling process with two or more sampling stages.

Consider the problem of determining public opinion before a state-wide election in a large state like New York. The state is subdivided into counties that are relatively heterogeneous in population. A single-stage cluster random sample might begin with a random sample of the counties, followed by an opinion poll of every registered voter within the selected counties. Much more likely would be a multistage cluster random sample, in which random samples of smaller subdivisions within the counties are taken before the polling is begun.

3.22   NONRANDOM SAMPLING

In Section 3.17 we said "Random sampling, then, is any sampling method in which every sampling unit in the population has a known nonzero probability of being included in the sample." Sampling designs that do not satisfy this definition are called *nonrandom sampling designs* or *nonprobability sampling designs*. They are also called *judgment sampling designs* because they typically involve personal, nonrandom judgments by the investigator with regard to which units to include in the sample. Another term for them is *biased sampling designs*, because they generally lead to some type of *sampling bias* (systematic sampling error), either a *selection bias* or a *response bias*. [Recall that systematic errors of measurement are also called bias (see Section 2.13).] A selection bias occurs when sampling procedure tends to select certain units from the population and to exclude others. A response bias occurs when the design leads to a great deal of missing data. Thus, for example, in a mailed survey a response bias is produced by a low rate of return of the survey questionnaires.

EXAMPLE 3.4 Why is the following true story an example of nonrandom sampling, with both selection bias and response bias?

The 1936 presidential election in the United States had two major candidates: the Republican, Alfred M. Landon, and the Democrat, the incumbent president, Franklin D. Roosevelt. Several weeks before the election, *Literary Digest* magazine tried to predict the outcome by mailing 10 million questionnaires to people selected from three sources: the subscription list for the magazine, telephone directories, and automobile registration records. The magazine received back approximately 2.3 million answers, and of these some 57% favored Landon. From these results the magazine predicted a landslide victory for Landon. A few weeks later, however, in the actual election, it was Roosevelt who got the majority of the votes (62%).

Solution

This is an example of nonrandom sampling because by limiting the sample to magazine subscribers and to owners of telephones and automobiles, most of the voting population had a zero probability of being included in the sample. The time was 1936, in the depths of the Depression, and the judgement-selection limited the sample to a relatively prosperous stratum of the population. Besides this severe selection bias, produced by a discrepancy between the target population and the sampling frame, there was also a response bias. This response bias, called *self-selection bias*, occurred because only about 25% of the selected sample returned their questionnaires. Thus even for this chosen stratum of the population, the probabilities for inclusion in the sample were unknown before sampling.

3.23 TABLES OF RANDOM NUMBERS

The most commonly used inanimate device for introducing chance into the sampling process is a *table of random numbers* (or *table of random digits*). Such a table, which typically has been created with a computer random-number-generating function, consists of thousands of digits, each of which is any one of the ten numbers from 0 to 9. Every digit has, in essence, been selected by a simple random sample from the numbers 0 to 9. Consequently, the numbers 0 to 9 are equally likely to appear in any digit-position in the table, and there are no systematic connections between digits. Table A.1 (*Random Numbers*) in the Appendix is such a table of random numbers; it consists of 6,000 digits arranged in two-digit pairs, in 40 rows and 75 columns.

EXAMPLE 3.5 A 64-student statistics class is shown, in their seats, in Fig. 3-5. Each bracket in the figure ([ ]) is an individual seat, containing the student's initials and whether the student is female (**f**) or male (**m**). There are 16 females (25%) and 48 males (75%). Using Table A.1 in the Appendix, select a simple random sample of 16 from this class.

Solution

To select such a sample, we must meet the conditions of random sampling (see Section 3.17) and simple random sampling (see Section 3.18). To do this here, we consider the class to be a population from which we create a sampling frame by giving each student a unique two-digit number from 01 to 64 (see Fig. 3-6). We then take student numbers from Table A.1, without replacement (see Section 3.16), in such a way that: all students have a $\frac{1}{64}$ probability of being the first selection; all remaining students have a $\frac{1}{63}$ probability of being the second selection; and so on until all remaining students have a $\frac{1}{49}$ probability of being the 16th selection. With Table A.1, this sampling can be accomplished by simply taking a sequence of 16 two-digit numbers between 01 and 64 without accepting any repetitions.

	1	2	3	4	5	6	7	8
1	[CA-**m**]	[FE-**m**]	[LB-**f**]	[HE-**m**]	[LW-**m**]	[OA-**m**]	[PS-**m**]	[OF-**m**]
2	[AA-**f**]	[HC-**m**]	[EB-**m**]	[MA-**m**]	[ME-**m**]	[HK-**m**]	[AD-**m**]	[RE-**m**]
3	[AE-**f**]	[FA-**m**]	[CE-**m**]	[BP-**m**]	[EO-**m**]	[RA-**m**]	[DA-**m**]	[GK-**m**]
4	[JA-**m**]	[GB-**m**]	[MJ-**f**]	[NO-**f**]	[JW-**m**]	[LT-**f**]	[HO-**m**]	[WA-**m**]
5	[JD-**f**]	[DD-**f**]	[NA-**m**]	[SM-**m**]	[MQ-**m**]	[JT-**m**]	[TS-**m**]	[MU-**m**]
6	[GM-**m**]	[BC-**m**]	[CI-**m**]	[EF-**f**]	[JL-**m**]	[JQ-**m**]	[FV-**m**]	[DW-**m**]
7	[AC-**f**]	[DM-**m**]	[JH-**f**]	[BF-**m**]	[AH-**f**]	[NP-**m**]	[GT-**m**]	[GY-**f**]
8	[MZ-**f**]	[BJ-**f**]	[LR-**m**]	[CR-**m**]	[PB-**m**]	[EJ-**m**]	[AT-**m**]	[TM-**f**]

Fig. 3-5

The first step in using Table A.1 is to randomly determine a *starting place* in the table for taking the sequence of two-digit numbers. There are many techniques for doing this, such as the following: first, blindly touch a sharp point to any place in the table and take the nearest two-digit number as the column number for the starting place; then, repeat the blind-touching process elsewhere in the table, now taking the nearest two-digit number for the row of the starting place. When we actually did this, the first touch yielded 59 and the second 24, which gives a starting place at the intersection of column 59 and row 24: the number 68.

From such a starting place, two-digit numbers can be collected in any direction: up, down, diagonally, sideways to the right or left. In this case we went downward from the starting place in column 59, taking all nonrepeating numbers from 01 to 64. By the bottom of the column we had collected these numbers: 59, 64, 17, 22, 07, 39, 44, 32, 26, 53, 45, 38, and 13. As these are only 13 of the required 16, we then went to the top of column 60 and proceded downward again, collecting these numbers: 63, 27, and 37. Therefore, our simple random sample of 16 from this class is:

[LR-**m**], [TM-**f**], [AE-**f**], [RA-**m**], [PS-**m**], [TS-**m**], [EF-**f**], [WA-**m**],

[GB-**m**], [AH-**f**], [JL-**m**], [JT-**m**], [ME-**m**], [AT-**m**], [MJ-**f**], [MQ-**m**].

	1	2	3	4	5	6	7	8
1	[CA-**m**] 01	[FE-**m**] 02	[LB-**f**] 03	[HE-**m**] 04	[LW-**m**] 05	[OA-**m**] 06	[PS-**m**] 07	[OF-**m**] 08
2	[AA-**f**] 09	[HC-**m**] 10	[EB-**m**] 11	[MA-**m**] 12	[ME-**m**] 13	[HK-**m**] 14	[AD-**m**] 15	[RE-**m**] 16
3	[AE-**f**] 17	[FA-**m**] 18	[CE-**m**] 19	[BP-**m**] 20	[EO-**m**] 21	[RA-**m**] 22	[DA-**m**] 23	[GK-**m**] 24
4	[JA-**m**] 25	[GB-**m**] 26	[MJ-**f**] 27	[NO-**f**] 28	[JW-**m**] 29	[LT-**f**] 30	[HO-**m**] 31	[WA-**m**] 32
5	[JD-**f**] 33	[DD-**f**] 34	[NA-**m**] 35	[SM-**m**] 36	[MQ-**m**] 37	[JT-**m**] 38	[TS-**m**] 39	[MU-**m**] 40
6	[GM-**m**] 41	[BC-**m**] 42	[CI-**m**] 43	[EF-**f**] 44	[JL-**m**] 45	[JQ-**m**] 46	[FV-**m**] 47	[DW-**m**] 48
7	[AC-**f**] 49	[DM-**m**] 50	[JH-**f**] 51	[BF-**m**] 52	[AH-**f**] 53	[NP-**m**] 54	[GT-**m**] 55	[GY-**f**] 56
8	[MZ-**f**] 57	[BJ-**f**] 58	[LR-**m**] 59	[CR-**m**] 60	[PB-**m**] 61	[EJ-**m**] 62	[AT-**m**] 63	[TM-**f**] 64

Fig. 3-6

Solved problems

POPULATIONS

3.1   For the following populations, indicate whether they are finite or infinite physical populations or measurement populations: (*a*) the current ages of all living women who have swum the English Channel,   (*b*) the number of trials it takes in an experiment for a Norway rat to learn a specific maze under specific conditions,   (*c*) all the body cells in a living 20-year old man,   (*d*) all chocolate cakes baked from a specific recipe,   (*e*) the body weights of all current and past citizens of the United States.

Solution

(*a*)   Finite measurement population

(*b*)   The measurement results of an experiment that hypothetically could be repeated an infinite number of times is an infinite measurement population.

(*c*)   Finite physical population

(*d*)   Infinite physical population

(*e*)   Finite measurement population

SAMPLES

3.2   For the following samples, indicate whether they are physical samples or measurement samples: (*a*) the surface areas of four oil paintings by Rembrandt,   (*b*) six living men who have run a mile in less than four minutes,   (*c*) daily measurements for 20 days of the water temperature (°C) of a lake.

Solution

(*a*)   Measurement sample

(*b*)   Physical sample

(*c*)   Measurement sample

3.3   For the following, indicate first whether they are a population, a sample, or whether they could be either, and then indicate whether they are physical or measurement: (*a*) current weights of all living former presidents of the United States,   (*b*) attitudes of 50 Americans toward immigrants as measured on a five-point scale from 1 (= unfavorable) to 5 (= favorable),   (*c*) number of new television sets purchased in one month in each store of a chain of stores,   (*d*) all remaining 132 butterflies in a species that is going extinct.

Solution

(*a*)   Population, measurement

(*b*)   Sample, measurement

(*c*)   Could be either, measurement

(*d*)   Population, physical

3.4   When is a physical sample measured instead of a physical population?

Solution

While the interest that motivates data collection and analysis in any field is almost always in the characteristics of an entire physical population, it is rarely practical or even possible to measure entire populations. For example, it is impossible to measure an infinite physical population (see Section 3.2). Even a finite population cannot be entirely measured if it is very large or widely spread out in time or space. Finally, even if the physical population is finite and measurable, if the measurement process destroys the item being measured then, typically, the population will not be measured. Thus, for example, if an apple grower tests all of his apples for sugar content, then he will have no apples to sell. In most cases, physical samples of a population are taken and measured, and techniques from inferential statistics are used to generalize from the sample to the unavailable population.

PARAMETERS AND STATISTICS

3.5   Calculate arithmetic means for the following: (a) measurement sample $x_1 = 4$, $x_2 = 2$, $x_3 = 3$, (b) measurement population $x_1 = 5$, $x_2 = 5$, $x_3 = 5$.

Solution

(a) $\bar{x} = \dfrac{\sum_{i=1}^{n} x_i}{n} = \dfrac{\sum_{i=1}^{3} x_i}{3} = \dfrac{4+2+3}{3} = 3$

(b) $\mu = \dfrac{\sum_{i=1}^{N} x_i}{N} = \dfrac{\sum_{i=1}^{3} x_i}{3} = \dfrac{5+5+5}{3} = 5$

ESTIMATION PROBLEMS AND HYPOTHESIS-TESTING PROBLEMS

3.6   Why is inferential statistics also called *inductive statistics?*

Solution

The field of logic divides forms of human reasoning into two broad categories: *inductive reasoning* (also called *inductive logic*), and *deductive reasoning* (also called *deductive logic*). Inductive reasoning is the process of forming generalizations. The following is an example of inductive reasoning:

All cows are mammals and have brains.

All humans are mammals and have brains.

All dogs are mammals and have brains.

Therefore, *probably* all mammals have brains.

This argument, characteristic of inductive reasoning, goes from specific examples to a general (or universal) conclusion. But the word "probably" in the conclusion shows that it is an uncertain conclusion, really only a guess. In essence, what is concluded is: From what we know, it is probably true that all mammals have brains.

Deductive reasoning goes in the opposite direction—from general conclusions to specific examples. The following is a comparable example of deductive reasoning:

All mammals have brains.

Humans are mammals.

Therefore, humans have brains.

The deductive argument goes from general statements (called *premises*) to a particular conclusion. If the premises are true, then the conclusion must also be true. A deductive conclusion is not an uncertain probability statement, but rather a clear declaration of what must be true. Deductive reasoning investigates what is implied by the premises; what must also be true if the premises are true.

Inferential statistics in both estimation and hypothesis-testing problems uses inductive reasoning. It draws conclusions (statistical inferences) about an entire measurement population from specific and limited sample information, and these conclusions are uncertain probability statements. However, unlike nonstatistical inductive conclusions, infererential statistics gives a quantitative estimate of the probable truth of its conclusions.

EXPERIMENTS AND OBSERVATIONAL STUDIES

3.7   Indicate whether each of the following is an exploratory experiment, an experiment with a control group, or an observational study: (*a*) asking a sample of 1,500 registered voters across the U.S. which candidate they prefer in the next presidential election, (*b*) determining by timing surgeons in two hospitals the average time it takes in each hospital to do an appendectomy, (*c*) testing whether vitamin C prevents colds by giving the vitamin to one group of women and not to another group of women and then counting the number of colds in each group over a subsequent time period.

Solution

(*a*)   Observational study

(*b*)   Observational study

(*c*)   Experiment with a control group

3.8   Indicate whether each of the following is an exploratory experiment, an experiment with a control group, or an observational study: (*a*) determining by weighing samples of babies in two countries, the average birth weights in the two countries, (*b*) growing samples of new hybrid flower in four different percentages (levels) of nitrogen in the soil to determine the optimal level for growth, (*c*) testing whether a new gasoline additive for increasing miles-per-gallon actually works, by comparing average mpg from samples of cars driven with the additive in the gasoline and without it.

Solution

(*a*)   Observational study

(*b*)   Exploratory experiment

(*c*)   Experiment with a control group

PROBABILITIES FOR SAMPLING: WITH AND WITHOUT REPLACEMENT

3.9   A coin has two surfaces, a *head* surface and a *tail* surface. Using equation (3.3), what is the probability of the head surface landing upwards with: (*a*) the first flip of the coin, (*b*) the second flip of the coin?

Solution

(*a*) $P(\text{head}) = \frac{1}{2}$

(*b*) $P(\text{head}) = \frac{1}{2}$

3.10   There is a suit of 13 *heart* cards in a standard 52-card deck of playing cards. Using equation (3.3), what is the probability in picking a card from the deck of: (*a*) picking the 10 of hearts, (*b*) picking any heart card?

Solution

(*a*) $P(\text{10 of hearts}) = \dfrac{1}{52}$

(*b*) $P(\text{heart card}) = \dfrac{13}{52} = \dfrac{1}{4}$

3.11 If in Problem 3.10 you picked a heart card, then, using equation (3.3), what is the probability of picking a heart card on a second pick if you use: (a) sampling with replacement,   (b) sampling without replacement?

Solution

(a) $P(\text{heart card}) = \dfrac{13}{52} = \dfrac{1}{4}$

(b) $P(\text{heart card}) = \dfrac{12}{51} = \dfrac{4}{17}$

RANDOM AND NONRANDOM SAMPLING

3.12 There are two gender strata in the statistics class shown in Fig. 3-6: females (25%) and males (75%). For these strata, use Table A.1 and simple random sampling to take a proportional stratified random sample (see Section 3.19) of 16 from this class.

Solution

With regard to gender, a proportional stratified random sample of 16 from the class would have four females (25%) and 12 males (75%). We can get such a sample by taking separate simple random samples from each strata. Doing this for the females, we first use the technique from Example 3.5 to get a starting place (the intersection of column 37 and row 40: the number 42), and then go upward from there in column 37, collecting these nonrepeating female numbers: 49, 17, 27, and 51. Then, for the 12 males, we first get a starting place (the intersection of column 30 and row 29: the number 34), and then go downward from there in column 30 and, if necessary, upward from the bottom of column 31. In this way, we collect the following nonrepeating male numbers: 06, 48, 42, 29, 41, 55, 46, 52, 43, 07, 47, and 01. Therefore, our proportional stratified random sample of 16 from this class is:

[AC-**f**], [AE-**f**], [MJ-**f**], [JH-**f**], [OA-**m**], [DW-**m**], [BC-**m**], [JW-**m**],

[GM-**m**], [GT-**m**], [JQ-**m**], [BF-**m**], [CI-**m**], [PS-**m**], [FV-**m**], [CA-**m**].

3.13 In Fig. 3-7, the class in Fig. 3-5 has been arbitrarily subdivided into 16 clusters of four students each. Using Table A.1 and simple random sampling, take a single-stage cluster random sample (see Section 3.21) of 16 students from these 16 clusters.

Solution

To get this single-stage cluster random sample, we first take a simple random sample of four from the 16 clusters, and then take all students from each selected cluster. Using the technique from Section 3.5, we first get a starting place (the intersection of column 3 and row 6: the number 57), and then go across row 6 to the right, collecting the first four nonrepeating numbers between 01 and 16. We get 07 and 11. Needing two more, we then go down to row 7, and go to the right from column 1, collecting 06 and 01. Taking all students from clusters 1, 6, 7, and 11, we get this single-stage cluster random sample:

[CA-**m**], [FE-**m**], [AA-**f**], [HC-**m**], [CE-**m**], [BP-**m**], [MJ-**f**], [NO-**f**],

[EO-**m**], [RA-**m**], [JW-**m**], [LT-**f**], [MQ-**m**], [JT-**m**], [JL-**m**], [JQ-**m**].

3.14 You are a psychologist who wants to do a maze-learning experiment with rats. You purchase 20 male rats of the same species from an animal dealer. Is this a simple random sample of these rats?

Solution

In the pure and abstract world of mathematical statistics, this would not be a simple random sample. In that world, as we indicated in Sections 3.17 and 3.18, to be a simple random sample: (1) every sampling unit

	1	2	3	4	5	6	7	8
1	[CA-**m**]	[FE-**m**]	[LB-**f**]	[HE-**m**]	[LW-**m**]	[OA-**m**]	[PS-**m**]	[OF-**m**]
		01		02		03		04
2	[AA-**f**]	[HC-**m**]	[EB-**m**]	[MA-**m**]	[ME-**m**]	[HK-**m**]	[AD-**m**]	[RE-**m**]
3	[AE-**f**]	[FA-**m**]	[CE-**m**]	[BP-**m**]	[EO-**m**]	[RA-**m**]	[DA-**m**]	[GK-**m**]
		05		06		07		08
4	[JA-**m**]	[GB-**m**]	[MJ-**f**]	[NO-**f**]	[JW-**m**]	[LT-**f**]	[HO-**m**]	[WA-**m**]
5	[JD-**f**]	[DD-**f**]	[NA-**m**]	[SM-**m**]	[MQ-**m**]	[JT-**m**]	[TS-**m**]	[MU-**m**]
		09		10		11		12
6	[GM-**m**]	[BC-**m**]	[CI-**m**]	[EF-**f**]	[JL-**m**]	[JQ-**m**]	[FV-**m**]	[DW-**m**]
7	[AC-**f**]	[DM-**m**]	[JH-**f**]	[BF-**m**]	[AH-**f**]	[NP-**m**]	[GT-**m**]	[GY-**f**]
		13		14		15		16
8	[MZ-**f**]	[BJ-**f**]	[LR-**m**]	[CR-**m**]	[PB-**m**]	[EJ-**m**]	[AT-**m**]	[TM-**f**]

Fig. 3-7

(here individual male rat) must have a known nonzero probability of inclusion, and (2) at every selection, all remaining sampling units must have the same probability of inclusion. Clearly for this rat population, and for that matter for any other large, widely dispersed, or hypothetical population, these conditions cannot be met. Yet it is common practice in applied statistics to take such samples from such populations and to accept them as satisfying the simple-random-sampling requirements of inferential statistics. Why?

This is one example of many that we will present in the book of the conflict between the rigid, abstract properties of mathematical models and the practical necessities of doing statistical analyses on messy real-world data. Here, unless there is some reason to believe the sample is unrepresentative of the population [large discrepancy between sampling frame and target population as in Example 3.4; presence of some form of systematic bias (e.g., the dealer has provided only very old rats); etc.], the sample is accepted as a simple random sample. In essence, you do the best you can to meet the requirements of the mathematical models. Fortunately, most statistical models are "robust," which means that even if assumptions have been violated, inferential techniques will give valid results if the violations are within certain limits.

Supplementary Problems

POPULATIONS

3.15 For the following populations, indicate whether they are finite or infinite physical populations or measurement populations: (a) all men who have stood on the surface of the moon, (b) blood-sugar levels in all living women with diabetes.

Ans. (a) Finite physical population, (b) finite measurement population

3.16 For the following populations, indicate whether they are finite or infinite physical populations or measurement populations: (a) the diameters in inches of all circles, (b) the results (heads or tails) of flips of a coin.

Ans. (a) Infinite measurement population, (b) infinite measurement population

SAMPLES

3.17 For the following, indicate first whether they are a population, a sample, or whether they could be either, and then indicate whether they are physical or measurement: (a) classifying, as to type, all new cases of cancer that occur within a fifty-mile radius of a nuclear plant from the time the plant was built, (b) miles/gallon/month calculated for one year for all of the taxis in one company's fleet of taxis.

Ans. (a) Could be either, measurement, (b) could be either, measurement

3.18 For the following, indicate first whether they are a population, a sample, or whether they could be either, and then indicate whether they are physical or measurement: (a) all the apples in an orchard, (b) 10 white rats being used in a psychology experiment.

Ans. (a) Could be either, physical, (b) sample, physical

PARAMETERS AND STATISTICS

3.19 Calculate arithmetic means for the following: (a) measurement sample $x_1 = 0$, $x_2 = 0$, $x_3 = 3$, $x_4 = 5$, (b) measurement population $x_1 = 10,000$, $x_2 = 5,000$.

Ans. (a) 2, (b) 7,500

EXPERIMENTS AND OBSERVATIONAL STUDIES

3.20 Indicate whether each of the following is an exploratory experiment, an experiment with a control group, or an observational study: (a) determining the average number of vehicles that pass a potential site for a supermarket between 8 a.m. and 10 a.m. weekdays, (b) determining the effectiveness of a new "memory-improvement drug" by comparing the performances of two groups of rats learning to run a maze: each member of group (1) is injected with the drug in a saline solution before starting the learning trials, each member of group (2) is injected with the saline solution without the drug before the trials, (c) determining how many adults live in each U.S. household.

Ans. (a) Observational study, (b) experiment with a control group, (c) observational study

3.21 Indicate whether each of the following is an exploratory experiment, an experiment with a control group, or an observational study: (a) comparing the average maximum lung capacities of twenty-year-old males and females, (b) testing three new methods for teaching reading by using each with a different third-grade class and comparing subsequent scores, (c) determining how all the shareholders of a company feel about a proposed merger.

Ans. (a) Observational study, (b) exploratory experiment, (c) observational study

PROBABILITIES FOR SAMPLING: WITH AND WITHOUT REPLACEMENT

3.22 Half of the cards in a standard 52-card deck of playing cards are *red* cards. Using equation (3.3), what is the probability in picking a card from the deck of getting: (a) a red card, (b) a black card?

Ans. (a) $\dfrac{26}{52} = \dfrac{1}{2}$, (b) $\dfrac{26}{52} = \dfrac{1}{2}$

3.23 If in Problem 3.22 you picked a red card, then, using equation (3.3), what is the probability of picking a black card on a new pick if you use: (a) sampling with replacement, (b) sampling without replacement?

Ans. (a) $\dfrac{26}{52} = \dfrac{1}{2}$ (b) $\dfrac{26}{51}$

RANDOM AND NONRANDOM SAMPLING

3.24 Using Table A.1 and simple random sampling to get the starting unit, take a systematic random sample (see Section 3.20) of 16 from the class in Figure 3-6 by taking every 4th student from the starting unit.

Ans. We first use the technique from Example 3.5 to get a starting place in the table (the intersection of column 16 and row 26: the number 65), and then go downward in column 16, taking the first two-digit number we find that is between 01 and 04—03. This is the starting unit. This means that in the sampling frame in Fig. 3-6 we take assigned numbers: 03, 07, 11, 15, 19, 23, 27, 31, 35, 39, 43, 47, 51, 55, 59, and 63. Therefore, our systematic random sample of 16 is:

[LB-**f**], [PS-**m**], [EB-**m**], [AD-**m**], [CE-**m**], [DA-**m**], [MJ-**f**], [HO-**m**],

[NA-**m**], [TS-**m**], [CI-**m**], [FV-**m**], [JH-**f**], [GT-**m**], [LR-**m**], [AT-**m**]

3.25 A radio talk-show host in a large city runs a pre-election call-in survey to determine voter preference between two candidates for mayor. Of the 800 listeners who call the radio station, 500 prefer candidate A, 250 prefer candidate B, and 50 have no preference. Is this a random sample?

Ans. Phone-in surveys such as this one cannot be accepted as random sampling because the sample is too unrepresentative of the target population (see Example 3.4 and Problem 3.14). It was judgment-selected in that the talk-show host invited only his listeners to participate, which produced a selection bias by excluding most potential voters. There is also a self-selection response bias as only the most motivated listeners made the call. There are telephone-survey techniques that can produce random samples, such as the technique called *random-digit dialing*. In this technique a random sample is taken from a population list of telephone numbers (the sampling frame) and the selected numbers are then called repeatedly until there is a response.

Chapter 4

Descriptive Statistics: Organizing the Data into Summary Tables

4.1 ARRAYS AND RANGES

An *array* (or *data array*) is an arrangement of a set of measurements into either an ascending or a descending order. In an *ascending array*, the measurements go from smallest value to largest value; in a *descending array*, they go from largest to smallest.

The *range* of a set of measurements is the difference (or distance on the measurement scale) between the largest and smallest measurements. If these measurements are of some variable X, then

$$\text{Range} = x_l - x_s \tag{4.1}$$

where x_l is the symbol for the largest value, and x_s is the symbol for the smallest. If the measurements have been placed in an ascending array, then x_s and x_l are, respectively, the first and last measurements in the array. If the measurements have been placed in a descending array, the ordering of x_s and x_l is reversed.

EXAMPLE 4.1 This is a sample of length measurements: 5.1 mm, 2.9 mm, 6.4 mm, 9.2 mm, 7.7 mm. (*a*) Place the sample into an ascending array. (*b*) Find the range of the sample.

Solution

(*a*) 2.9 mm, 5.1 mm, 6.4 mm, 7.7 mm, 9.2 mm

(*b*) Range $= x_l - x_s$
$$= 9.2 \text{ mm} - 2.9 \text{ mm} = 6.3 \text{ mm}$$

EXAMPLE 4.2 Table A.2 in the Appendix (*Statistics Class Data*) is a summary of a variety of measurements taken from the statistics class introduced in Example 3.5. Place the height measurements (column 4) for the 16 females in a descending array, and then find the range for this data.

Solution

Taken directly from the table, the female heights (measured to the nearest $\frac{1}{4}$ inch) are: 67.75, 60.25, 63.75, 65.25, 62.00, 63.50, 65.25, 65.50, 65.25, 64.75, 67.00, 64.25, 69.25, 66.25, 63.00, 64.75. The descending array for these measurements is: 69.25, 67.75, 67.00, 66.25, 65.50, 65.25, 65.25, 65.25, 64.75, 64.75, 64.25, 63.75, 63.50, 63.00, 62.00, 60.25. The range for these measurements is: 69.25 in $-$ 60.25 in $= 9.00$ in.

4.2 FREQUENCY DISTRIBUTIONS

A *frequency distribution* is a summary of how many times (how *frequently*) each category on a measurement scale occurs within a set of measurements. It shows how the measurements are *distributed* (or spread) across the used part of the measurement scale. Frequency distributions are presented either in summary tables, called *frequency tables*, or in various types of graphs (see Chapter 5).

EXAMPLE 4.3   Place the following sample of 20 weight measurements (in kilograms) into an ascending array and then into a frequency distribution: 1.0, 1.5, 1.3, 1.3, 1.3, 1.4, 1.4, 1.0, 1.2, 1.2, 1.3, 1.2, 1.3, 1.3, 1.4, 1.3, 1.5, 1.2, 1.4, 1.3.

Solution

The ascending array is: 1.0, 1.0, 1.2, 1.2, 1.2, 1.2, 1.3, 1.3, 1.3, 1.3, 1.3, 1.3, 1.3, 1.3, 1.4, 1.4, 1.4, 1.4, 1.5, 1.5. The frequency distribution for this data is presented in a frequency table in Table 4.1.

There are three columns in the table: *weight, tally,* and *frequency*. The weight column shows all categories (unit steps) in the part of the measurement scale that was used, with the symbol x_i representing the ith category of variable X (weight). There are $k = 6$ categories, from $x_i = x_1 = 1.0$ to $x_k = x_6 = 1.5$.

The *tally column* shows a *tallying* (or counting) of the number of measurements in each category. This is done by placing a mark, called a *tally mark*, next to the appropriate category each time that category appears in a set of measurements. There are several types of tally marks, but the most common (shown in Table 4.1) are single vertical lines to represent individual measurements and four vertical lines crossed by a diagonal to represent a unit of five measurements.

The *frequency column* is simply a numerical representation of the tally count for each category. The symbol f_i represents the frequency in each of the k categories on this part of the scale.

The symbol $\sum$ at the bottom-left of Table 4.1 indicates that the boxed-value 20 in the row to its right is a summation of the frequency values in the column above it, from the first to the kth category. This is stated symbolically

$$\sum_{i=1}^{k} f_i = \sum_{i=1}^{6} f_i = n = 20$$

Note that as the set of measurements represents a sample, the sum of the values equals the sample size n (number of measurements in the sample). If the set of measurements had represented an entire population, then

$$\sum_{i=1}^{k} f_i = N$$

where N is the population size (the number of measurements in the population).

4.3   RELATIVE FREQUENCY DISTRIBUTIONS AND PERCENTAGE DISTRIBUTIONS

The *relative frequency* (or *proportion*) of each measurement category in a sample is the frequency of that category divided by the sample size. It is symbolized by f_i/n. For a population, it is the frequency in each category divided by the population size: f_i/N. The percentage for each category is the percent of the total frequency (n or N) that is found in that category. This percentage is achieved by multiplying the relative frequency by 100, which is symbolized for a sample by $[(f_i/n) \times (100)]\%$ and for a population by $[(f_i/N) \times (100)]\%$.

Relative frequency distributions and *percentage distributions* are summaries of how relative frequency and percentage are distributed across the used part of the measurement scale. As with frequency distributions, they are presented either in a summary table or in various types of graphs (see Chapter 5).

EXAMPLE 4.4   Add columns that show relative frequency and percentage to Table 4.1.

Solution

Table 4.1 is shown, with relative frequency and percentage columns added, in Table 4.2.

Note: The tally column from Table 4.1 has not been included in Table 4.2. This column is only useful in the development of a frequency distribution, and is rarely seen in summaries of statistical information.

Table 4.1

Weight (kg) x_i	Tally	f_i
1.0	||	2
1.1		0
1.2	||||	4
1.3	ⅣⅠ |||	8
1.4	||||	4
1.5	||	2
$\sum$		20

Table 4.2

Weight (kg) x_i	Frequency f_i	Relative frequency f_i/n	Percentage $[(f_i/n) \times (100)]\%$
1.0	2	2/20 = 0.1	10
1.1	0	0/20 = 0.0	0
1.2	4	4/20 = 0.2	20
1.3	8	8/20 = 0.4	40
1.4	4	4/20 = 0.2	20
1.5	2	2/20 = 0.1	10
$\sum$	20		

EXAMPLE 4.5   The following ascending array shows the finishing times (in minutes; continuous ratio measurement) for the first 30 male runners in a Boston Marathon: 129, 130, 130, 133, 134, 135, 136, 136, 138, 138, 138, 141, 141, 141, 142, 142, 142, 142, 143, 143, 143, 143, 143, 144, 144, 145, 145, 145, 145, 145. Put this array into a summary table with columns for time, frequency, relative frequency, and percentage.

Solution

The completed summary table is shown in Table 4.3. The symbol f_i/N is used for relative frequency because here we are considering these first 30 times to be a population: the first 30 finishing times for males in this specific running of the Boston Marathon. They could just as legitimately have been considered a sample of world-class marathon times.

4.4   GROUPED FREQUENCY DISTRIBUTIONS

In Table 4.4, the frequency distribution in Table 4.3 (Boston-Marathon times) has been converted into a grouped frequency distribution by condensing the steps on the time scale into six groups of three-unit steps each. The first group, 128–130, starts one step below the fastest time and extends across the next two steps; the next group extends from 131 minutes to 133 minutes; and so on to the last group that extends from 143 minutes to the slowest time—145 minutes—for the first 30 male finishers. In such a grouped frequency distribution, each group is called a *class* and the symbol used to represent the class (e.g., 128–130) is called the *class interval*. In Table 4.4, there are six columns: class intervals [here labeled *time (min)*], *class limits, class boundaries, class mark, tally,* and *frequency.*

The class limits (also called *stated class limits*) are the smallest (*lower class limit*) and largest (*upper class limit*) measurements in the class. They define the class interval.

Table 4.3

Time (min) x_i	Frequency f_i	Relative frequency f_i/n	Percentage $[(f_i/n) \times (100)]\%$
129	1	0.033	3.3
130	2	0.067	6.7
131	0	0.000	0.0
132	0	0.000	0.0
133	1	0.033	3.3
134	1	0.033	3.3
135	1	0.033	3.3
136	2	0.067	6.7
137	0	0.000	0.0
138	3	0.100	10.0
139	0	0.000	0.0
140	0	0.000	0.0
141	3	0.100	10.0
142	4	0.133	13.3
143	5	0.167	16.7
144	2	0.067	6.7
145	5	0.167	16.7
$\sum$	30		

Table 4.4

Time (min)	Class limits lower–upper	Class boundaries lower–upper	Class mark m_i (min)	Tally	Frequency f_i
128–130	128–130	127.5–130.5	129	|||	3
131–133	131–133	130.5–133.5	132	|	1
134–136	134–136	133.5–136.5	135	||||	4
137–139	137–139	136.5–139.5	138	|||	3
140–142	140–142	139.5–142.5	141	‖‖ ||	7
143–145	143–145	142.5–145.5	144	‖‖ ‖‖ ||	12
$\sum$					30

The class boundaries (also called *true class limits*) represent the *implied class interval*. Recall from Section 2.10 that the last digit of an approximate measurement is actually considered to be an interval called the implied range. Therefore, 128 minutes is actually somewhere between 127.5 and 128.5 minutes, and 130 minutes is somewhere between 129.5 and 130.5 minutes. As these two measurements define the first class interval (128–130), this class has a *lower class boundary* of 127.5 minutes and an *upper class boundary* of 130.5 minutes. Similarly, the next class (131–133) has a lower class boundary of 130.5 minutes (also the upper class boundary of the first class) and an upper class boundary of 133.5 minutes, and so on to the sixth class (143–145) which has class boundaries of 142.5 minutes and 145.5 minutes.

The class marks are the exact middles (*midpoints*) of the classes, which can be determined by adding the lower and upper class limits and dividing by two. As here the variable i represents the ith class, we denote the class mark by m_i. Thus, for this first class, the class mark is

$$m_1 = \frac{128 + 130}{2} = 129$$

The tally and frequency columns are essentially the same as they were for ungrouped frequency distributions, but now they show the number of measurements in each class rather than each category.

Not shown but easily calculated from Table 4.4 are the *class widths* (also called *between-class widths*). These are calculated by finding the distances between successive class limits, either between successive lower limits or successive upper limits. Thus in this table, calculating for successive lower limits, the first class width is: $131 - 128 = 3$. Actually, for this table it turns out that all the class widths are 3; the table has *equal class widths*. Such equal widths can also be calculated by finding the distance between successive class marks. Here, using the first two marks, all widths are: $132 - 129 = 3$.

4.5 GROUPED RELATIVE FREQUENCY AND PERCENTAGE DISTRIBUTIONS

For each class in a grouped distribution, relative frequencies are calculated as they were for ungrouped distributions: f_i/n or f_i/N. Also as before, to find percentage for each class, relative frequencies are multiplied by 100.

EXAMPLE 4.6   After first removing the class boundaries, class mark, and tally columns from Table 4.4, add columns for relative frequency and percentage.

Solution
The requested summary table is presented in Table 4.5.

Table 4.5

Time (min)	Class limits lower–upper	Frequency f_i	Relative frequency f_i/N	Percentage $[(f_i/N) \times (100)]\%$
128–130	128–130	3	0.1000	10.00
131–133	131–133	1	0.0333	3.33
134–136	134–136	4	0.1333	13.33
137–139	137–139	3	0.1000	10.00
140–142	140–142	7	0.2333	23.33
143–145	143–145	12	0.4000	40.00
Σ		30		

4.6 GUIDELINES FOR TRANSFORMING UNGROUPED DISTRIBUTIONS INTO GROUPED DISTRIBUTIONS

In transforming an ungrouped distribution into a grouped distribution, the challenge is to group the data in such a way that the most significant trends become sharply visible. There is no single best solution to this problem, as so much of it involves personal insights and judgments, but statistics does offer some guidelines.

(1)   Use no fewer than five classes and no more than 20.

(2)   If possible, use the same class width for all classes.

(3)   Class widths can be odd or even numbers, but an odd number is preferable as then the class marks will be one of the unit steps on the measurement scale.

(4)   Class widths can be any number, but for ease of comprehension it is recommended that they be some multiple of 5, 10, 50, 100, 500, and so on.

(5)   If the class widths are equal, make sure that

$$(\text{range} = x_l - x_s) < [(\text{number of classes used}) \times (\text{class width})]$$

(6) Make sure that the class with the smallest lower class limit includes x_s and the class with the largest upper class limit includes x_l.

(7) To avoid emphasizing measurement categories below x_s and above x_l, x_s should be as near to the lower limit of its class as possible, and x_l should be as near to the upper limit of its class as possible.

(8) In general, the larger the amount of data (n or N) the larger the number of classes that should be used.

EXAMPLE 4.7 Use the above transformation guidelines to show how the frequency distribution in Table 4.3 was transformed into the grouped frequency distribution in Table 4.4.

Solution

The first step in using these guidelines was to determine the range of the 30 marathon times

$$\text{Range} = x_l - x_s = 145 \text{ min} - 129 \text{ min} = 16 \text{ min}$$

Next, several factors had to be considered simultaneously.

(a) We want equal class widths (guideline 2) and would prefer that the width be an odd number (guideline 3), possibly a multiple of 5 (guideline 4).

(b) There must be at least five classes (guideline 1) and it must be true (guideline 5) that

$$16 < [(\text{number of classes used}) \times (\text{class width})]$$

(c) The class with the smallest lower class limit must include 129 min and the class with the largest upper class limit must include 145 min (guideline 6).

(d) 129 min should be near to its lower class limit and 145 min should be near to its upper class limit (guideline 7).

(e) The *subjective factor*: Select the arrangement that most clearly shows the important trends in the data.

Only the following combinations of (number of classes used) and (class width) satisfied most of these specifications:

	Number of classes used		Class width
16 < [20 =	5	×	4]
16 < [18 =	6	×	3]
16 < [21 =	7	×	3]
16 < [18 =	9	×	2]

After investigating the grouping possibilities using these combinations, we decided that the best choice was the grouped distribution used in Table 4.4. This distribution (six classes, each of width 3) was selected because it: has odd-numbered class widths and thus class marks identical to scale units (e.g., 129, 132), extends from 128 min (one unit below the fastest time) to 145 min (the slowest time for the first 30 males), and gives a clear impression of increasing frequency with increasing times.

4.7   OPEN-ENDED GROUPED DISTRIBUTIONS AND UNEQUAL CLASS WIDTHS

An *open-ended class* has only one class limit, an upper class limit or a lower class limit. Any grouped distribution that has an open-ended class at either or both extremes is called an *open-ended grouped distribution*.

Open-ended classes are typically used when there are a few extremely large or small measurements, far from where most of the data are concentrated. Another reason for open-ended classes is to keep information confidential. Thus, for example, in presenting the results of the second examination to the statistics class [see Table A.2, column (3)], to avoid humiliating the students who did badly you may want to use an open-ended distribution with a lower open-ended class of, say, 64 or less.

A variety of class widths are often used between the open-ended classes. Such unequal widths are used either to isolate and give emphasis to certain groupings that have political or economical significance, or to join areas within the distribution that have relatively few measurements.

It is generally advised that open-ended classes should be avoided if possible. Without defined properties (limits, boundaries, widths, marks), such classes make it difficult to graph the data (see Chapter 5) and impossible to calculate most descriptive statistical measurements from the data (see Chapters 6 and 7).

EXAMPLE 4.8 The grouped frequency distribution shown in Table 4.6 was taken from page 74 of the *Statistical Abstract of the United States: 1995* (published by the U.S. Bureau of the Census). It shows, for the physical population of babies born alive in the United States in 1992, the ages of their mothers in that year, grouped into six age groups. The number of babies (column labeled 1992) is given in units of 1,000, which means that the actual number of babies can be obtained by multiplying the given frequency by 1,000. Using the information given, add the following columns to the table: class limits, class boundaries, class width, and class mark.

Table 4.6

Age of mother	1992 (births in thousands)
Under 20 years old	518
20–24 years old	1,070
25–29 years old	1,179
30–34 years old	895
35–39 years old	345
40 years old or more	58

Solution

The distribution in Table 4.6 illustrates open-ended classes and open-ended grouped frequency distributions. The "Under 20 years old" class is an open-ended class because it has an upper class limit of 19 but no lower class limit. Similarly, the "40 years old or more" class is open-ended because it has a lower class limit of 40 but no upper class limit. The completed summary table with all the information requested is shown in Table 4.7. A question mark has been used instead of a number where numerical values cannot be determined because of the nature of open-ended classes.

Table 4.7

Age of mother	Class limits	Class boundaries	Class width	Class mark m_i (years)	1992 (births in thousands) f_i
Under 20 years old	?–19	?–19.5	?	?	518
20–24 years old	20–24	19.5–24.5	5	22	1,070
25–29 years old	25–29	24.5–29.5	5	27	1,179
30–34 years old	30–34	29.5–34.5	5	32	895
35–39 years old	35–39	34.5–39.5	5	37	345
40 years old or more	40–?	39.5–?	?	?	58
$\sum$					4,065

EXAMPLE 4.9 The open-ended grouped frequency distribution shown in Table 4.8 was taken from *USA Statistics in Brief: 1992*, a supplement to the *Statistical Abstract of the United States: 1992*. It shows the age-group frequency distribution for the resident population of the United States for the year 1991. The frequency for each age group (column labeled 1991) is given in units of 1,000,000. Therefore, the actual number of people in each

Table 4.8

Population	1991 (people in millions)
Under 5 years old	19.2
5–17 years old	45.9
18–24 years old	26.4
25–34 years old	42.9
35–44 years old	39.3
45–64 years old	46.7
65 years old and over	31.8

age group is obtained by multiplying the given frequency by 1,000,000. Using the information provided, add the following completed columns to the summary table: class limits, class boundaries, class width, and class mark.

Solution

The completed summary table is given in Table 4.9, where a (?) symbol is again used to indicate that numerical values cannot be determined. This distribution illustrates that open-ended grouped frequency distributions can be constructed with unequal class widths. Here, there are two open-ended classes and four different class widths between the open ends.

Table 4.9

Population	Class limits	Class boundaries	Class width	Class mark m_i (years)	1991 (births in thousands) f_i
Under 5 years old	?–4	?–4.5	?	?	19.2
5–17 years old	5–17	4.5–17.5	13	11	45.9
18–24 years old	18–24	17.5–24.5	7	21	26.4
25–34 years old	25–34	24.5–34.5	10	29.5	42.9
35–44 years old	35–44	34.5–44.5	10	39.5	39.3
45–64 years old	45–64	44.5–64.5	20	54.5	46.7
65 years old or more	65–?	64.5–?	?	?	31.8
Σ					252.2

4.8 "LESS THAN" CUMULATIVE DISTRIBUTIONS

"Less than" cumulative frequency distributions show how many values in a data set are less than any given value. To produce such a distribution from a nongrouped frequency distribution, the frequencies are *cumulated* (added to the total) as one goes across the used part of the measurement scale from the smallest category (x_s) to the largest (x_l).

Where the sample data consist of continuous and therefore approximate measurements, the "less than" cumulation is to the upper boundary of the implied range of a measurement category (see Section 2.10). At each upper boundary the cumulation is the total number of measurements in the sample that are less than that boundary value.

If the sample data is discrete ratio and therefore exact measurement (see Section 2.10), its values do not have implied ranges, and thus "less than" cumulation of frequencies must be obtained differently.

Again cumulation is from category-to-category from x_s to x_l, but now the cumulation indicates for each category how many values are less than the category value itself.

EXAMPLE 4.10 Convert the frequency distribution in Table 4.1 into a "less than" cumulative frequency distribution.

Solution

The requested distribution is shown in Table 4.10, in a summary table called a "*less than*" *cumulative frequency table*. The top value in Table 4.1 (x_s) is 1.0 kg, an approximate measurement, which has an implied range of 0.95 kg to 1.05 kg. Therefore the cumulation in Table 4.10 starts at the upper boundary of the category below x_s, 0.95 kg, and indicates there are no numbers less than 0.95 kg. Next, as there are two numbers in the 1.0 kg category lying somewhere between 0.95 kg and 1.05 kg, the cumulation indicates that there are two numbers less than 1.05 kg. As there are no values in the 1.1 kg category (1.05 kg to 1.15 kg), the cumulation states that there are two values less than 1.15 kg. As there are four values in the next category, 1.2 kg (1.15 kg to 1.25 kg), the cumulation indicates there are now six values less than 1.25 kg. Continuing the process to its completion at 1.5 kg (x_l, 1.45 kg to 1.55 kg), the cumulation indicates that all 20 values in the sample are less than 1.55 kg.

Table 4.10

Weight (kg)	Cumulative frequency
Less than 0.95	0
Less than 1.05	2
Less than 1.15	2
Less than 1.25	6
Less than 1.35	14
Less than 1.45	18
Less than 1.55	20

4.9 "OR MORE" CUMULATIVE DISTRIBUTIONS

In the "*or more*" *cumulative frequency distribution*, the questions that can be answered are of the form: How many values in a data set are equal to or more than a given value? Working with approximate measurements, the cumulation now goes from x_l to x_s, asking for each measurement category: How many values are equal to or greater than the lower boundary of the implied range for the category? Working with exact measurements, the cumulation determines how many values are equal to or greater than a given category.

"Less than" and "or more" distributions are the most common versions of cumulative frequency distributions, but it is also legitimate to construct an "*or less*" *cumulative frequency distribution* (for approximate measurements, all values are equal to or less than the upper boundary of a category's implied range; for exact measurements, all values are equal to or less than the category itself) or a "*more than*" *cumulative frequency distribution* (for approximate measurements, all values are more than the lower boundary of a category's implied range; for exact measurements, all values are more than the category itself).

EXAMPLE 4.11 Convert the frequency distribution in Table 4.1 into an "or more" cumulative frequency distribution.

Solution

The requested "or more" distribution is shown in Table 4.11, in a summary table called an "*or more*" *cumulative frequency table*. In converting the distribution in Table 4.1 to the distribution in Table 4.11, the cumulation starts at the lower boundary of the next larger category to x_l. Since x_l equals 1.5 kg, the lower boundary of the next larger category (1.6 kg) equals 1.55 kg. This start indicates that no values in the sample are equal to or more than 1.55 kg. As there are then two values in x_l, lying between 1.45 kg and 1.55 kg, the cumulation indicates there are two values equal to or more than 1.45 kg. As there are four values in the next category, 1.4 kg (1.35 kg to 1.45 kg), the cumulation indicates that there are six values equal to or more than 1.35 kg. This process continues to completion at x_s, 1.0 kg (0.95 kg to 1.05 kg), where the cumulation indicates that all 20 values in the sample are equal to or more than 0.95 kg.

Table 4.11

Weight (kg)	Cumulative frequency
0.95 or more	20
1.05 or more	18
1.15 or more	18
1.25 or more	14
1.35 or more	6
1.45 or more	2
1.55 or more	0

4.10 GROUPED CUMULATIVE DISTRIBUTIONS

Grouped "less than" cumulative distributions are typically cumulated to an upper class boundary, but it is also legitimate to cumulate to a lower class limit. Similarly, for a grouped "or more" cumulative distribution, it is legitimate to cumulate from either a lower class boundary or a lower class limit.

EXAMPLE 4.12 Convert the grouped frequency distribution of marathon times in Table 4.4 into a grouped "less than" cumulative frequency distribution, using the upper class boundary for the cumulation.

Solution

The requested grouped cumulative distribution is shown in Table 4.12. It is seen that there are no values less than 127.5 min, the upper boundary for the class below the class containing x_s; and there are three values less than 130.5 min, the upper boundary of the class containing x_s; and so on until all 30 values are less than 145.5 min, the upper boundary of the class containing x_l.

Table 4.12

Time (min)	Cumulative frequency
Less than 127.5	0
Less than 130.5	3
Less than 133.5	4
Less than 136.5	8
Less than 139.5	11
Less than 142.5	18
Less than 145.5	30

Solved Problems

ARRAYS AND RANGES

4.1 From Table A.2, find ranges for the weights (in column 5) of the 16 females as well as the 16 students (both males and females) in the simple random sample from Example 3.5 (identified by the * symbol in column 9). Which of the two ranges is larger and why is this so?

Solution

For the 16 females

$$\text{Range} = 136 \text{ lb} - 105 \text{ lb} = 31 \text{ lb}$$

For the simple random sample

$$\text{Range} = 186 \text{ lb} - 115 \text{ lb} = 71 \text{ lb}$$

The range for the simple random sample is more than twice as large as the range for the females, because males tend to be heavier than females and this simple random sample happens to include both one of the four lightest females and one of the four heaviest males.

4.2 For both the proportional stratified random sample from Problem 3.12 and the systematic random sample from Problem 3.24 (identified in Table A.2 by the * in columns 10 and 11 respectively), put the term paper grades (column 8) into ascending arrays and find their ranges.

Solution

The assignment of letter grades (from A for excellent to B, C, D, and finally F for failure) to term papers is ordinal-level measurement (see Section 2.5). The measurements can be ordered (ranked) from smallest-to-largest or from largest-to-smallest, but it is not legitimate to determine exact distances between measurements by addition or subtraction. Therefore, for these sets of ordinal measurements we can determine their ascending arrays but not their ranges.

These are the grades for the proportional stratified random sample, taken in order of appearance from Table A.2: F, B, C, A, B, B, A, B, C, A, A, A, B, B, B, A, and this is their ascending array: F, C, C, B, B, B, B, B, B, B, A, A, A, A, A, A.

These are the grades for the systematic random sample, taken in order of appearance from Table A.2: B, B, A, A, B, C, B, C, D, B, C, A, B, A, F, B, and this is their ascending array: F, D, C, C, C, B, B, B, B, B, B, B, A, A, A, A.

Note: While the statistical measure of a range (range $= x_l - x_s$) is not a legitimate calculation for ordinal-level data, there is a nonstatistical meaning for the term range that could be applied here. In everyday English the term means the extent or scope of a group of things, and therefore it could be said here that "the grades range in both samples from Fs to As."

4.3 For the single-stage cluster random sample (see Problem 3.13) in Table A.2 (identified by the * in column 12), put the second lecture exam scores (column 3) into a descending array and then find the range.

Solution

The second lecture exam scores are the number of points each student earned out of a possible 100 points. This is discrete ratio-level measurement (see Section 2.8), and at this level both arrays and ranges are legitimate. The scores for the single-stage cluster random sample taken directly from the table are: 88, 78, 97, 82, 90, 94, 79, 90, 74, 88, 86, 81, 85, 64, 64, 92. The descending array for this sample is: 97, 94, 92, 90, 90, 88, 88, 86, 85, 82, 81, 79, 78, 74, 64, 64; and the range is: $97 - 64 = 33$.

4.4 Why were the female heights in Table A.2 measured to the nearest $\frac{1}{4}$ inch rather than to the nearest inch?

Solution

In order to provide sufficient separation between measurements in a data set, the part of the measurement scale being used should have between 30 and 300 unit steps. If the heights had been taken to the nearest inch (i.e., with unit steps of one inch), then the scale would extend from 60 to 69, covering only ten unit steps. By taking the measurements to the nearest $\frac{1}{4}$ inch (i.e., unit steps of $\frac{1}{4}$ inch), the number of usable steps increases to an acceptable 37.

DISTRIBUTIONS: FREQUENCY, RELATIVE FREQUENCY, AND PERCENTAGE

4.5 For Table 4.2, what are: (a) $\sum_{i=1}^{k}(f_i/n)$,   (b) $\sum_{i=1}^{k}\left[(f_i/n) \times (100)\right]\%$?

Solution

(a) Within the limits of rounding errors, the sum of the relative frequency column will always be 1.0.

$$\sum_{i=1}^{k}(f_i/n) = 1.0$$

(b) Again, within the limits of rounding errors, the sum of the percentage column will always be 100%.

$$\sum_{i=1}^{k}\left[(f_i/n) \times (100)\right]\% = 100\%$$

4.6 You are studying the characteristics of a type of pig that has been imported from China because its sows are said to produce an unusually large litter (the number of offspring produced in a single birth; discrete ratio-level measurement). The typical American sow has 10 to 12 piglets in a litter, and you have been told that these Chinese pigs can have 16 to 20 piglets per litter. Your results for 50 litters ($n = 50$) from these Chinese pigs are summarized in Table 4.13. Complete this table by filling in the missing values.

Table 4.13

Litter size x_i	Frequency f_i	Relative frequency f_i/n
15		0.02
16		0.04
17		0.10
18		0.40
19		0.24
20		0.12
21		

Solution

There is one missing value in the relative frequency column, for a litter size of 21. To find this value we utilize two facts: the sum of the entire relative frequency column is always 1.00 [see Problem 4.5(a)], and the sum of the relative frequencies for all litter sizes but 21 is 0.92. Therefore, the relative frequency for a litter of 21 is $1.00 - 0.92 = 0.08$.

With the relative frequency column completed, we can now calculate the frequencies for each litter size using the relationship $f_i = (f_i/n)(n)$. Thus, for a litter of 17 ($i = 3$ in the table), $f_3 = (0.10)(50) = 5$. The completed table, with frequencies calculated for all litter sizes, is shown in Table 4.14.

Table 4.14

Litter size x_i	Frequency f_i	Relative frequency f_i/n
15	1	0.02
16	2	0.04
17	5	0.10
18	20	0.40
19	12	0.24
20	6	0.12
21	4	0.08
$\sum$	50	1.00

4.7   From Table A.2, place the term-paper grades (column 8) into separate female and male summary tables, each with columns for grade, frequency, relative frequency, and percentage.

Solution

The summary table for females is shown in Table 4.15 and the summary table for males in Table 4.16.

Table 4.15   Females

Grade x_i	Frequency f_i	Relative frequency f_i/n	Percentage $[(f_i/n) \times (100)]\%$
F	1	0.0625	6.25
D	1	0.0625	6.25
C	3	0.1875	18.75
B	6	0.3750	37.50
A	5	0.3125	31.25
$\sum$	16	1.0000	100.00%

4.8   From Table A.2, arrange the hair colors (column 7) into separate female and male tables, each with columns for color, frequency, relative frequency, and percentage.

Solution

The summary table for females is shown in Table 4.17 and the summary table for males is shown in Table 4.18.

GROUPED DISTRIBUTIONS

4.9   When, for the same set of data, would a grouped distribution be preferred over an ungrouped distribution?

Solution

The function of a distribution, grouped or ungrouped, is to present summaries of quantitative information. Therefore, in deciding which type of distribution to use, the goal is to provide the most concise and immediately understandable summary—the one that most clearly shows what you want to emphasize. To achieve this goal, a grouped distribution is often preferable to an ungrouped distribution.

Table 4.16   Males

Grade x_i	Frequency f_i	Relative frequency f_i/n	Percentage $[(f_i/n) \times (100)]\%$
F	2	0.0417	4.17
D	4	0.0833	8.33
C	10	0.2083	20.83
B	20	0.4167	41.67
A	12	0.2500	25.00
$\sum$	48	1.0000	100.00%

Table 4.17   Females

Color x_i	Frequency f_i	Relative frequency f_i/n	Percentage $[(f_i/n) \times (100)]\%$
Black	3	0.1875	18.75
Blonde	8	0.5000	50.00
Brown	4	0.2500	25.00
Red	1	0.0625	6.25
$\sum$	16	1.0000	100.00%

Table 4.18   Males

Color x_i	Frequency f_i	Relative frequency f_i/n	Percentage $[(f_i/n) \times (100)]\%$
Black	10	0.2083	20.83
Blonde	14	0.2917	29.17
Brown	20	0.4167	41.67
Red	4	0.0833	8.33
$\sum$	48	1.0000	100.00%

The ungrouped distribution must include categories for all unit steps on the used part of the measurement scale, and recall (see Problem 4.4) that 30–300 unit steps are recommended. With so many unit steps, the data are often spread far apart, with many empty or low-frequency categories. The grouped distribution, by condensing these categories into a relatively few classes, provides a much clearer summary of the important information in the data. In the example given in Table 4.4, by condensing the 17 categories of the original frequency distribution (Table 4.3) into six classes, many empty and low-frequency categories were eliminated. This made the important trend in the data more visible: the rapid increase in frequency with longer finishing times.

4.10   From the information given, fill in the empty spaces in Table 4.19.

Solution

Table 4.20 is the completed table. Using the top row as an example of the calculations: the class limits are 1.1–1.2, which is the same as the class interval [column labeled Length (mm)]; the lower class boundary is

Table 4.19

Length (mm)	Class limits	Class boundaries	Class mark m_i (mm)	Class width	Frequency f_i	Relative frequency f_i/n
1.1–1.2					5	
1.3–1.4					6	
1.5–1.6					6	
1.7–1.8					8	
1.9–2.0					25	
Σ					50	

1.05, which is the lower end of the implied range for the approximate measurement 1.1 mm, and the upper class boundary is 1.25, which is the upper end of the implied range for the approximate measurement 1.2 mm; the class mark is $m_1 = \dfrac{1.1 + 1.2}{2} = 1.15$; and the class width is $1.3 - 1.1 = 0.2$. The relative frequency for this class is $f_1/n = 5/50 = 0.10$.

Table 4.20

Length (mm)	Class limits	Class boundaries	Class mark m_i (mm)	Class width	Frequency f_i	Relative frequency f_i/n
1.1–1.2	1.1–1.2	1.05–1.25	1.15	0.2	5	0.10
1.3–1.4	1.3–1.4	1.25–1.45	1.35	0.2	6	0.12
1.5–1.6	1.5–1.6	1.45–1.65	1.55	0.2	6	0.12
1.7–1.8	1.7–1.8	1.65–1.85	1.75	0.2	8	0.16
1.9–2.0	1.9–2.0	1.85–2.05	1.95	0.2	25	0.50
Σ					50	1.00

4.11   In Example 4.5 we gave the top 30 finishing times for male runners in a Boston Marathon. We now expand this set of data, giving, in an ascending array, the top 90 finishing times (in minutes) for males in this race:

129, 130, 130, 133, 134, 135, 136, 136, 138, 138, 138, 141, 141, 141, 142, 142, 142, 142, 143, 143, 143, 143, 143, 144, 144, 145, 145, 145, 145, 145, 146, 146, 146, 146, 147, 147, 147, 148, 148, 148, 148, 148, 148, 149, 149, 149, 149, 149, 150, 150, 150, 150, 151, 151, 151, 151, 152, 152, 152, 152, 152, 152, 152, 152, 153, 153, 153, 153, 153, 153, 153, 153, 153, 153, 154, 154, 154, 154, 154, 154, 154, 154, 155, 155, 155, 155, 155.

Using the guidelines from Section 4.6, put the fastest 85 times into a grouped distribution with nine classes, and place it in a summary table with columns for class intervals, class boundaries, class mark, frequency, and percentage.

Solution

For these 85 times, the range $= 154$ min $- 129$ min $= 25$ min. Therefore, as we prefer equal and odd class widths, and as it must be true that $25 < [(9) \times \text{(class width)}]$, the optimal class width is again 3. From these decisions and the other guidelines, we selected the grouped distribution in Table 4.21 as the best choice for the data.

Table 4.21

Time (min)	Class boundaries	Class mark m_i (min)	Frequency f_i	Percentage $[(f_i/n) \times (100)]\%$
128–130	127.5–130.5	129	3	3.53
131–133	130.5–133.5	132	1	1.18
134–136	133.5–136.5	135	4	4.71
137–139	136.5–139.5	138	3	3.53
140–142	139.5–142.5	141	7	8.24
143–145	142.5–145.5	144	12	14.12
146–148	145.5–148.5	147	13	15.29
149–151	148.5–151.5	150	13	15.29
152–154	151.5–154.5	153	29	34.12
Σ			85	100.01%

4.12 From Table A.2, using the guidelines from Section 4.6 and making the class width a multiple of 5, put all 64 of the second lecture exam scores (column 3) into a grouped distribution in a summary table with columns for class intervals, class boundaries, class mark, tally, frequency, and percentage.

Solution

Calculated from Table A.2, the range $= 99 - 49 = 50$. Using a class width of 5, we want: $50 < $ [(number of classes used) $\times$ (5)]. Therefore, the optimal number of classes is 11. Finally, with a class width of 5 and a highest score of 99 out of a possible 100, there are only two choices for the class with the largest upper class limit: (95–99) or (96–100). To emphasize the fact that no one in the statistics class achieved the maximum possible score, we have chosen (95–99). These decisions determined the eleven class intervals in the completed summary shown in Table 4.22 (column labeled "2d Exam"). Because the data is discrete ratio level, the class intervals, class limits, and class boundaries are all the same. The class mark for each class is the sum of the class limits divided by two. For the tally, a mark was placed next to the appropriate class mark for each of the 64 scores in Table A.2. Finally, tally was converted to frequency and percentage.

Table 4.22

2d Exam	Class boundaries	Class mark m_i	Tally	Frequency f_i	Percentage $[(f_i/n) \times (100)]\%$
45–49	45–49	47	|	1	1.5625
50–54	50–54	52		0	0.0000
55–59	55–59	57	||||	4	6.2500
60–64	60–64	62	|||	3	4.6875
65–69	65–69	67	|| 5	7.8125
70–74	70–74	72	|||	3	4.6875
75–79	75–79	77	|| |	6	9.3750
80–84	80–84	82	|| || |	11	17.1875
85–89	85–89	87	|| ||||	9	14.0625
90–94	90–94	92	|| || || ||	17	26.5625
95–99	95–99	97	||	5	7.8125
Σ				64	100.0000%

4.13   From Table A.2, using the guidelines from Section 4.6, put the 64 weights (column 5) into a grouped frequency distribution. Use a summary table showing the usual columns for class intervals and class mark, but then use separate columns of tally and frequency for females, males, and totals.

Solution

From Table A.2, for the 64 weights: the range = 194 lb − 105 lb = 89 lb. For this range, nine classes each of width 10 seem optimal. The only possible grouped frequency distribution for this combination is shown in Table 4.23.

Table 4.23

Weight (1b)	Class mark m_i (1b)	Females		Males		Totals	
		Tally	Frequency f_i	Tally	Frequency f_i	Tally	Frequency f_i
105–114	109.5	⦀	3		0	⦀	3
115–124	119.5	ℍ ⦀	8	⦀	3	ℍ ℍ ⎮	11
125–134	129.5	⦀⦀	4	⎮	1	ℍ	5
135–144	139.5	⎮	1	ℍ ⎮	6	ℍ ⎮⎮	7
145–154	149.5		0	ℍ ℍ ⎮	11	ℍ ℍ ⎮	11
155–164	159.5		0	ℍ ℍ ⎮	11	ℍ ℍ ⎮	11
165–174	169.5		0	ℍ ⦀	8	ℍ ⦀	8
175–184	179.5		0	⦀⦀	4	⦀⦀	4
185–194	189.5		0	⦀⦀	4	⦀⦀	4
$\sum$			16		48		64

4.14   Can grouped distributions be formed for nominal- and ordinal-level data?

Solution

No. The term grouped distribution is restricted to distributions of either interval- or ratio-level measurements. As defined (see Section 4.4), a grouped distribution is formed by grouping the unit steps of the original measurement scale into ordered classes with clearly defined and nonsubjective limits, boundaries, widths, and marks. Such grouping is not possible with nominal or ordinal data (see Sections 2.4 and 2.5).

It is not uncommon, however, for nominal and ordinal data to be reorganized for presentation or analysis by forming larger and more comprehensive categories. Thus, for example, the frequency distributions of the term paper grades (see Tables 4.15 and 4.16) may be reorganized to emphasize good and poor performance with these categories: (A), (B), and (C or below); or the hair color distributions (see Tables 4.17 and 4.18) may be reorganized to emphasize one hair color, say blonde and nonblonde. Reorganized nominal- and ordinal-level distributions, however, are not called grouped distributions.

OPEN-ENDED GROUPED DISTRIBUTIONS AND UNEQUAL CLASS WIDTHS

4.15   From examination of the open-ended grouped frequency distribution in Table 4.6, can you say whether the principles used in constructing this distribution are in agreement with the guidelines from Section 4.6?

Solution

It would seem that most of the guidelines were followed. Guideline 1 suggests 5–20 classes, and 6 were used. Guideline 2 says: "If possible, use the same class widths for all classes." The open-ended classes have no known width, but the other four classes each have a width of 5. This width (odd and a multiple of 5) agrees with guidelines 3 and 4. With open-ended classes, guidelines 5 and 7 do not apply, but we can assume that guideline 6 was followed.

4.16 The open-ended grouped frequency distribution shown in Table 4.24 was taken from page 471 of the *Statistical Abstract of the United States: 1995*. It shows the number of families in the United States that were in each of nine income groups in 1993. The number of families (column labeled Number) is in units of 1,000. Using the information provided, add the following completed columns to the summary table: class boundaries, class width, percentage.

Table 4.24

Household income	Number (families in thousands)
Under $5,000	4,407
$5,000–$9,999	9,467
$10,000–$14,999	8,956
$15,000–$19,999	8,319
$20,000–$24,999	8,103
$25,000–$34,999	14,318
$35,000–$49,999	15,791
$50,000–$74,999	15,632
$75,000 and over	12,114

Solution

The completed table is shown in Table 4.25. While household income is discrete ratio-level measurement (see Section 2.8), the class boundaries are not identical to the class limits. It is discrete ratio because income in the United States cannot be subdivided further than the level of cents, but the boundaries and limits are not identical because the incomes used for the table were rounded off to the nearest dollar. Thus, it is legitimate to calculate a boundary halfway between adjacent limits.

Table 4.25

Household income	Class boundaries	Class width	Number (families in thousands) f_i	Percentage $[(f_i/n) \times (100)]\%$
under $5,000	?–4,999.5	?	4,407	4.5383
$5,000–$9,999	4,999.5–9,999.5	5,000	9,467	9.7490
$10,000–$14,999	9,999.5–14,999.5	5,000	8,956	9.2228
$15,000–$19,999	14,999.5–19,999.5	5,000	8,319	8.5668
$20,000–$24,999	19,999.5–24,999.5	5,000	8,103	8.3444
$25,000–$34,999	24,999.5–34,999.5	10,000	14,318	14.7446
$35,000–$49,999	34,999.5–49,999.5	15,000	15,791	16.2614
$50,000–$74,999	49,999.5–74,999.5	25,000	15,632	16.0977
$75,000 and over	74,999.5–?	?	12,114	12.4749
Σ			97,107	100.0000%

4.17   The following descending array shows the money (in dollars) that was won by each of 70 golfers in a recent tournament [amount won followed by (number of golfers who won that amount)]:

200,000 (1); 83,333 (3); 45,000 (1); 40,000 (1); 36,250 (2); 30,000 (3); 21,900 (5); 15,000 (7); 10,000 (3); 7,535 (7); 5,750 (7); 4,260 (5); 3,220 (5); 2,750 (2); 2,490 (5); 2,380 (3); 2,330 (2); 2,290 (2); 2,260 (1); 2,240 (1); 2,220 (2); 2,180 (1); 2,170 (1).

You are a newspaper sportswriter doing a story about the tournament, and you want to use a grouped frequency distribution of the winnings to illustrate the following point: While a few golfers make an enormous amount of money from such a tournament, most make relatively little. After first rounding off the winnings to the nearest $1,000, illustrate this point with a grouped frequency distribution that has ten classes, none of which is open-ended. Use the following as the first of the ten classes: $2,000–$4,000. While in illustrating your story you would use only two columns for the distribution (class intervals and frequency), here, for your summary table, show: class intervals, class boundaries, class width, tally, and frequency.

Solution

The grouped frequency distribution shown in Table 4.26 is one possible solution to this problem. It has the ten classes, none of which is open-ended. Seven different class widths are used to illustrate the rapid decline and spread of the frequencies as winnings increased, and also to clearly show the largest prizes that were won.

Table 4.26

Winnings ($)	Class boundaries	Class width	Tally	Frequency f_i
2,000–4,000	1,500–4,500	3,000	|||| |||| |||| |||| |||| ||||	30
5,000–10,000	4,500–10,500	6,000	|||| |||| |||| ||	17
11,000–20,000	10,500–20,500	10,000	|||| ||	7
21,000–30,000	20,500–30,500	10,000	|||| |||	8
31,000–40,000	30,500–40,500	10,000	|||	3
41,000–45,000	40,500–45,500	5,000			1
46,000–81,000	45,500–81,500	36,000		0
82,000–83,000	81,500–83,500	2,000	|||	3
84,000–198,000	83,500–198,500	115,000		0
199,000–200,000	198,500–200,500	2,000			1
$\sum$				70

CUMULATIVE DISTRIBUTIONS

4.18   To the "less than" cumulative frequency table in Table 4.10, add columns showing the *"less than" cumulative relative frequency distribution* and the *"less than" cumulative percentage distribution*.

Solution

The completed summary table with the requested columns is shown in Table 4.27. Again considering only approximate measurements in nongrouped frequency distributions, a "less than" cumulative relative frequency distribution shows the proportion of the values in the data set that are less than the implied upper boundary of a measurement category, and a "less than" cumulative percentage distribution shows the percentage of the values less than that boundary.

All that was required to convert the "less than" cumulative frequency distribution in Table 4.10 to the "less than" cumulative relative frequency distribution in Table 4.27 was to divide each cumulative frequency

Table 4.27

Weight (kg)	Cumulative frequency	Cumulative relative frequency	Cumulative percentage
Less than 0.95	0	0.0	0
Less than 1.05	2	0.1	10
Less than 1.15	2	0.1	10
Less than 1.25	6	0.3	30
Less than 1.35	14	0.7	70
Less than 1.45	18	0.9	90
Less than 1.55	20	1.0	100

value by the sample size ($n = 20$). Thus, $0/20 = 0.0$, $2/20 = 0.1$, and so on to $20/20 = 1.0$. Next, to convert these relative frequency values to percentages and produce a "less than" cumulative percentage distribution, it is only necessary to multiply each "less than" cumulative relative frequency value by 100. Thus: $0.0 \times 100 = 0$, $0.1 \times 100 = 10$, and so on to $1.0 \times 100 = 100$.

4.19 Convert the frequency and relative frequency distributions for litter size in Table 4.14 to "less than" cumulative frequency and relative frequency distributions.

Solution

Litter size is exact measurement (see Section 2.10). Therefore, to obtain the "less than" cumulative frequency distribution, the cumulation from category-to-category from x_s to x_l must now indicate for each category how many values are less than the category itself.

Then, to obtain the "less than" cumulative relative frequency distribution, each cumulative frequency value is again divided by sample size. Using these techniques, the requested cumulative distributions for this problem are shown in Table 4.28. Thus, from Table 4.14, $x_s = 15$ with a frequency of 1, so the top row in Table 4.28 indicates that there are no values less than 15, and that the proportion of values less than 15 is $0/50 = 0.0$. Then, in the next row, the cumulation indicates there is one value less than 16 (proportion $= 1/50 = 0.02$). Continuing this process to $x_l = 21$, we find that 46 values are less than 21 (proportion $= 46/50 = 0.92$). To complete the cumulation, we include the next category above x_l and say that there are 50 values less than 22 (proportion $= 50/50 = 1.00$).

4.20 To the "or more" cumulative frequency table in Table 4.11, add columns showing the *"or more"* *cumulative relative frequency distribution* and the *"or more"* *cumulative percentage distribution*.

Table 4.28

Litter size	Cumulative frequency	Cumulative relative frequency
Less than 15	0	0.00
Less than 16	1	0.02
Less than 17	3	0.06
Less than 18	8	0.16
Less than 19	28	0.56
Less than 20	40	0.80
Less than 21	46	0.92
Less than 22	50	1.00

Solution

The completed summary table with the requested columns is shown in Table 4.29. As with the conversion from frequency to relative frequency in Problem 4.18, to convert the "or more" cumulative frequency in Table 4.11 to an "or more" cumulative relative frequency, it is simply a matter of dividing the cumulative frequency by the sample size ($n = 20$). Thus, as 20 values are 0.95 kg or more, the proportion of values that are 0.95 kg or more is $20/20 = 1.0$. And, at the other extreme, as 0 values are 1.55 kg or more, the proportion of values that are 1.55 kg or more is $0/20 = 0.0$. Now to convert an "or more" cumulative relative frequency distribution to an "or more" cumulative percentage distribution, it is simply a matter of multiplying each cumulative relative frequency value by 100. Thus: $1.0 \times 100 = 100$, $0.9 \times 100 = 90$, and so on to $0.0 \times 100 = 0$.

Table 4.29

Weight (kg)	Cumulative frequency	Cumulative relative frequency	Cumulative percentage
0.95 or more	20	1.0	100
1.05 or more	18	0.9	90
1.15 or more	18	0.9	90
1.25 or more	14	0.7	70
1.35 or more	6	0.3	30
1.45 or more	2	0.1	10
1.55 or more	0	0.0	0

4.21 Convert the grouped frequency distribution of the second lecture exam scores in Table 4.22 into grouped *"or more" cumulative frequency* and *percentage distributions*.

Solution

The requested grouped cumulative distributions are shown in Table 4.30. In this problem, as the number of points is exact discrete ratio measurement (see Section 2.8), the class boundaries and limits are identical. Whichever is considered here, 100% of the scores are 45 or more; 98.4% are 50 or more; and so on until 0.0% are equal to 100, the maximum possible score for the exam.

4.22 Convert the grouped frequency distribution for golf winnings shown in Table 4.26 into a grouped "less than" cumulative percentage distribution. Use upper class boundaries in the "less than" cumulation.

Solution

The requested grouped cumulative distribution is shown in Table 4.31.
Note: This is an example of a cumulative distribution involving unequal class widths, and the same techniques used for equal-width distributions were applied without change.

4.23 A test of reaction times was given to 50 people who want to join a police force. The results, measured to the nearest tenth of a second, are summarized in the "or more" cumulative percentage distribution shown in Table 4.32, where the lower implied boundary of a category was used in the cumulation. If the police department requires a reaction time of 0.5 seconds or less on this test, how many of the 50 candidates passed the test? How many were only 1/10 of a second too slow?

Table 4.30

2d Exam	Cumulative frequency	Cumulative percentage
45 or more	64	100.0
50 or more	63	98.4
55 or more	63	98.4
60 or more	59	92.2
65 or more	56	87.5
70 or more	51	79.7
75 or more	48	75.0
80 or more	42	65.6
85 or more	31	48.4
90 or more	22	34.4
95 or more	5	7.8
100	0	0.0

Table 4.31

Winnings ($)	Cumulative percentage
Less than 1,500	0.0
Less than 4,500	42.9
Less than 10,500	67.1
Less than 20,500	77.1
Less than 30,500	88.6
Less than 40,500	92.9
Less than 45,500	94.3
Less than 81,500	94.3
Less than 83,500	98.6
Less than 198,500	98.6
Less than 200,500	100.0

Table 4.32

Reaction time (sec)	Cumulative percentage
0.15 or more	100
0.25 or more	98
0.35 or more	94
0.45 or more	86
0.55 or more	76
0.65 or more	50
0.75 or more	30
0.85 or more	18
0.95 or more	8
1.05 or more	0

Solution

We can see from Table 4.32 that 76% of the candidates had times of 0.55 seconds or more, so we know that (100% − 76% = 24%) of the candidates had times of 0.5 seconds or less. Thus, (0.24 × 50 = 12) of the candidates passed the test.

For the second question, we know that while 76% were 1/10 of a second or more slower than required (0.55 or more), 50% were 2/10 of a second slower or more. Therefore, (76% − 50% = 26%) of the candidates, or 13 people, were only 1/10 of a second too slow.

4.24 A car dealership has 20 salespeople assigned to selling new cars. The grouped "or more" cumulative percentage distribution shown in Table 4.33 summarizes how many cars were sold by each of the salespeople during a three-month interval. Thus, 100% of them sold "0 or more," 90% of them sold "3 or more," and so on.   (a) If the numbers given in the "or more" column are the lower class boundaries, then for the class that has a lower boundary of 0, what are the class limits, class mark, and class width?   (b) How many of the salespeople sold exactly 7 cars?   (c) How many of the salespeople sold 7 cars or less?   (d) What proportion of the salespeople sold at least 9 cars?   (e) What is the total number of new cars that were sold by the 20 salespeople during this three-month interval?

Table 4.33

Cars sold by each salesperson	Cumulative percentage
0 or more	100
3 or more	90
6 or more	75
9 or more	75
12 or more	45
15 or more	20
18 or more	5
21 or more	0

Solution

(a)   Class limits: 0–2; class mark: 1; class width: 3

(b)   We know that 75% of the salespeople sold "6 or more" cars, and that 75% also sold "9 or more." As this is an "or more" distribution, we are cumulating from large to small; and as there was no change in percentage from "9 or more" to "6 or more" we can deduce that no new sales were cumulated in the (6–8) class. Therefore, none of the salespeople sold exactly 6, 7, or 8 cars.

(c)   We know that (100% − 75% = 25%) of the salespeople, or five of them, sold 5 cars or less, and that none sold 6 or 7 cars. Therefore, we can conclude that five salespeople sold 7 cars or less.

(d)   The phrase "at least 9" is the same as "9 or more." Therefore, to find the proportion of "at least 9" we divide the percentage for "9 or more" by 100: 75/100 = 0.75.

(e)   Because the data have been grouped, it is not possible to calculate from Table 4.33 an exact answer to this question. Thus, for example, while we can say that 5% of the 20, or one person, sold between 18 and 20 cars, we cannot say whether it was exactly 18, 19, or 20. While it is not possible to give an exact total from the information given in Table 4.33, we can calculate a reasonable approximation of the totals by assuming that all the values in the class are concentrated at the class mark. Then, after determining the frequency for each class, we can get the approximate total by taking the sum over all the classes of this product: [(class mark) × (class frequency)]. Therefore:

$$\text{Approximate total} = (1 \times 2) + (4 \times 3) + (7 \times 0) + (10 \times 6) + (13 \times 5)$$

$$+ (16 \times 3) + (19 \times 1)$$

$$= 206$$

Note: You will see in Chapters 6 and 7 that this technique of assuming all values in a class are concentrated at the class mark is used to approximate other descriptive measures when exact calculations are not possible because only grouped data are available.

Supplementary problems

ARRAYS AND RANGES

4.25   In a figure skating competition, a male skater received this set of marks for "artistic impression" from the eight judges: 6.0, 5.5, 5.6, 5.9, 6.0, 5.9, 5.7, 5.7. Place these marks in an ascending array and calculate their range.

   Ans.   Such "artistic impression" scores are ordinal-level measurement; the judges have rated a performance on a subjective scale of aesthetic quality ranging from absolutely terrible (0.0) to perfection (6.0). Therefore, while it is not legitimate to calculate their range (see Problem 4.2), they can be placed in this ascending array: 5.5, 5.6, 5.7, 5.7, 5.9, 5.9, 6.0, 6.0.

4.26   For the females in Table A.2, place the hair color data (column 7) into a descending array and calculate the range.

   Ans.   Hair color is nominal-level measurement, which means (see Section 2.4) that it is not legitimate to either order the measurements from large to small (place them in a descending array) or calculate distances between them (calculate their range).

4.27   The map of the state of Colorado in Fig. 4-1 shows the maximum temperatures (°F) recorded in 12 cities on a day in July. If it is legitimate, put these measurements into an ascending array and calculate their range.

   Ans.   Temperature measurement on the Fahrenheit scale is an example of interval-level measurement (see Section 2.6), and at this level both arrays and ranges are legitimate. The ascending array is: 74, 76, 79, 80, 81, 83, 84, 85, 85, 88, 92, 93, and the range is: $93°F - 74°F = 19°F$.

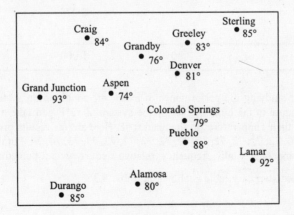

Fig. 4-1

4.28   For a men's downhill ski race, these are the times it took (minutes: seconds. hundredths of a second) for the ten fastest racers to cover the course, recorded in the order in which they raced: 1:24.77, 1:26.23, 1:26.03, 1:26.82, 1:25.61, 1:24.91, 1:23.38, 1:25.80, 1:26.65, 1:24.07. If it is legitimate, after first converting these times to the nearest second, place them in an ascending array and calculate their range.

   Ans.   These times are an example of continuous ratio-level measurement (see Section 2.8), in which both arrays and ranges are legitimate. The ascending array for the times converted to seconds is: 83, 84, 85, 85, 86, 86, 86, 86, 87, 87; and the range is: $87 \sec - 83 \sec = 4 \sec$.

DISTRIBUTIONS: FREQUENCY, RELATIVE FREQUENCY, AND PERCENTAGE

4.29 For Table 4.3, what are: (a) $\sum_{i=1}^{k}(f_i/N)$, (b) $\sum_{i=1}^{k}[(f_i/N) \times (100)]\%$?

Ans. (a) 1.0, (b) 100%

4.30 An economist studying trends in gasoline prices within a city takes a simple random sample of 40 of the city's gas stations, determining for each station the price per gallon (in dollars; discrete ratio-level measurement) of unleaded regular gasoline. Here are the results, presented in an ascending array: 1.05, 1.07, 1.07, 1.07, 1.08, 1.10, 1.10, 1.10, 1.10, 1.10, 1.10, 1.10, 1.10, 1.10, 1.10, 1.10, 1.10, 1.11, 1.11, 1.11, 1.11, 1.11, 1.11, 1.11, 1.11, 1.11, 1.11, 1.12, 1.12, 1.12, 1.12, 1.12, 1.13, 1.13, 1.13, 1.13, 1.14, 1.14, 1.14, 1.15. Put this array into a summary table with columns for price, tally, frequency, relative frequency, and percentage.

Ans. The completed summary table is shown in Table 4.34.

Table 4.34

Price ($) x_i	Tally	Frequency f_i	Relative frequency f_i/n	Percentage $[(f_i/n) \times (100)]\%$
1.05	|	1	0.025	2.5
1.06		0	0.000	0.0
1.07	|||	3	0.075	7.5
1.08	|	1	0.025	2.5
1.09		0	0.000	0.0
1.10	⫦ ⫦ ||	12	0.300	30.0
1.11	⫦ ⫦	10	0.250	25.0
1.12	⫦	5	0.125	12.5
1.13	||||	4	0.100	10.0
1.14	|||	3	0.075	7.5
1.15	|	1	0.025	2.5
$\sum$		40	1.000	100.0%

4.31 You are an engineer studying the performance of a new air conditioning system in a 60-story building. To evaluate the temperature in the building, set by the system at 72°F you take a simple random sample of 20 rooms and measure their temperatures to the nearest °F. Here are the results presented in an ascending array: 69, 69, 70, 70, 70, 70, 72, 72, 72, 72, 72, 72, 72, 72, 72, 72, 73, 73, 74, 74. Put this array into a summary table with columns for temperature, tally, frequency, relative frequency, and percentage.

Ans. The completed summary table is shown in Table 4.35.

GROUPED DISTRIBUTIONS

4.32 In reading a journal article, you find the grouped frequency distribution presented in Table 4.36. The first column [Weight (kg)] gives the class mark for each of the five classes in the table, and the second column (Frequency) gives the frequency for each class. Using this information, add completed columns to the table for class limits, class boundaries, class width, and percentage.

Ans. The completed summary table is shown in Table 4.37.

Table 4.35

Temperature (°F) x_i	Tally	Frequency f_i	Relative frequency f_i/n	Percentage $[(f_i/n) \times (100)]\%$
69	⋅‖	2	0.1	10
70	‖‖	4	0.2	20
71		0	0.0	0
72	⊮ ⊮	10	0.5	50
73	‖	2	0.1	10
74	‖	2	0.1	10
$\sum$		20	1.0	100%

Table 4.36

Weight (kg) m_i	Frequency f_i
0.3	3
0.8	13
1.3	42
1.8	15
2.3	7
$\sum$	80

Table 4.37

Weight (kg) m_i	Class limits	Class boundaries	Class width	Frequency f_i	Percentage $[(f_i/n) \times (100)]\%$
0.3	0.1–0.5	0.05–0.55	0.5	3	3.75
0.8	0.6–1.0	0.55–1.05	0.5	13	16.25
1.3	1.1–1.5	1.05–1.55	0.5	42	52.50
1.8	1.6–2.0	1.55–2.05	0.5	15	18.75
2.3	2.1–2.5	2.05–2.55	0.5	7	8.75
$\sum$				80	100.00%

4.33   The results of the second exam for the statistics class are summarized in Table 4.22. Before the exam, the class was given the following equivalents of letter grades for exam scores: (90–99) is an A, (80–89) is a B, (70–79) is a C, (60–69) is a D, and (below 60) is an F. What percentage of the class got each letter grade for the exam?

Ans. As to $(26.5625 + 7.8125 = 34.3750)\%$, Bs to $(17.1875 + 14.0625 = 31.2500)\%$, Cs to $(4.6875 + 9.3750 = 14.0625)\%$, Ds to $(4.6875 + 7.8125 = 12.5000)\%$, and Fs to $(1.5625 + 6.2500 = 7.8125)\%$.

4.34   From Table A.2, using the guidelines from Section 4.6 and making the class width 1.00 and the first class mark 60.00, put all 64 height measurements (column 4) into a grouped frequency distribution. In your summary table, show: class intervals, class boundaries, class mark, and separate columns for tally and frequency for females and males.

Ans.   The completed summary is shown in Table 4.38.

Table 4.38

Height (in)	Class boundaries	Class mark m_i (in)	Females		Males	
			Tally	Frequency f_i	Tally	Frequency f_i
59.50–60.49	59.495–60.495	60.00	|	1		0
60.50–61.49	60.495–61.495	61.00		0		0
61.50–62.49	61.495–62.495	62.00	|	1		0
62.50–63.49	62.495–63.495	63.00	|	1		0
63.50–64.49	63.495–64.495	64.00	|||	3	|	1
64.50–65.49	64.495–65.495	65.00	ℍ	5	||||	4
65.50–66.49	65.495–66.495	66.00	||	2	|||	3
66.50–67.49	66.495–67.495	67.00	|	1	||||	4
67.50–68.49	67.495–68.495	68.00	|	1	ℍ |	6
68.50–69.49	68.495–69.495	69.00	|	1	ℍ ℍ ||	12
69.50–70.49	69.495–70.495	70.00		0	ℍ ||	7
70.50–71.49	70.495–71.495	71.00		0	||||	4
71.50–72.49	71.495–72.495	72.00		0	|||	3
72.50–73.49	72.495–73.495	73.00		0	|||	3
73.50–74.49	73.495–74.495	74.00		0	|	1
$\sum$				16		48

OPEN-ENDED GROUPED DISTRIBUTIONS AND UNEQUAL CLASS WIDTHS

4.35   The open-ended grouped frequency distribution shown in Table 4.39 was taken from page 671 of the *Statistical Abstract of the United States: 1995*. It shows the frequency distribution of farms of different sizes (acreage) in the United States in 1992. The number of farms in each size group (column labeled 1992) is in units of 1,000. Using the information given in the distribution, add the following completed columns to the summary table: class limits, class boundaries, class width, class mark.

Ans.   The completed summary table is shown in Table 4.40.

4.36   From Table A.2, place all 64 household incomes (column 6) in the same nine classes used in Table 4.24. In the completed summary table, give: class intervals, tally, frequency, and percentage.

Ans.   The completed summary table is shown in Table 4.41.

Table 4.39

Size of farm	1992 (farms in thousands)
Under 10 acres	166
10–49 acres	388
50–99 acres	283
100–179 acres	301
180–259 acres	172
260–499 acres	255
500–999 acres	186
1,000–1,999 acres	102
2,000 acres and over	71

Table 4.40

Size of farm	Class limits	Class boundaries	Class width	Class mark m_i (acres)	1992 (farms in thousands) f_i
Under 10 acres	?–9	?–9.5	?	?	166
10–49 acres	10–49	9.5–49.5	40	29.5	388
50–99 acres	50–99	49.5–99.5	50	74.5	283
100–179 acres	100–179	99.5–179.5	80	139.5	301
180–259 acres	180–259	179.5–259.5	80	219.5	172
260–499 acres	260–499	259.5–499.5	240	379.5	255
500–999 acres	500–999	499.5–999.5	500	749.5	186
1,000–1,999 acres	1,000–1,999	999.5–1,999.5	1,000	1,499.5	102
2,000 acres and over	2,000–?	1,999.5–?	?	?	71
$\sum$					1,924

Table 4.41

Household income	Tally	Frequency f_i	Percentage $[(f_i/n) \times (100)]\%$
Under $5,000		0	0.000
$5,000 to $9,999		0	0.000
$10,000 to $14,999	‖	2	3.125
$15,000 to $19,999	‖‖	4	6.250
$20,000 to $24,999	ɴ̄ ɴ̄	10	15.625
$25,000 to $34,999	ɴ̄ ɴ̄ ɴ̄ ɴ̄ ‖‖	24	37.500
$35,000 to $49,999	ɴ̄ ɴ̄ ‖‖	14	21.875
$50,000 to $74,999	ɴ̄ ‖	6	9.375
$75,000 and over	‖‖	4	6.250
$\sum$		64	100.000%

CUMULATIVE DISTRIBUTIONS

4.37   Convert the frequency distribution of temperatures (°F) in Table 4.35 into both a "less than" cumulative frequency distribution and a "less than" cumulative percentage distribution.

   Ans.   The requested distributions are shown in Table 4.42.

4.38   Convert the frequency and relative frequency distributions for litter size in Table 4.14 to "or more" cumulative frequency and percentage distributions.

   Ans.   The completed table is shown in Table 4.43.

4.39   Convert the grouped frequency distributions for students weights in Table 4.23 into grouped "less than" cumulative frequency distributions.

   Ans.   The requested cumulative distributions are shown in Table 4.44. The cumulation was done to the upper class boundaries.

Table 4.42

Temperature (°F)	Cumulative frequency	Cumulative percentage
Less than 68.5	0	0
Less than 69.5	2	10
Less than 70.5	6	30
Less than 71.5	6	30
Less than 72.5	16	80
Less than 73.5	18	90
Less than 74.5	20	100

Table 4.43

Litter size	Cumulative frequency	Cumulative percentage
15 or more	50	100
16 or more	49	98
17 or more	47	94
18 or more	42	84
19 or more	22	44
20 or more	10	20
21 or more	4	8
22 or more	0	0

Table 4.44

Weight (lb)	Cumulative frequency		
	Females	Males	Totals
Less than 104.5	0		0
Less than 114.5	3	0	3
Less than 124.5	11	3	14
Less than 134.5	15	4	19
Less than 144.5	16	10	26
Less than 154.5		21	37
Less than 164.5		32	48
Less than 174.5		40	56
Less than 184.5		44	60
Less than 194.5		48	64

4.40 Convert the grouped frequency distributions for student weights in Table 4.23 into grouped "or more" relative frequency distributions.

Ans. The requested cumulative distributions are shown in Table 4.45. First the group cumulative frequencies were calculated from the lower class boundaries, and then to get cumulative relative frequencies these cumulative frequencies were divided by sample size $(n_f = 16$, $n_m = 48$, and $n_t = 64)$.

Table 4.45

Weight (lb)	Cumulative relative frequency		
	Females	Males	Totals
104.5 or more	1.00		1.00
114.5 or more	0.81	1.00	0.95
124.5 or more	0.31	0.94	0.78
134.5 or more	0.06	0.92	0.70
144.5 or more	0.00	0.79	0.59
154.5 or more		0.56	0.42
164.5 or more		0.33	0.25
174.5 or more		0.17	0.12
184.5 or more		0.08	0.06
194.5 or more		0.00	0.00

4.41 Convert the open-ended grouped frequency distribution of farm sizes shown in Table 4.40 into grouped "or more" cumulative frequency and percentage distributions. Use lower class limits in the "or more" cumulation.

Ans. A possible solution to this problem is given in Table 4.46. Note that because there are open-ended classes at both ends of the distribution, it is not possible to establish either the 0% or 100% boundaries. Therefore, the frequencies and percentages for the open-ended classes must in some way be identified separately. Because of these complexities, some statistics books recommend that cumulative distributions not be used for open-ended distributions.

Table 4.46

Size of farm (acres)	Cumulative frequency	Cumulative percentage
(under 10)	(166)	(8.6 under 10)
10 or more	1,758	91.4
50 or more	1,370	71.2
100 or more	1,087	56.5
180 or more	786	40.9
260 or more	614	31.9
500 or more	359	18.7
1,000 or more	173	9.0
(2,000 and over)	(71)	(3.7 2,000 and over)

Chapter 5

Descriptive Statistics: Graphing the Data

5.1 BAR GRAPHS, LINE GRAPHS, AND PIE GRAPHS

A *graph* is a diagram that shows relationships between variables: how changes in one variable are related to changes in another, how one variable (the dependent variable) is a function of another (the independent variable) (see Section 1.17). In this chapter we concentrate on graphs that show how changes in a measurement variable X are related to changes in either frequency, relative frequency, or percentage. Three types of graphs are used: *bar graphs, line graphs*, and *pie graphs*.

A bar graph is constructed on a rectangular coordinate system (see Section 1.20), where by tradition the X axis represents the independent variable, the Y axis the dependent variable, and rectangles (or bars) show the relationship between the variables. An example of such a bar graph used to illustrate a frequency distribution is presented in Fig. 5-1(a). It shows how the frequency of occurrence of measurement values in a sample changes as one goes across the measurement scale, and thus how frequency (the Y axis) could be a function of the measurement variable (the X axis). The frequency of each measurement value is proportional to the height of the vertical rectangle above that value. Note that, by convention, when presenting empirical distributions in graphs the axes are not identified with X and Y symbols.

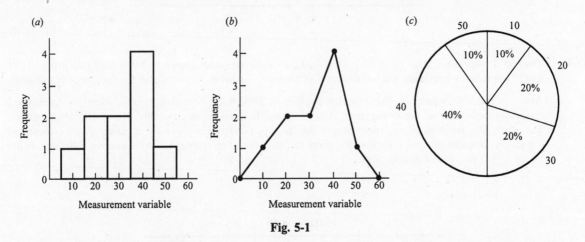

Fig. 5-1

A line graph, also constructed on a rectangular coordinate system, shows the relationships between variables by means of dots connected by lines or by continuous lines. A line graph version of the frequency distribution in Fig. 5-1(a) is shown in Fig. 5-1(b). Now the frequency of each measurement value is represented by the height of a dot above the X axis, and the dots are connected by straight lines.

Bar graphs and line graphs can also be used to display relative frequency or percentage distributions. All that is required is to convert the Y axis of the coordinate system to relative frequency or percentage.

Pie graphs, unlike bar graphs and line graphs, are not constructed on rectangular coordinate systems. They show the relationship between variables by dividing a circle (or pie) into appropriately sized sectors (or slices). In this chapter they are used to display relative frequency and percentage distributions. In the pie graph shown in Fig. 5-1(c), the frequency distribution from Figs. 5-1(a) and 5-1(b) has been converted to a percentage distribution, with the percentage for each measurement value proportional to the area of the sector for that value.

5.2   BAR CHARTS

A *bar chart* is a bar graph for nominal-level, ordinal-level, and discrete ratio-level data (see Sections 2.4, 2.5, and 2.8) that shows frequency, relative frequency, or percentage by the height of the bar and not by the bar's area, and shows vagueness or discontinuities in the measurement scale by spaces between the bars.

EXAMPLE 5.1   Construct a *frequency bar chart* for the frequency distribution of male hair color shown in Table 4.18.

Solution

The frequency bar chart (also called a *frequency bar graph* or a *frequency bar diagram*) for this distribution is shown in Fig. 5-2. In this chart, the categories of the measurement variable (hair color) are evenly spaced along the X axis. If a measurement scale has an order (small to large, low to high, etc.), this order should be preserved on the X axis, with the smallest (or lowest) category placed near the Y axis. Here, because the variable is nominal level, and thus unordered, the categories could have been arranged in any order on the X axis. The Y axis has a frequency scale marked off in equal units of five. Above each category on the X axis, each measurement in the sample in that category is represented by a small rectangle that has a standard base width and a height equal to one on the frequency scale. The frequency of a category, therefore, is represented by the total number of small rectangles piled vertically above it. The ten measurements of black hair, for example, are represented by ten rectangles in the bar labeled "black." While piles of component rectangles are shown in Fig. 5-2 to illustrate how bar charts are constructed, they are rarely shown in finished presentations. Instead, the bars are left clear or are shaded in some fashion, and the frequency of the category is read by projecting the top line of the bar across to the vertical axis.

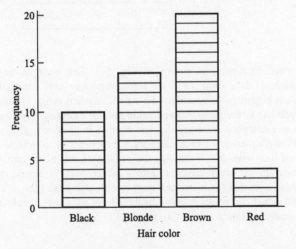

Fig. 5-2

The spacing between the bars—the fact that they are not touching—has significance. Such spacing indicates that the measurement variable is not continuous with uniform and standard reference units. The opposite also has meaning: When the bars are shown to be touching, in a form of bar graph called a *histogram* (see Section 5.3), it generally indicates that the measurement variable is continuous interval-level or ratio-level measurement.

There are no fixed rules for constructing the frequency scale on the Y axis, but the following guidelines are suggested by statistics books.

(1)   The length of the Y axis should be roughly 60%–75% of the length of the X axis.

(2)   The frequency scale should start at the X axis at zero and extend slightly above the highest category frequency.

(3)   It is arbitrary how many numbers are used for the frequency scale and how it is subdivided, but generally there are anywhere from 2 to 20 equal-length subdivisions.

5.3 HISTOGRAMS: UNGROUPED DATA

As indicated in Example 5.1, a *histogram* is a bar graph for continuous interval-level and ratio-level data. (Its use for discrete ratio-level data is discussed in Problem 5.9.) A histogram differs from a bar chart in that the histogram shows frequency, relative frequency, or percentage by the area of the bar and not always by its height, and shows the high level of measurement being graphed by bars that touch each other to form a continuous structure. All suggestions and guidelines from Example 5.1 for the construction of bar charts apply to the construction of histograms.

EXAMPLE 5.2 Construct a *frequency histogram* for the data summarized in Table 5.1.

Table 5.1

Length (cm) x_i	Frequency f_i
1.2	2
1.3	7
1.4	10
1.5	12
1.6	10
1.7	7
1.8	2
$\sum$	50

Solution

The requested histogram is shown in Fig. 5-3. For such an ungrouped frequency distribution of continuous ratio-level data, each bar in the histogram represents a measurement category with a base width that extends from boundary to boundary across the implied range for that category. The number below the midpoint of each bar is the measurement value for the category. This construction assumes that all sample measurements in a category are evenly spread over its implied range. If equal base widths are used for each category, then the frequency for each category is proportional to both the height and area of its bar.

If a vertical line were drawn through the middle of this histogram (above 1.5 on the X axis) it would divide the histogram into two identical, mirror-image parts. Because of this the histogram is said to be *symmetrical*. Also note that the histogram rises to only one peak (above 1.5), and because of this it is said to be *unimodal*. (If it had two peaks it would be called *bimodal*, three peaks *trimodal*, and so on; there is further discussion of *modes* and *modality* in Chapter 6.)

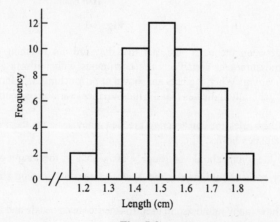

Fig. 5-3

5.4   HISTOGRAMS: GROUPED DATA

As grouped data, by definition (see Problem 4.14), is interval or ratio level, histograms should be used for continuous measurement variables. There is the usual problem with discrete ratio-level data (see Problem 5.9), but because the data is grouped it is suggested that histograms be used.

Each bar in a frequency histogram for grouped data represents frequency of a class. The base width of the bar, extending from the lower to upper boundaries of the class, represents class width, and if equal base widths (classes) are used, then the class frequency is proportional to both the area and the height of the bar. It is assumed that the values within a class are uniformly distributed across the bar width from boundary to boundary.

EXAMPLE 5.3   Construct a frequency histogram for the grouped frequency distribution of weights presented in Table 4.36. Show class marks (m_i) along the X axis.

Solution

The requested histogram is shown in Fig. 5-4. The bars of the histogram are centered on class marks along the X axis, but it is equally appropriate for such grouped data to show along the X axis: class boundaries, class limits, or the used part of the measurement scale.

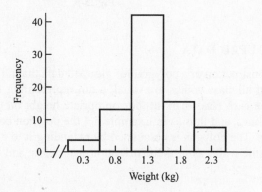

Fig. 5-4

5.5   POLYGONS: UNGROUPED DATA

A *polygon* is simply a line graph (see Section 5.1) of a frequency, relative frequency, or percentage distribution. In constructing a polygon for ungrouped continuous data, it is assumed that all measurements in a category are at the midpoint of the implied range for that category. For a *frequency polygon*, for example, a dot representing frequency for each category is placed above the category midpoint at the heights indicated by the vertical frequency scale. The category dots are then connected by straight lines. It is customary to extend the polygon to the X axis at both ends, by connecting it to the category midpoints just below and above the used part of the measurement scale. Attaching the polygon in this fashion to the X axis produces a closed plane figure bounded by straight lines, which is the definition of a polygon.

EXAMPLE 5.4   Construct a frequency polygon for the data presented in Table 5.1. In your graph, show how the frequency histogram for this data (Fig. 5-3) is related to the frequency polygon.

Solution

The requested combination of polygon and histogram is shown in Fig. 5-5. The polygon connects the midpoints of the tops of adjacent histogram bars. While these two figures are closely related, they are rarely presented together. When the polygon is constructed independently, the coordinates of each dot (x = measurement value, y = frequency) are used to plot them directly onto the coordinate system (see Section 1.20), and then the dots are connected by straight line segments. The same descriptions apply to polygons that are used for histograms, so this polygon is unimodal and symmetrical.

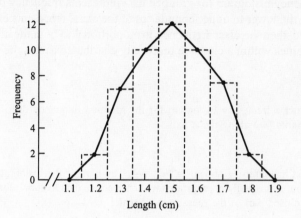

Fig. 5-5

5.6 POLYGONS: GROUPED DATA

It is assumed in the construction of a polygon for grouped data that all measurements within a class are at the class mark. Then, if all class widths are equal, a dot representing frequency, relative frequency, or percentage is placed above each class mark at the appropriate height on the vertical scale. The class dots are connected by straight lines, and thus as in Example 5.4 the polygon connects the midpoints of the tops of adjacent histogram bars. The polygon is completed by extending it to the X axis at both ends, assuming there are classes of the same width as in the distribution just below and above the histogram.

EXAMPLE 5.5 Construct a frequency polygon and histogram for the grouped data presented in Table 4.21. Show class marks (m_i) along the X axis.

Solution

The requested combination of polygon and histogram is shown in Fig. 5-6.

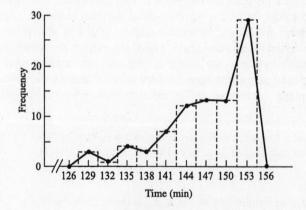

Fig. 5-6

5.7 FREQUENCY CURVES, RELATIVE FREQUENCY CURVES, AND PERCENTAGE CURVES

A frequency polygon for grouped data typically represents the frequency distribution of a sample that has been grouped into 5 to 20 classes. However, consider what would happen to the polygon if sample sizes were steadily increased, more and more classes were used in grouping the distribution, and class widths were steadily decreased. Under these conditions, as sample size n approached population size N the polygon would become a smooth curve called a *frequency curve* (or *smooth-curve frequency polygon*). Similarly, a relative frequency polygon would become a *relative frequency curve* and a percentage polygon would become a *percentage curve*. Such smooth curves for populations can be approximated (or *fit*) by theoretical mathematical curves. These theoretical curves, as you will see in later chapters, are the basic tools that allow probability decisions in inferential statistics.

5.8 PICTOGRAPHS

A *pictograph* (or *pictogram*) is a form of bar graph in which stylized, easily recognizable figures are used in place of rectangular bars.

EXAMPLE 5.6 The open-ended grouped frequency distribution shown in Table 5.2 was taken from page 471 of the *Statistical Abstract of the United States: 1995*. It shows, for males and females separately, the number of single-person households in the United States that were in each of nine income groups in 1993. You want to present these data in a *consecutive-parts relative frequency pictograph*. To simplify the presentation, you first condense the nine income groups into three arbitarily selected groups: lower income (under \$25,000), middle income (\$25,000 to \$49,999), and upper income (\$50,000 and over). Convert the frequency distributions in Table 5.2 into relative frequency distributions that have these three groups, and then graph the distributions with a consecutive-parts pictograph.

Table 5.2

	Single-person households (in thousands)	
Household income	Male householder f_i	Female householder f_i
Under \$5,000	704	1,386
\$5,000–\$9,999	1,257	4,013
\$10,000–\$14,999	1,324	2,439
\$15,000–\$19,999	1,106	1,678
\$20,000–\$24,999	1,080	1,279
\$25,000–\$34,999	1,639	1,578
\$35,000–\$49,999	1,217	1,083
\$50,000–\$74,999	698	520
\$75,000 and over	414	196
Σ	9,439	14,172

Solution

The requested relative frequency distributions are shown in Table 5.3, and the requested pictograph is shown in Fig. 5-7. Each complete small figure in the pictograph, male or female, represents a relative frequency of 0.1.

Table 5.3

| Household income | Single-person households (in thousands) | | | |
| | Male householder | | Female householder | |
	Frequency f_i	Relative frequency f_i/N	Frequency f_i	Relative frequency f_i/N
Lower income	5,471	0.5796	10,795	0.7617
Middle income	2,856	0.3026	2,661	0.1878
Upper income	1,112	0.1178	716	0.0505
$\sum$	9,439	1.0000	14,172	1.0000

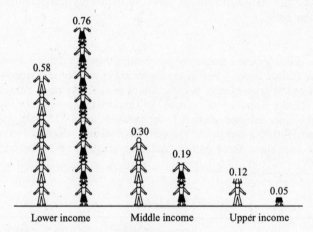

Fig. 5-7

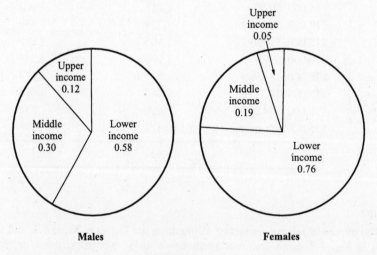

Fig. 5-8

5.9 PIE GRAPHS

As indicated in Section 5.1, *pie graphs* (or *pie charts*, or *circle diagrams*) represent relative frequency or percentage distributions with circles divided into sectors that are proportional in areas to the relative frequency or percentage values. If there is an ordering to the categories (or classes) of the distribution, it is typically preserved in a clockwise sequence starting at the 12 o'clock position.

EXAMPLE 5.7   From the data in Table 5.3, construct separate *relative frequency pie graphs* for males and females.

Solution

The requested pie graphs are shown in Fig. 5-8.

5.10 STEM-AND-LEAF DISPLAYS

Stem-and-leaf displays are techniques that allow rapid and informal exploration of the characteristics of a set of data. Among other things, they give a first view of symmetry and modality, and help determine the most appropriate class widths for grouping data in distributions. In a typical display there is a vertical line of numbers called *starting parts*, and for each starting part there is horizontal line of numbers called *leaves*. Each complete horizontal line (starting part plus leaves) is called a *stem*. Every number in the data set being displayed has both a starting part and a leaf.

The *stem width* (or *category interval*) determines which numbers in the data set are recorded on a given stem. Similar to a class width, it is the distance between the lowest value that is recorded on the stem in question and the lowest value that is recorded on the stem just beneath it.

A *simple stem-and-leaf display* has the following characteristics: (1) each stem has a different starting part, and (2) while each starting part can have more than one digit, each leaf of a stem must be only one digit.

EXAMPLE 5.8   Arrange the following set of numbers in a simple stem-and-leaf display that has single-digit starting parts and leaves, and a stem width of 10: 46, 35, 37, 20, 43, 15, 15, 26, 45, 25, 29, 13, 39, 44, 21, 24, 16, 40, 19, 45, 30, 34, 17, 39, 16, 40, 31, 21, 14, 42, 16, 43, 22, 11, 24, 25, 31, 27, 40, 33.

Solution

The requested simple stem-and-leaf display is shown in Fig. 5-9. The starting parts are the vertical numbers (1, 2, 3, 4) to the left of the vertical line, and the leaves are the numbers extending horizontally to the right. For starting-part 2, for example, the leaves are 06591412457. As we are asked to use single-digit starting parts and leaves, the first value in the set of numbers, 46, has a starting part of 4 and a leaf of 6; the second value, 35, has a starting part of 3 and a leaf of 5; and so on. Each value in a set of numbers is recorded in the display as it is encountered in the raw data, by writing its leaf to the right of its starting part. Thus, for the first value in the set, 46, the 6 is written to the immediate right of its starting part 4; for the second value encountered, 35, the 5 is written to the immediate right of starting part 3; then, for the next value, 37, the 7 is written to the right of starting part 3 but after leaf 5; and so on. As we are asked to form a stem width of 10, the stem with starting part 1 includes the numbers 10 through 19, and the next stem beneath it starts at the number 20.

When all the data is in the display, a *check count* is made in which the number of leaves on each stem is recorded in parentheses at the right of the stem, and the stem counts are then totaled at the bottom. This serves as a quick check on whether all the data have been recorded in the display.

```
1 | 5536976461   (10)
2 | 06591412457  (11)
3 | 579049113      (9)
4 | 6354050230    (10)
                  ----
                  (40)
```

Fig 5-9

5.11 GRAPHS OF CUMULATIVE DISTRIBUTIONS

An *ogive* is a graphical representation of a continuous cumulative frequency, relative frequency, or percentage distribution (see Sections 4.8, 4.9, and 4.10). Considering frequency as an example, a *"less than" frequency ogive* (or *"less than" cumulative frequency polygon*) is a line graph (see Section 5.1) constructed on a rectangular coordinate system from continuous data organized into a grouped or nongrouped "less than" cumulative frequency distribution.

For nongrouped data, where the "less than" cumulation is to the upper implied boundary of a measurement category (see Section 4.8), these upper boundaries are plotted along the X axis. The Y axis is cumulative frequency, which goes from zero at the X axis to the completed cumulation. The "less than" cumulative frequencies at each upper boundary are plotted as dots directly onto the coordinate system using the coordinates: $x =$ upper implied boundary of a category, $y =$ "less than" cumulative frequency at that boundary. The dots are connected by straight lines, and the completed "less than" frequency ogive rises to the right, from zero at the upper boundary of the category below x_s to the completed sample size at the upper boundary of x_l.

An *"or more" frequency ogive* (or *"or more" cumulative frequency polygon*) is a line graph constructed on a rectangular coordinate system from continuous data organized into a grouped or nongrouped "or more" cumulative frequency distribution. For nongrouped data, where the cumulation is from the lower implied boundary of a measurement category (see Section 4.9), all lower implied boundaries are plotted along the X axis. The Y axis is again cumulative frequency going from zero at the X axis to the completed cumulation. The "or more" cumulative frequencies at each lower boundary are plotted as dots directly onto the coordinate system using the coordinates: $x =$ lower implied boundary of a category, $y =$ "or more" cumulative frequency at that boundary. Again the dots are connected by straight lines, and the completed "or more" frequency ogive falls to the right from the complete sample size at the lower boundary of x_s to zero at the lower boundary of the category above x_l.

EXAMPLE 5.9 Construct a "less than" frequency ogive for the "less than" cumulative frequency distribution shown in Table 4.10.

Solution

The requested ogive is shown in Fig. 5-10. Reading from it, the height of the dot above 1.25 kg, for example, tells you there are six measurements in the sample that are less than 1.25 kg; and, from the dot above 1.55 kg, that all 20 measurements in the sample are less than 1.55 kg. Also, the fact that there is no change in "less than" cumulative frequency from 1.05 kg to 1.15 kg indicates that as there were two numbers less than both 1.05 kg and 1.15 kg, none of the measurements was 1.1 kg.

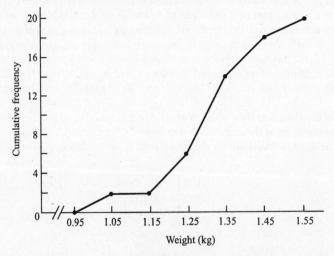

Fig. 5-10

EXAMPLE 5.10 Construct an "or more" frequency ogive for the "or more" cumulative frequency distribution shown in Table 4.29.

Solution

The requested ogive is shown in Fig. 5-11. Reading from it, the height of the dot above 1.25 kg tells us there are 14 measurements in the sample equal to or more than 1.25 kg; and the dot on the X axis at 1.55 kg indicates there are no measurements in the sample equal to or more than 1.55 kg. Again, the lack of change in the "or more" cumulative frequency from 1.05 kg to 1.15 kg indicates there were no 1.1 kg measurements in the sample.

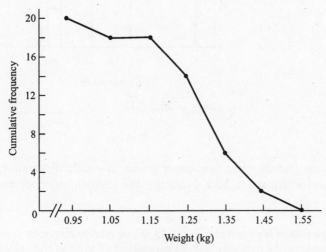

Fig. 5-11

Solved Problems

BAR CHARTS

5.1 The frequency distribution for male term-paper grades presented in Table 4.16 is shown in the bar chart in Fig. 5-12. What mistakes in technique were made in constructing this bar chart?

Solution

There are five mistakes (see Example 5.1):

(1) In Table 4.16, the term-paper grades are ordered from low-to-high, and thus the categories should be arranged left-to-right: *F, D, C, B, A*.

(2) As the measurement scale is ordinal and thus "not a continuous variable with uniform and standard reference units," there should be spaces between the vertical bars.

(3) The vertical scale is improperly marked off to show the frequency for each category. It should be marked off in equal intervals.

(4) The bars do not have equal base widths. Instead, each bar has a base width that is 1/3 its height, and because of this the bar areas magnify and distort the frequency differences.

(5) The recommendation that the vertical axis be 60%–75% of the horizontal axis has not been followed. Instead, the two axes have the same length.

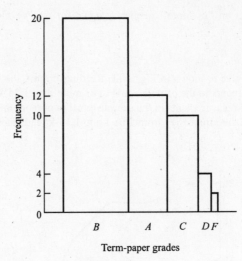

Fig. 5-12

5.2   Construct a bar chart for the male term-paper grades in which all the mistakes in Fig. 5-12 have been corrected, and which shows both frequency and relative frequency on the same graph.

Solution

The requested bar chart is presented in Fig. 5-13, where relative frequency is shown by a second vertical axis that has been added to the right side, with a scale marked off in equal units of 0.1. This relative frequency scale was coordinated with the frequency scale by first converting the desired relative frequency scale values to their equivalent frequencies; then finding the location of these frequencies on the frequency scale; and finally, marking off the relative frequency values at the same location (height above the X axis) on the right-hand scale. Thus, for example, a relative frequency of 0.1 is equivalent to a frequency of $(0.1 \times 48 = 4.8)$, and therefore the line marking 0.1 on the relative frequency axis was drawn at the same height as 4.8 on the frequency axis. Bar charts can be constructed with only frequency, relative frequency, or percentage on the left-hand axis, or with some combination of these variables on two parallel vertical axes.

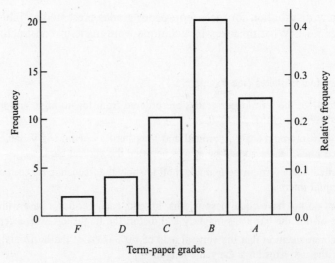

Fig. 5-13

5.3   From data for the female hair color in Table 4.17, construct a *horizontally oriented, percentage bar chart*.

Solution

The requested bar chart is shown in Fig. 5-14. In a horizontally oriented version of a bar chart for nominal data, the categories on the measurement scale are typically arranged such that either the longest bar is on top with a smooth progression downward to the shortest bar (as here), or the longest bar is on the bottom with a smooth progression upward to the shortest bar. The amount represented by the length of the bar can be indicated (as here) by both a number to the right of the bar and a scale at the base of the graph, or by either of these techniques alone. Note that the spacing between the bars, indicating noncontinuity and vague category boundaries, is also used in the horizontal version.

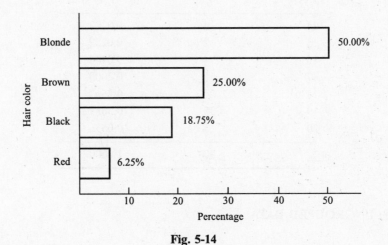

Fig. 5-14

5.4   Place the data for female (Table 4.17) and male (Table 4.18) hair color together in a *component-parts frequency bar chart* and a *consecutive-parts frequency bar chart*.

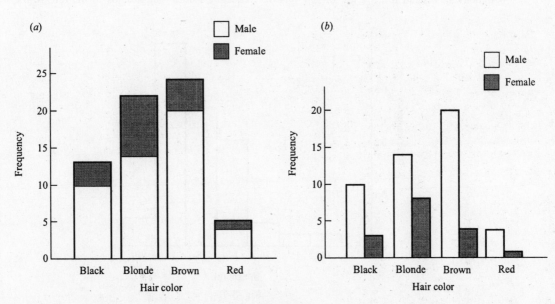

Fig. 5-15

Solution

The component-parts chart (also called a *component bar graph*) is shown in Fig. 5-15(*a*) and the consecutive-parts chart is shown in Fig. 5-15(*b*). It can be seen in the components-parts frequency bar chart that the total length of each bar represents the total frequency of a hair-color category, and that each bar has component lengths representing male and female contributions to the total frequency. In the consecutive-parts frequency bar chart, the male and female frequencies are shown as paired vertical bars for each category—two separate bar charts on the same axes.

Table 5.4

Weight (g) x_i	Frequency f_i	Relative frequency f_i/n
14	2	0.0222
15	2	0.0222
16	4	0.0444
17	18	0.2000
18	24	0.2667
19	35	0.3889
20	5	0.0556
$\sum$	90	1.0000

HISTOGRAMS: UNGROUPED DATA

5.5   Construct a *relative frequency histogram* for the data summarized in Table 5.4.

Solution

The requested histogram is shown in Fig. 5-16. Again, as in Example 5.2, a histogram is used because the measurements are continuous ratio, but now, while the histogram is unimodal as in Fig. 5-3, it is no longer symmetrical. Instead it has a *tail* of low relative frequency values that extends to the left, in the negative

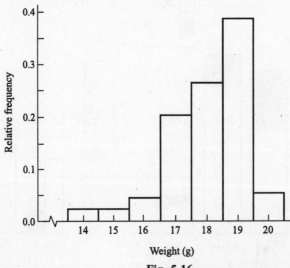

Fig. 5-16

Table 5-5

Temperature (°F) x_i	Frequency f_i	Relative frequency f_i/n
100	10	0.10
101	45	0.45
102	25	0.25
103	10	0.10
104	5	0.05
105	0	0.00
106	3	0.03
107	2	0.02
$\sum$	100	1.00

direction along the X axis. Because it is not symmetrical it is said to be *skewed*, and here it is said to be *skewed to the left* or *negatively skewed*.

5.6   Construct a histogram for the data summarized in Table 5.5, showing both frequency and relative frequency.

Solution

The requested histogram is shown in Fig. 5-17. As this data is continuous interval, a histogram is again appropriate. Note the gap in the histogram representing an absence of measurements for 105°F. This gap illustrates that while the histogram is for continuous measurements, it itself does not have to have a continuous structure. The techniques used to construct the coordinated frequency and relative frequency scales are those used for charts in Problem 5.2. This histogram is unimodal, and as it now has a tail extending to the right or in the positive direction, it is said to be *skewed to the right* or *positively skewed*.

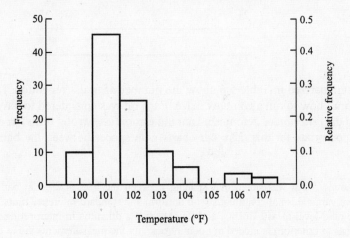

Fig. 5-17

5.7   In Figs. 5-3 and 5-16 there are breaks in the X axis near the Y axis that indicate a gap between zero and the first values on the used portion of the measurement scale. In Fig. 5-3 the symbol (∦) is used to indicate the gap and in Fig. 5-16 the symbol (∿) is used. There is a comparable gap in Fig. 5-17, for the lower range of temperatures, but no indication of it on the graph. Which of these techniques is the correct way to indicate such gaps?

Solution

Both types of symbols are commonly used to indicate such gaps as well as other symbols such as $(-\!/\!\!/-)$. It is also common practice, as in Fig. 5-17, to have the histogram start near the Y axis with no indication of a gap.

5.8   In Fig. 5-3 the bars in the histogram have equal base widths and therefore for each bar both height and area are proportional to frequency. Similarly, in Fig. 5-16 the equal base widths mean that for each bar both height and area are proportional to relative frequency. What are the total areas in the histograms (all the bar areas added together) proportional to in: (*a*) Fig. 5-3, (*b*) Fig. 5-16, and (*c*) Fig. 5-17?

Solution

(*a*)   The total area is proportional to the sum of the frequencies, or n (N if the data represent a population).

(*b*)   The total area is proportional to the sum of the relative frequencies, or 1.000.

(*c*)   With both frequency and relative frequency scales, the total area can be considered to be proportional to either n or 1.000.

Table 5.6

Number of trials to criterion x_i	Frequency f_i
15	1
16	3
17	9
18	15
19	8
20	2
21	2
Σ	40

5.9   The data summarized in Table 5.6 show the number of trials (repetitions) it took each of 40 white rats to "learn" how to run a complex maze. Each rat was considered to have learned the maze when it achieved the *criterion* of zero maze-running errors (i.e., wrong turns) in a trial. Which type of bar graph is appropriate for this data: bar charts with spaces between the bars or histograms?

Solution

In Example 5.1 it was stated that spacing between the bars is used in bar charts to indicate that the measurement variable is not continuous with uniform and standard reference units. Here, the data in Table 5.6 are discrete ratio-level measurements, and thus there is a dilemma in interpreting the rule. The measurement variable (trials to criterion) is indeed not continuous, but the measurements are at the highest level (ratio) and therefore the variable does have uniform and standard reference units (the number one). Because of this dilemma, both types of bar graph are used for discrete ratio data: bar charts to emphasize the discrete nature of the variable and histograms to emphasize that the data set is at a higher level than nominal-level or ordinal-level data.

There is another important reason why histograms might be used for this sort of data, and it has to do with the assumptions of parametric statistics. You will recall (see Section 3.13) that parametric statistical techniques are the preferred techniques for interval-level and ratio-level data, but only if the parametric

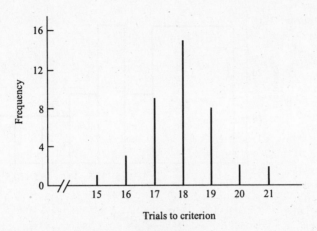

Fig. 5-18

assumptions are satisfied. One of these assumptions, the so-called *normal distribution assumption* includes the requirement that the measurement variable be continuous. It would thus seem that parametric techniques cannot be used with discrete ratio-level measurements. Fortunately, however, statisticians have determined that this discontinuity is not a severe problem for ratio-level data, so they suggest that discrete ratio-level data be treated as if they were continuous for the purpose of using parametric statistics. This, then, is a main reason why histograms are often used to display such data. Many statistics books attempt to resolve this contradiction by stating that while the measurement variable may be discrete, there is an underlying hypothetical variable (see Section 1.16) that is continuous. Here, for example, the measurement variable (trials to criterion) is discrete, but the underlying hypothetical variable of "learning" is continuous.

5.10   Construct a *rod graph* for the data summarized in Table 5.6.

Solution

The requested rod graph is shown in Fig. 5-18. Such a graph for frequency is a form of bar graph in which the height of a thin line is proportional to the frequency, and the line is assumed to have no width or area. It is another way to graph discrete ratio data to emphasize the discontinuities in a measurement variable.

HISTOGRAMS: GROUPED DATA

5.11   Construct a relative frequency histogram for the grouped data presented in Table 4.21. Show class boundaries along the X axis.

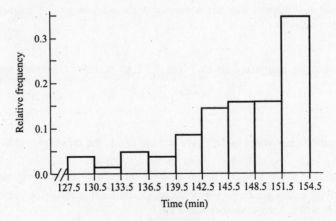

Fig. 5-19

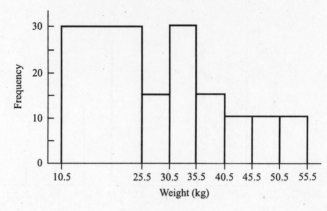

Fig. 5-20

Solution

The requested histogram is shown in Fig. 5-19. Having a maximum frequency at the right end, with a tapering downward of frequencies to the left, this negatively skewed distribution is described as *J-shaped*. If the characteristics were reversed (maximum at the left, tapering to the right), the positively skewed distribution would be described as *reversed-J-shaped*.

5.12   What is wrong with the frequency histogram presented in Fig. 5-20, in which class boundaries are shown along the X axis and class frequencies are proportional to bar heights?

Solution

This is an example of grouped data with unequal class widths. Thus, while the first class to the left has boundaries from 10.5 kg to 25.5 kg, and therefore a width of 15, the other six classes each have a width of 5. With such unequal class widths, if the bar heights are made proportional to frequency, then the bar areas will not be proportional to frequency, and area must always be proportional to frequency in a frequency histogram. It is this disproportionality that is wrong with the graph. Thus, for example, while the ratio of frequencies for the two classes having the boundaries (10.5 kg to 25.5 kg) and (30.5 kg to 35.5 kg) is 30/30 or 1/1, their bar areas are in the ratio 3/1. There is an accepted technique for correcting this problem (see how it is used to correct this graph in Problem 5.13).

(1)   Pick a *standard reference width*; typically the smallest class width in the grouped distribution.

(2)   Divide each class width by the reference width, and take the reciprocal of the resulting number.

(3)   Multiply each class frequency by the reciprocal for that class, achieving a product called the *frequency per reference width* (or the *frequency density*).

(4)   Construct a histogram that has a vertical scale marked off for frequency per reference width (or frequency density).

5.13   Use the technique suggested in Problem 5.12 to make a corrected version of the histogram in Fig. 5-20.

Solution

(1)   The smallest class width in the figure is 5, so this is the reference width.

(2)   Dividing each class width by 5 gives $15/5 = 3$, with a reciprocal of $1/3$, for the first class (10.5 kg to 25.5 kg), and $5/5 = 1$, with a reciprocal of 1, for each of the other six classes.

(3)   Multiplying each class frequency by the reciprocal for that class (see Table 5.7) gives the frequency per reference width; here, frequency per 5 kg.

(4)   The resulting histogram, with a vertical scale marked off in frequency per 5 kg, is shown in Fig. 5-21.

Table 5.7

Class boundaries	(Class frequency	×	Reciprocal for class)	=	Frequency per 5 kg
10.5–25.5	(30	×	1/3)	=	10
25.5–30.5	(15	×	1)	=	15
30.5–35.5	(30	×	1)	=	30
35.5–40.5	(15	×'	1)	=	15
40.5–45.5	(10	×	1)	=	10
45.5–50.5	(10	×	1)	=	10
50.5–55.5	(10	×	1)	=	10

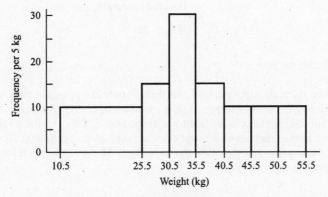

Fig. 5-21

Now the area of each bar is proportional to class frequency, while the height is proportional to frequency per 5 kg. To see the relationship between area and frequency, consider again the two classes (10.5 kg to 25.5 kg) and (30.5 kg to 35.5 kg), which have the ratio of frequencies of 30/30 or 1/1. In Fig. 5-21 the ratio of their bar areas is 1/1 instead of 3/1 as in Fig. 5-20. Note further that while the original histogram is bimodal and positively skewed, the corrected version in Fig. 5-21 is unimodal and symmetrical.

5.14 Construct a percentage histogram for the grouped household income data in Table 4.25, showing the used part of the measurement scale on the *X* axis.

Table 5-8

Household income	(Class percentage	×	Reciprocal for class)	=	Percentage per $5,000
Under $5,000	(4.5383	×	undefined)	=	?
$5,000 –$9,999	(9.7490	×	1/1)	=	9.7490
$10,000 –$14,999	(9.2228	×	1/1)	=	9.2228
$15,000 –$19,999	(8.5668	×	1/1)	=	8.5668
$20,000 –$24,999	(8.3444	×	1/1)	=	8.3444
$25,000 –$34,999	(14.7446	×	1/2)	=	7.3723
$35,000 –$49,999	(16.2614	×	1/3)	=	5.4199
$50,000 –$74,999	(16.0977	×	1/5)	=	3.2195
$75,000 and over	(12.4749	×	undefined)	=	?

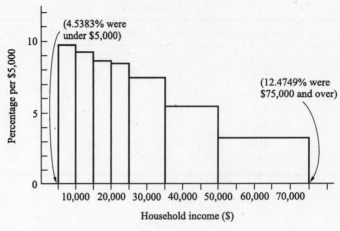

Fig. 5-22

Solution

    This is an example of graphing grouped data that not only have unequal class widths but also open-ended classes. In constructing the correct histogram, it is necessary to convert percentage to percentage per standard reference width, but only for the classes that have defined boundaries. As the smallest defined class width is $5,000, we use this as the reference width, and calculate percentage per $5,000. These calculations are shown in Table 5.8, and the requested histogram is shown in Fig. 5-22. Note that the information for the open-ended classes is written above the histogram.

POLYGONS: UNGROUPED DATA

5.15 Showing the related polygon and histogram, construct a *relative frequency polygon* for the data presented in Table 5.4.

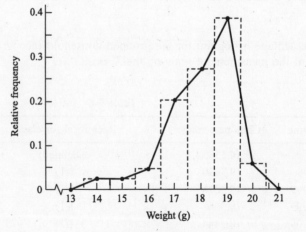

Fig. 5-23

Solution

    The requested combination of polygon and histogram is shown in Fig. 5-23, where you can see that a relative frequency polygon is a line graph of a relative frequency distribution. Again all values for a category are assumed to be at the midpoint of the implied range for that category, but now the height of the dot above the midpoint represents relative frequency. As before, the dots are connected by straight-line segments, and the polygon is extended to the X axis. As with the frequency polygon, the relative frequency polygon can be constructed from its histogram or by plotting the dots directly onto the coordinate system. This polygon, like its histogram, is unimodal and negatively skewed.

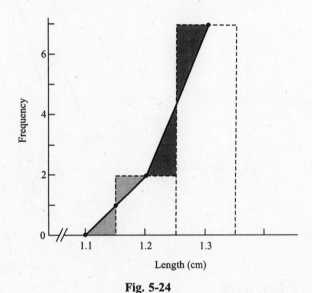

Fig. 5-24

5.16 We know from Example 5.2 and Problem 5.8 that the total area of the histogram in Fig. 5-5 is proportional to $n = 50$, and the total area of the histogram in Fig. 5-23 is proportional to 1.000. What are the total areas enclosed by the polygons in these figures proportional to?

Solution

In Fig. 5-5 the histogram and the polygon have identical areas, and thus the area enclosed by the polygon is proportional to $n = 50$. This equality of areas between polygon and histogram is also true in Fig. 5-23, so the area enclosed by the polygon is proportional to 1.000. In fact, it is always true that related histograms and polygons have identical areas if they are constructed from the same distribution onto the same coordinate system. To see why, we have magnified in Fig. 5-24 the first three categories in Fig. 5-5.

It can be seen in Fig. 5-24 that constructing the left side of the polygon onto the left side of the histogram has produced pairs of identical triangles along the left walls of the histogram bars. Examining the first pair (light shading) shows that one of the pair is under the polygon but outside the histogram, while the other is under the histogram but above the polygon. This same relationship to the histogram and polygon is true for the second pair (dark shading), and in examining Fig. 5-5 you can see that it is true for all left-side pairs in the graph, and that the mirror image of this pattern is true for all right-side pairs. Thus, for each triangular area in the histogram that is lost to the polygon, an identical triangular area is added to the polygon. This is why the total areas of the histogram and the polygon are identical. You can also see why, to achieve these identical areas, it is always necessary to extend the polygon to the X axis at both ends.

5.17 Can polygons be constructed for distributions of nominal, ordinal, or discrete ratio data?

Solution

Bar charts with spacings between the bars are used for nominal (Fig. 5-2) and ordinal (Fig. 5-13) data because their measurement variables are not continuous with uniform and standard reference units. For the same reason, it is not legitimate to represent such data with a continuous line graph. For discrete ratio data, on the other hand, it is allowable to use continuous polygons by the same arguments that permitted histograms for such data (see Problem 5.9).

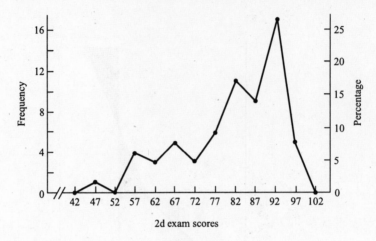

Fig. 5-25

POLYGONS: GROUPED DATA

5.18 Construct a polygon for the grouped discrete ratio data presented in Table 4.22, showing both frequency and percentage. Do not show the related histogram, but instead plot the dots directly onto the coordinate system. Show class marks (m_i) along the X axis.

Solution

The requested polygon is shown in Fig. 5-25.

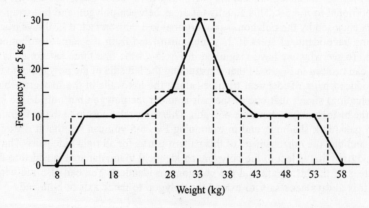

Fig. 5-26

5.19 Using information from Problem 5.12, convert the frequency per 5 kg histogram (Fig. 5-21) into a frequency per 5 kg polygon. Show the related histogram and polygon, and place class marks along the X axis.

Solution

The requested combination of histogram and polygon is shown in Fig. 5-26. The problem in constructing this polygon is in making it identical in area to the histogram shown in Fig. 5-21. This is done in Fig. 5-26 by ignoring the fact that the first histogram bar is three times as wide as the others, and instead treating it as if it were composed of three bars, each 5 kg in width. Then, the same techniques used for constructing polygons from equal-width distributions are used to construct the frequency per 5 kg polygon.

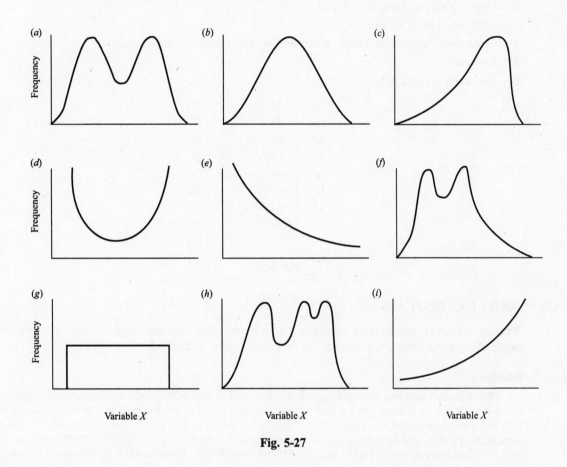

Fig. 5-27

5.20 Can the percentage per \$5,000 histogram shown in Fig. 5-22 be converted to a percentage per \$5,000 polygon?

Solution

No. A percentage polygon is a closed-plane figure. Thus, it is not legitimate to construct such a polygon for grouped data with open-ended classes.

FREQUENCY CURVES, RELATIVE FREQUENCY CURVES, AND PERCENTAGE CURVES

5.21 Theoretical frequency curves for a variety of distributions are shown in Fig. 5-27. For each curve, select from the following list the term (or terms) that describes the illustrated distribution: symmetrical, positively skewed, negatively skewed, J-shaped, reversed-J-shaped, unimodal, bimodal, trimodal.

Solution

(a) Symmetrical, bimodal

(b) Symmetrical, unimodal

(c) Negatively skewed, unimodal

(d) Symmetrical, bimodal (also called a *U-shaped distribution*)

(e) Positively skewed, reversed-J-shaped

(f) Positively skewed, bimodal

(g) Symmetrical (also called a *uniform distribution* or a *rectangular distribution*)

(h) Trimodal

(i) Negatively skewed, J-shaped

1a	341	(3)
1b	5569766	(7)
2a	014124	(6)
2b	65957	(5)
3a	04113	(5)
3b	5799	(4)
4a	3400230	(7)
4b	655	(3)
		(40)

Fig. 5-28

STEM-AND-LEAF DISPLAYS

5.22 Arrange the set of numbers in Example 5.8 in a *stretched stem-and-leaf display* that again has single-digit starting parts and leaves, but now has a stem width of 5.

Solution

The requested stretched stem-and-leaf display is shown in Fig. 5-28. While the simple display shown in Fig. 5-9 has a unique starting part for each stem and a stem width of 10, the stretched display shown in Fig. 5-28 has a stem width that has been halved, "stretching" the stem into two stems each of width 5, and the same starting part is used for both stems. Thus while in Fig. 5-9 the first stem was 1|5536976461, this has now been stretched into two stems: 1a|341 and 1b|5569766. Note that the letters a and b to the right of the starting parts indicate the first and second stems for each starting part. Also note that while a simple stem-and-leaf display has stem widths such as 0.1, 1, 10, 100, etc., the stretched stem-and-leaf display has stem widths such as 0.05, 0.5, 5, 50, etc.

While in Fig. 5-9 the frequency distribution of the sample appears to be rectangular in shape [see Fig. 5-27(g)], when "stretched" in Fig. 5-28 it becomes bimodal with peaks at starting parts 1b and 4a. This suggests that if we want to develop a grouped distribution for this data we should use class widths of 5 or less.

5.23 Arrange the set of numbers shown in Example 5.8 in a *squeezed stem-and-leaf display* that again has single-digit starting parts and leaves, but now has a stem width of 2.

Solution

The requested squeezed stem-and-leaf display is shown in Fig. 5-29, where it can be seen that each original stem of width 10 in Fig. 5-9 has now been subdivided ("squeezed") into five stems that each have a width of 2. All five stems have the same starting part as the original stem but are differentiated by the letters a, b, c, d, and e. Thus, 1a includes the values 10 and 11, 1b the values 12 and 13, 1c the values 14 and 15, 1d the values 16 and 17, and 1e the values 18 and 19. Squeezed stem-and-leaf displays always have stem widths such as 0.02, 0.2, 2, 20, etc.

1a	1	(1)
1b	3	(1)
1c	554	(3)
1d	6766	(4)
1e	9	(1)
2a	011	(3)
2b	2	(1)
2c	5445	(4)
2d	67	(2)
2e	9	(1)
3a	011	(3)
3b	3	(1)
3c	54	(2)
3d	7	(1)
3e	99	(2)
4a	000	(3)
4b	323	(3)
4c	545	(3)
4d	6	(1)
		(40)

Fig. 5-29

	Females		Males	
(2)	59	10		
(4)	1597	11	9	(1)
(7)	1143937	12	44	(2)
(3)	164	13	881	(3)
		14	82707394	(8)
		15	287622092181	(12)
		16	340900651	(9)
		17	2052360	(7)
		18	046	(3)
		19	014	(3)
(16)				(48)

Fig. 5-30

The stretched display in Fig. 5-28 is an improvement over the simple display in Fig. 5-9 in that it reveals the bimodality of the data and indicates that class widths for grouping the data should be restricted to 5 or less. But, in "squeezing" the stems we have spread the data too thin and lost sight of the bimodality. We have thus determined with three quick displays that class widths for grouping this data should be 5 or less, but greater than 2.

5.24 Directly from Table A.2, arrange the weight data (column 5) for the entire class in a *two-sided simple stem-and-leaf display* (females to the left, males to the right) that has two-digit starting parts, single-digit leaves, and a stem width of 10.

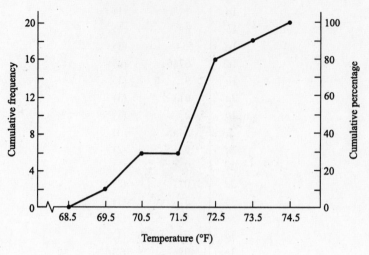

Fig. 5-31

Solution

The requested two-sided stem-and-leaf display is shown in Fig. 5-30. Such a display is used when it is known, or suspected, that two populations have been included in the same sample; it enables a rapid and informal comparison of the two distributions. The same starting parts are used for both distributions, with lines of leaves extending to the left and right of the starting parts.

Table 5.9

Number of spots x_i	Frequency f_i
1	2
2	3
3	4
4	3
5	2
$\sum$	14

GRAPHS OF CUMULATIVE DISTRIBUTIONS

5.25 For the distribution shown in Table 4.42, construct on the same coordinate system both a "less than" frequency ogive and a *"less than" percentage ogive* (or *"less than" cumulative percentage polygon*).

Solution

The techniques used in Section 5.11 are used to produce the line graph in Fig. 5-31. Now for the same dot, "less than" cumulative frequency can be read from the left-side vertical axis and "less than" cumulative percentage can be read from the right-side vertical axis.

5.26 As part of a genetics study, you count the number of body spots on each beetle in a sample of 14. The results are shown in the frequency distribution in Table 5.9. First, convert this frequency distribution into both a "less than" cumulative frequency distribution and an "or less" cumulative frequency distribution (see Section 4.9). Then, graph these two cumulative distributions on the same coordinate system.

Table 5.10

Number of spots	Cumulative frequency
Less than 1	0
Less than 2	2
Less than 3	5
Less than 4	9
Less than 5	12
Less than 6	14

Table 5.11

Number of spots	Cumulative frequency
0 or less	0
1 or less	2
2 or less	5
3 or less	9
4 or less	12
5 or less	14

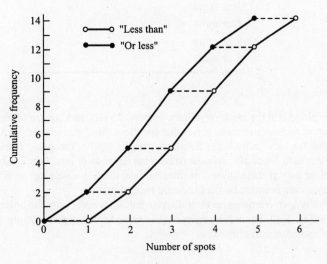

Fig. 5-32

Solution

The "less than" cumulative frequency distribution is shown in Table 5.10. Because these data are discrete ratio, cumulation indicates for each measurement category how many values are less than the category itself (see Problem 4.19). To achieve the "or less" cumulative frequency distribution shown in Table 5.11, the cumulation for each category was the number of values equal to or less than the category.

As the data are discrete ratio, these cumulative distributions can be graphed in either of two ways. They can be graphed to emphasize the discrete nature of the data or they can be graphed "as if they were continuous" (see Problem 5.9). If we consider them to be continuous, they can be graphed with ogives. This has been done in Fig. 5-32, where the cumulative distributions in Tables 5.10 and 5.11 have been graphed

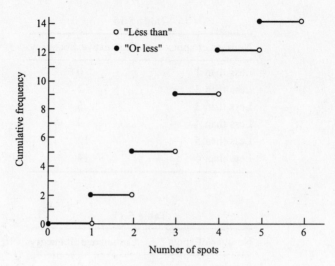

Fig. 5-33

Table 5.12

Number of spots	Cumulative frequency
1 or more	14
2 or more	12
3 or more	9
4 or more	5
5 or more	2
6 or more	0

with two ogives plotted on the same coordinate system. The dashed horizontal lines between the ogives show for two consecutive measurement categories that the "less than" cumulative frequency for the larger category is identical to the "or less" cumulative frequency for the smaller category. Thus, for example, there are five measurements less than 3 and also five measurements equal to or less than 2. The horizontal lines also show that as there are no intermediate values for discrete data, there is no change in either "less than" or "or less" cumulative frequencies between two consecutive measurements.

If we graph the data to emphasize their discrete nature, we plot only the horizontal lines in Fig. 5-32. This has been done in Fig. 5-33. There is no agreed upon name for this type of graph, so we will call them *discrete-data graphs for cumulative distributions*.

Table 5.13

Number of spots	Cumulative frequency
More than 0	14
More than 1	12
More than 2	9
More than 3	5
More than 4	2
More than 5	0

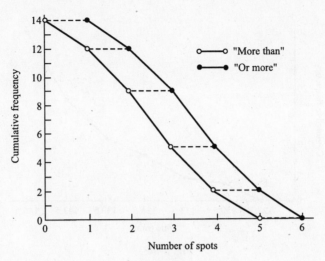

Fig. 5-34

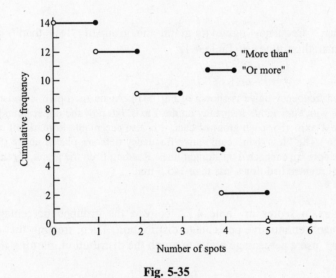

Fig. 5-35

5.27 Convert the frequency distribution in Table 5.9 into both an "or more" cumulative frequency distribution and a "more than" cumulative frequency distribution (see Section 4.9). Then, graph these two cumulative distributions on the same coordinate system.

Solution

The "or more" cumulative frequency distribution is shown in Table 5.12. Because these data are discrete ratio, the cumulation indicates how many values are equal to or more than a given measurement category. To construct the "more than" cumulative frequency distribution shown in Table 5.13, the cumulation for each category is the number of values that are more than the category.

As these data are discrete ratio, we can again graph them either as ogives or as discrete-data graphs for cumulative distributions. The two distributions are graphed as two ogives on the same coordinate system in Fig. 5-34 and as a discrete-data graph in Fig. 5-35. The dashed horizontal lines between the ogives in Fig. 5-34 show that for two consecutive measurement categories the "or more" cumulative frequency for the larger category is equal to the "more than" cumulative frequency for the smaller category, and also that there is no change in either "or more" or "more than" cumulative frequencies between the two consecutive categories.

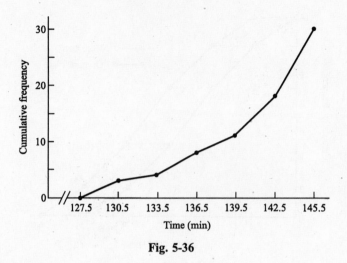

Fig. 5-36

Again, as in Fig. 5-33, the discrete-data graph of the data (Fig. 5-35) is simply the plot of the horizontal lines between the ogives.

5.28 Use a "less than" frequency ogive to graph the grouped "less than" cumulative frequency distribution of marathon times in Table 4.12.

Solution

The requested frequency ogive is shown in Fig. 5-36. Again the ogive is constructed on a rectangular coordinate system with cumulative frequency on the Y axis, but now the upper boundaries for the classes are marked off on the X axis. (For such grouped data, it is also acceptable to mark off class lower limits on the measurement scale.) The "less than" cumulative frequency dots are placed above the upper boundaries for their class and the dots are connected by straight lines. Reading from the graph, 8 runners had times less than 136.5 min, and all runners had times less than 145.5 min.

5.29 For the second exam scores in Table 4.22, convert the grouped percentage distribution into a grouped "less than" cumulative percentage distribution. Then, treating this discrete data "as if it were continuous," use a percentage ogive to graph the distribution, plotting the dots over the lower

Table 5.14

2d Exam score	Cumulative percentage
Less than 45	0.0000
Less than 50	1.5625
Less than 55	1.5625
Less than 60	7.8125
Less than 65	12.5000
Less than 70	20.3125
Less than 75	25.0000
Less than 80	34.3750
Less than 85	51.5625
Less than 90	65.6250
Less than 95	92.1875
Less than 100	100.0000

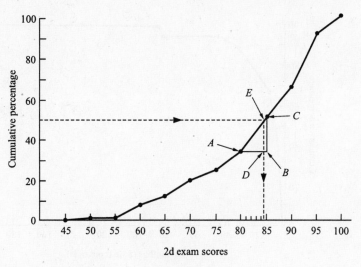

Fig. 5-37

limits for each class. Finally, using the graph, find the exam score below which are 50% of the scores.

Solution

The requested "less than" cumulative percentage distribution is shown in Table 5.14, and the requested percentage ogive is shown in Fig. 5-37. To use the graph to *interpolate* (or *approximate*) the requested score (the one that 50% of the scores are "less than"), the first step, as shown in Fig. 5-37, is to draw a horizontal line from the 50% mark on the *Y* axis to the percentage ogive. The *x* coordinate (or abscissa) of the point *E* where the horizontal line intersects the ogive is the required score. The next step is to drop a vertical line from *E* to the *X* axis. An approximate version of the score can then be read directly from where the vertical line crosses the *X* axis. For the scale of this graph it would seem to be roughly 84.5.

A more exact version of the score can be calculated using the similar triangles *ABC* and *ADE*. The sides of these triangles have the following relationship:

$$\frac{AD}{AB} = \frac{DE}{BC}$$

We know that: AB = class width = 5, BC = 51.5625 − 34.3750 = 17.1875, and DE = 50 − 34.3750 = 15.6250. So now we can solve for AD:

$$\frac{AD}{5} = \frac{15.6250}{17.1875}$$
$$AD = 5 \times 0.9091 = 4.5455$$

Therefore, the requested score is: 80.0 + 4.5455 = 84.5455, or 84.5.

This measurement, the one that 50% of the measurements in a sample are "less than," is called the *median* of the sample. It and other measures of *central tendency, average value*, and *location* are discussed in detail in Chapter 6.

5.30 Construct a "less than" percentage ogive for the "less than" cumulative percentage distribution of golf winnings shown in Table 4.31. Use the measurement scale on the *X* axis, and place the "less than" percentage dots over the upper boundaries of the classes.

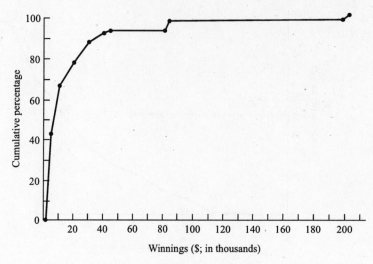

Fig. 5-38

Solution

The requested percentage ogive is shown in Fig. 5-38. Note that essentially the same graphing techniques are used for this grouped distribution with unequal class widths as are used for grouped distributions with equal class widths.

5.31   What is a *smooth-curve frequency ogive*?

Solution

In Section 5.7 we indicated that as sample size *n* approaches population size *N*, a frequency polygon becomes a smooth curve called a frequency curve. Similarly, as sample size approaches population size, a sample frequency ogive becomes a *smooth-curve frequency ogive* (as does a relative frequency ogive become a *smooth-curve relative frequency ogive* and a percentage ogive become a *smooth-curve percentage ogive*).

Supplementary Problems

BAR CHARTS

5.32   There will be three candidates (*A*, *B*, and *C*) for mayor in the next election. A poll is taken of 100 voters to determine which candidate each prefers. The results are shown in Fig. 5-39, in a frequency bar chart that is being distributed by candidate *C*. How many of the voters chose each of the candidates? Why is the chart a deceptive presentation of the results of this poll?

   Ans.   The results of the poll were actually fairly close: 31 for *A*, 33 for *B*, and 36 for *C*. The chart, however, makes it seem a decisive victory for *C* by three forms of deception: magnifying the height differences between the bars by starting the *Y* axis at 30 rather than zero, making the horizontal axis 75% of the vertical, and making the bar area for *C* disproportionately large by giving *C* twice the base width.

5.33   There are 200 balls in a jar; some are yellow (*Y*), some are red (*R*), some are blue (*B*), and the rest are white (*W*). The bar chart in Fig. 5-40 has a bar representing each color category and a relative frequency scale on the

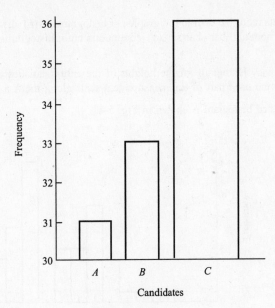

Fig. 5-39

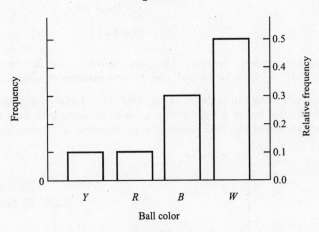

Fig. 5-40

right-hand axis. From the information given, determine the number of balls of each color in the jar and the numerical values for the five-line frequency scale marked off on the left-hand axis.

Ans. 20 *Y*, 20 *R*, 60 *B*, 100 *W*; the values are: 20, 40, 60, 80, and 100

HISTOGRAMS

5.34   For the following measurement variables, indicate first its level of measurement and then which form of bar graph is appropriate: (*a*) litter size (as in Table 4.13),   (*b*) attitudes of Americans toward immigrants as measured on a five-point scale from 1 (unfavorable) to 5 (highly favorable),   (*c*) body temperature (°F),   (*d*) diameter (in mm) of snail shells

Ans. (*a*) Discrete ratio; bar charts, rod graphs, or histograms,   (*b*) ordinal; bar charts,   (*c*) interval; histograms,   (*d*) continuous ratio; histograms

5.35   For the following measurement variables, indicate first its level of measurement and then which form of bar graph is appropriate: (*a*) words per minute in typing a sample section,   (*b*) price per gallon (in $) of gasoline in several areas,   (*c*) type of vitamin,   (*d*) miles driven per year by each truck in a fleet of trucks.

Ans. (*a*) Discrete ratio; bar charts, rod graphs, or histograms,   (*b*) discrete ratio; bar charts, rod graphs, or histograms,   (*c*) nominal; bar charts,   (*d*) continuous ratio; histograms

5.36 Construct a frequency histogram for the heights of the entire statistics class (males and females) shown in Table 4.38. Show the used part of the measurement scale along the *X* axis.

Ans. The requested histogram is shown in Fig. 5-41.

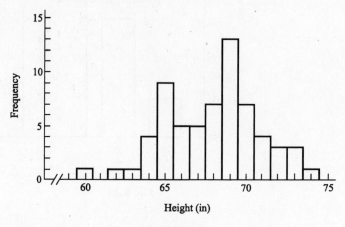

Fig. 5-41

5.37 Construct a *components-parts frequency histogram*, showing male and female components, for the data graphed in Fig. 5-41. Again show the used part of the measurement scale along the *X* axis.

Ans. The requested histogram is shown in Fig. 5-42. The version of this data in Fig. 5-41 is *bimodal* (has two peaks), which often means that the sample contains measurements from two different populations. The component-parts version in Fig. 5-42 confirms this interpretation, showing the two populations were males and females.

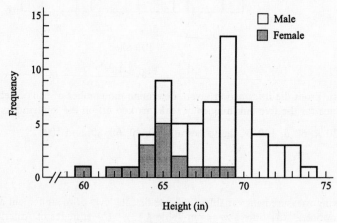

Fig. 5-42

POLYGONS

5.38 Construct a polygon for the data in Table 5.5, showing both frequency and relative frequency. Do not show the related histogram but instead plot the dots directly onto the coordinate system.

Ans. The requested polygon is shown in Fig. 5-43. Note for this unimodal and positively skewed polygon that it drops to the *X* axis for the zero-frequency category of 105°F.

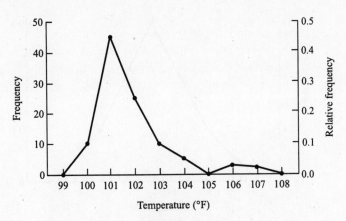

Fig. 5-43

5.39   You are the manager of a videotape rental store, and an employee has prepared a graph (Fig. 5-44) that summarizes for all customers who rented videotapes last week how many tapes each customer rented. (*a*) How many customers rented videotapes last week?   (*b*) What proportion of the customers rented two tapes?

Ans.   (*a*) 100,   (*b*) 0.5

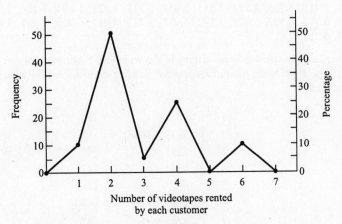

Fig. 5-44

5.40   From the information in Fig. 5-44, answer these questions.   (*a*) What percentage of the customers rented four tapes or fewer?   (*b*) How many customers rented three or more tapes?

Ans.   (*a*) 90%,   (*b*) 40

5.41   A fruit-importing company weighs a sample of melons for quality control. The results are shown in Fig. 5-45 in a polygon for grouped continuous ratio data, where the grouping employs equal class widths. The measurement scale is shown along the X axis.   (*a*) How many melons were weighed?   (*b*) The data was grouped into how many classes?

Ans.   (*a*) 40,   (*b*) 7

5.42   From the information in Fig. 5-45, answer these questions.   (*a*) What are the class marks for the seven classes?   (*b*) What are the class boundaries for these classes?

Ans.   (*a*) The class marks are 1.3, 1.6, 1.9, 2.2, 2.5, 2.8, 3.1,   (*b*) the class boundaries are: (1.15 to 1.45), (1.45 to 1.75), (1.75 to 2.05), (2.05 to 2.35), (2.35 to 2.65), (2.65 to 2.95), (2.95 to 3.25)

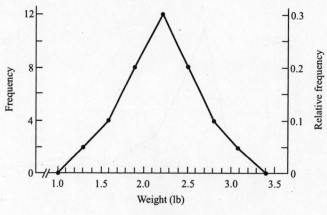

Fig. 5-45

STEM-AND-LEAF DISPLAYS

5.43 Arrange the following set of numbers in a simple stem-and-leaf display that has single-digit starting parts and leaves, and a stem width of 1:

3.792, 7.300, 1.419, 8.333, 3.212, 2.513, 2.937, 5.312, 4.821, 1.694, 2.100, 7.902, 9.111, 2.321, 2.119, 6.199, 8.774, 2.572, 3.999, 3.192, 5.988, 2.412, 4.911, 6.900, 7.297, 2.633, 4.431, 5.255, 6.591, 4.497, 6.511, 2.617.

Ans. The requested simple stem-and-leaf display is shown in Fig. 5-46, where it can be seen that while each original value has three digits to the right of its decimal, in this simple display only the first digit to the right of the decimal is recorded.

1	46	(2)
2	591315466	(9)
3	7291	(4)
4	8944	(4)
5	392	(3)
6	1955	(4)
7	392	(3)
8	37	(2)
9	1	(1)
		(32)

Fig. 5-46

5.44 Arrange the set of numbers in Problem 5.43 in a stem-and-leaf display that has single-digit starting parts, three-digit leaves, and a stem width of 1.

Ans. The requested stem-and-leaf display is shown in Fig. 5-47, where it can be seen that all the digits in the data can be presented in the leaves. If, as here, more than one digit is used in the leaves, then successive leaves are separated by commas.

GRAPHS OF CUMULATIVE DISTRIBUTIONS

5.45 In an international track meet, two semifinal heats were run in the men's 200-meter dash to determine the eight fastest men for the final. The results for the two heats are summarized in the two ogives in Fig. 5-48: an "or

1	419,694	(2)
2	513,937,100,321,119,572,412,633,617	(9)
3	792,212,999,192	(4)
4	821,911,431,497	(4)
5	312,988,255	(3)
6	199,900,591,511	(4)
7	300,902,297	(3)
8	333,774	(2)
9	111	(1)
		(32)

Fig. 5-47

more" frequency ogive for heat 1, where cumulation is from the lower implied boundary of a category; and a "less than" frequency ogive for heat 2, where cumulation is from the upper implied boundary of a category. (*a*) How many men ran in the two heats?   (*b*) How many men ran the 200 meters in 21.0 seconds?

Ans.   (*a*) 8 in heat 1, 7 in heat 2,   (*b*) 0

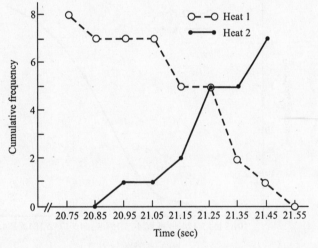

Fig. 5-48

5.46   From the information in Fig. 5-48, answer these questions. (*a*) What are the times of the eight fastest men?   (*b*) What are the slowest times in both heats?

Ans.   (*a*) [(heat 1) 20.8 sec], [(heat 2) 20.9 sec], [(heat 1) 21.1 sec, 21.1 sec], [(heat 2) 21.1 sec], [(heat 2) 21.2 sec, 21.2 sec, 21.2 sec],   (*b*) [(heat 1) 21.5 sec], [(heat 2) 21.4 sec]

5.47   Use a discrete-data graph to show the "or more" cumulative percentage distribution of litter size in Table 4.43.

Ans.   The requested discrete-data graph is shown in Fig. 5-49.

5.48   Construct separate female and male "less than" frequency ogives on the same coordinate system from the "less than" cumulative frequency distributions for weight in Table 4.44. Plot the upper boundaries on the *X* axis.

Ans.   The requested ogives are shown in Fig. 5-50.

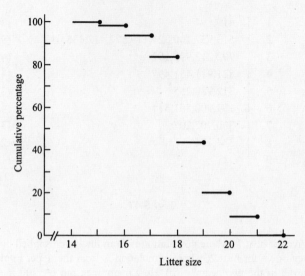

Fig. 5-49

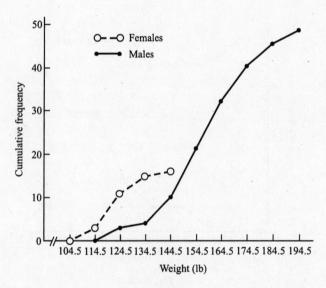

Fig. 5-50

Chapter 6

Descriptive Statistics: Measures of Central Tendency, Average Value, and Location

6.1 MEASURES OF CENTRAL TENDENCY, AVERAGE VALUE, AND LOCATION

The last two chapters dealt with two fundamental aspects of descriptive statistics: the organization of data into summary tables (Chapter 4) and the graphing of the organized data (Chapter 5). In this chapter and the next we go on to another aspect of descriptive statistics, the calculation of *descriptive measures* from the data: numerical values that summarize characteristics of the data, typically with a single number. In this chapter, we deal with the measures that describe *central tendency, average value*, and *location*, and then in Chapter 7 we deal with the measures that describe *dispersion* (the spread of data in a distribution). Other descriptive measures are introduced as they are needed throughout the book.

Descriptive statistical measures have two functions: they provide a mental image of a data distribution to someone with statistical training; and they are an essential component of inferential statistics, the basis of both estimation and hypothesis testing (see Section 3.6). They have this role in inferential statistics because most descriptive measures of samples have been developed as estimates of comparable population measures. As indicated in Section 3.4, the sample measure is called a statistic and the population measure that it is estimating is called a parameter.

To introduce the descriptive measures of this chapter, let us examine some of the characteristics of the symmetrical, unimodal frequency curve in Fig. 6-1. In this curve, the highest frequencies are found near the middle of the distance from x_s to x_l. This clustering of the measurements near the center of a distribution, typical of many types of data, is called *central tendency*, and the statistical measures that describe aspects of the "center" of a distribution are called *measures of central tendency*.

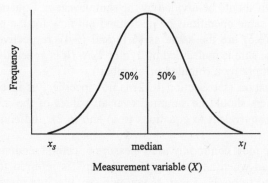

Fig. 6-1

The *average value* in a data set is the most typical, frequent, or representative measurement in the set. Because of the usual concentration of measurements in the center of a distribution, the various measures of central tendency are generally also called *measures of average value* (or *averages*).

Measures of location show where the characteristics of a distribution are located in relation to the measurement scale. Three measures of location that have already been introduced are shown in Fig. 6-1: x_s, and x_l, the minimum and maximum values in the data set, and the median (see Problem 5.29), which is shown here as the boundary point on the X axis to the left of which (and to the right of which) are 50% of the data. Because measures of central tendency and average value "locate" these characteristics relative to the measurement scale, some statistics books refer to all the measures described in this chapter as measures of location.

6.2   THE ARITHMETIC MEAN

All of the following formulas define the *arithmetic mean*:

$$\bar{x} = \frac{\sum_{i=1}^{n} x_i}{n} \tag{6.1}$$

$$\bar{x} = \frac{1}{n}\sum_{i=1}^{n} x_i \tag{6.2}$$

$$\bar{x} = \frac{\sum x_i}{n} \tag{6.3}$$

$$\mu = \frac{\sum_{i=1}^{N} x_i}{N} \tag{6.4}$$

$$\mu = \frac{1}{N}\sum_{i=1}^{N} x_i \tag{6.5}$$

$$\mu = \frac{\sum x_i}{N} \tag{6.6}$$

Equations (6.1) and (6.4) were introduced in Section 3.4. Equation (6.1) for the sample-statistic $\bar{x}$ states that to calculate $\bar{x}$ for a sample of measurements $x_1, x_2, \ldots, x_n$: the measurements should first be summed from x_1 to x_n, and then this sum should be divided by the sample size n. Equation (6.4) for the population-parameter μ instructs that the same operations be performed but now for the population $x_1, x_2, \ldots, x_N$ of size N. Equations (6.2) and (6.5) are the same as (6.1) and (6.4), respectively, except now instead of dividing the sum by n or N, the sum is multiplied by $1/n$ or $1/N$. (It is generally preferable to divide by an exact number rather than multiply by a rounded-off number.)

When the index of summation (see Section 1.22) is not specified, as in equations (6.3) and (6.6), it means the entire set of numbers should be summed over all values of the index variable (see Example 1.37). Thus, equation (6.3) is equivalent to equations (6.1) and (6.2), and equation (6.6) is equivalent to equations (6.4) and (6.5).

The arithmetic mean is the most commonly used measure of central tendency, average, and location. It is what is generally understood when an "average" or "mean" is referred to: batting average, average price, mean annual rainfall, and so on. However, as you will see, this interpretation may not be correct, as there are other measures called means and averages. The arithmetic mean is certainly the most important of these measures in inferential statistics, where the sample-statistic $\bar{x}$ is considered to be the most reliable and efficient estimate of its population-parameter μ. As to level of measurement, the arithemtic mean is really only legitimate for interval- and ratio-level measurements (continuous or discrete), but you will find it used for ordinal-level measurements as well.

EXAMPLE 6.1   Using equation (6.3), calculate the arithmetic means for the following samples: (*a*) $x_1 = 1$ g, $x_2 = 3$ g, $x_3 = 2$ g, $x_4 = 7$ g, $x_5 = 5$ g, $x_6 = 4$ g, $x_7 = 2$ g,   (*b*) 1 g, 3 g, 2 g, 7 g, 5 g, 4 g, 200 g.

Solution

(a) $\quad \bar{x} = \dfrac{\sum x_i}{n} = \dfrac{1\,g + 3\,g + 2\,g + 7\,g + 5\,g + 4\,g + 2\,g}{7} = \dfrac{24\,g}{7} = 3.4\,g$

(b) $\quad \bar{x} = \dfrac{\sum x_i}{n} = \dfrac{1\,g + 3\,g + 2\,g + 7\,g + 5\,g + 4\,g + 200\,g}{7} = \dfrac{222\,g}{7} = 31.7\,g$

Note: In (a) the values of x_i are identified: $x_1 = 1\,g$, $x_2 = 3\,g$, and so on. This is not done in part (b); instead it is assumed that the measurements are listed in the order $x_1, x_2, \ldots, x_n$. This ordering assumption holds true throughout this book wherever the specific values of the index of summation are not given. Note also in these calculations that the arithmetic means have the same units (g) as the measurements from which they were calculated. Also note how sensitive the arithmetic mean is to values that are quite different from the rest of their data set, values called *extreme values* or *outliers*. Thus, for example, between (a) and (b) when x_7 is changed from 2 g to 200 g the mean changes from 3.4 g to 31.7 g. The presence of extreme values in a data set often indicates some sort of procedural error or equipment failure. They can, however, be real, indicating the influence of some extraneous variable (see Section 3.10).

6.3   ROUNDING-OFF GUIDELINES FOR THE ARITHMETIC MEAN

In calculating descriptive measures like the arithmetic mean, there are two related but different rounding-off problems: (1) when and how to round off at the different steps in the calculations, and (2) how many digits are to be reported in the final answer.

Most descriptive measures presented in this book are defined by a formula, and the measure is *exactly equal* to the end result of the sequence of calculations specified by the formula. Thus, for example, $\bar{x}$ is exactly equal to the specific fraction: (sum of data) ÷ (sample size). While this is true in the abstract, in practice it is typically impossible to achieve the exact value—to do so may require that an infinite number of digits be retained throughout the steps of the calculations. While this is not possible, in order to get as close as you can to the exact value, *it is recommended that rounding off be kept to a minimum throughout the calculations*. There are no agreed-upon rules as to how many digits should be retained throughout the calculations, but some books suggest at least six.

Once the descriptive measure has been calculated, if it is to be used in further calculations then again many digits should be retained. If, instead, it is to be reported to an audience, then we have the second rounding-off question: How many digits should be reported in the final answer?

In reporting a descriptive measure, we are no longer concerned with the multiple digits of the exact value, but instead with conveying information. There are also no absolute rules for this sort of rounding off, only a variety of guidelines. One of the most commonly accepted guidelines for reporting arithmetic means, the ones we used in rounding off the answers in Example 6.1, can be stated as follows:

If the data are all at the same level of precision (see Section 2.15), then the mean should be reported at the next level of precision.

Thus, in Example 6.1(a) all of the original data are at the same level of precision, at the *units digit*, and therefore the arithmetic mean should be reported to the *tenths digit*. The calculated result for $\bar{x}$, using a calculator with a 12-digit display, is: 3.42857142857 g, and rounding this off to the tenths digit gives the reported answer: $\bar{x} = 3.4\,g$.

While this guideline for reporting the arithmetic mean is commonly stated in statistics books, there are other common guidelines: Report the arithmetic mean at either the next level of precision or at the same level as the data, coordinate the number of digits reported with the number reported for either the *standard deviation* (see Chapter 7) or the *standard error of the mean* (see Chapter 13). We present other important guidelines where they are relevant, but when rounding-off guidelines are not given for a descriptive measure, you can assume we are using the most typical procedure.

Finally, it should be mentioned that all of the calculations in this book were done with a calculator that allows a maximum display of 12 digits, of which 11 can be *decimal places* (digit positions to the right of the decimal point), and that this calculator is programmed to retain 15 digits during the steps of its calculations. If you are putting fewer digits in your calculator and the calculator is carrying fewer digits in its calculations, then you may get an answer to a problem that differs somewhat from the given answer.

EXAMPLE 6.2 Using equation (6.3), calculate the arithmetic means for the following samples: (a) $-3°C$, $+2°C$, $-1°C$, $+4°C$, $-6°C$, $-5°C$,   (b) 2.002 g, 3.7 g, 2.963 g, 3.5041 g, 2.737 g, 1.99999 g.

Solution

The suggested rounding-off guideline is used on part (a) data because they are all at the same level of precision. The data in part (b), however, should be treated differently because they are not at a uniform level of precision. For such data, we have to go back to the basic rounding-off rules for algebraic operations (see Section 2.15).

(a) $\bar{x} = \dfrac{\sum x_i}{n} = \dfrac{(-3°C) + 2°C + (-1°C) + 4°C + (-6°C) + (-5°C)}{6} = \dfrac{-9°C}{6} = -1.5°C$

(b) $\bar{x} = \dfrac{\sum x_i}{n} = \dfrac{2.002\,g + 3.7\,g + 2.963\,g + 3.5041\,g + 2.737\,g + 1.99999\,g}{6}$

$= \dfrac{16.9\,g}{6}$,   after rounding off the numerator

$= 2.81666666667\,g$,   or 2.82 g after rounding off the answer.

6.4 DEVIATIONS FROM AN ARITHMETIC MEAN AND THE CENTER OF GRAVITY OF A DISTRIBUTION

The difference (or distance) between any measurement in a population and the arithmetic mean of the population is called the measurement's *deviation from the population's arithmetic mean* (or simply *deviation from the mean*), and it is defined by the quantity $x_i - \mu$. Similarly, for any measurement in a sample, its *deviation from the sample's arithmetic mean* (or again *deviation from the mean*) is defined by $x_i - \bar{x}$. In a frequency histogram, deviations of measurements to the left of the mean (smaller than the mean) have negative signs and are called *negative deviations*, whereas deviations of measurements to the right of the mean (larger than the mean) have positive signs and are called *positive deviations*. It is a property of the arithmetic mean, for both populations and samples, that the negative and positive deviations exactly balance. This property is proven mathematically for a population by the following demonstration that $\sum_{i=1}^{N}(x_i - \mu) = 0$.

Given that the population mean μ is a constant for any given population, therefore (see Problem 1.45)

$$\sum_{i=1}^{N}(x_i - \mu) = \sum_{i=1}^{N}[x_i + (-\mu)] = \sum_{i=1}^{N}x_i + \sum_{i=1}^{N}-\mu = \sum_{i=1}^{N}x_i - N\mu$$

Substituting $\dfrac{\sum_{i=1}^{N}x_i}{N}$ for μ,

$$\sum_{i=1}^{N}(x_i - \mu) = \sum_{i=1}^{N}x_i - N\left(\frac{\sum_{i=1}^{N}x_i}{N}\right) = \sum_{i=1}^{N}x_i - \sum_{i=1}^{N}x_i = 0$$

An equivalent proof for samples can be used to show that $\sum_{i=1}^{n}(x_i - \bar{x}) = 0$.

As it is thus true that the sum of the deviations from an arithmetic mean will always be zero, it follows that for any data set the sum of the positive deviations will always equal the sum of the negative deviations. If we consider frequency histograms of such data sets, and consider deviations to be distances, then the sum of the positive distances from the mean (of measurements to its right) will equal the sum of the negative distances. If a frequency histogram were constructed from this data using a solid material, it would balance along its horizontal axis (X axis) exactly at the arithmetic mean. This is why the arithmetic mean is called a measure of the *center of gravity* of a distribution. Because measurements tend to cluster near the center of a distribution, the measure of the center of gravity is also a measure of central tendency and of central location (see Section 6.1).

6.5 THE ARITHMETIC MEAN AS A MEASURE OF AVERAGE VALUE

We stated in Section 6.1 that "the average value in a data set is the most typical, frequent, or representative measurement in the set." We can show how this statement describes the arithmetic mean, why it is called a *measure of average value*, by the following manipulation.

We know that

$$\mu = \frac{\sum_{i=1}^{N} x_i}{N} \quad \text{and that} \quad \bar{x} = \frac{\sum_{i=1}^{n} x_i}{n}$$

Therefore

$$N\mu = \sum_{i=1}^{N} x_i \tag{6.7}$$

and

$$n\bar{x} = \sum_{i=1}^{n} x_i \tag{6.8}$$

Thus, if all the measurements in a data set were replaced by the arithmetic mean of the data set, the sum of the measurements would remain the same. This is only true for the arithmetic mean, and therefore in this sense it is the most representative (or average) value for the data set.

6.6 CALCULATING ARITHMETIC MEANS FROM NONGROUPED FREQUENCY DISTRIBUTIONS

The arithmetic mean of the sample that is summarized in the frequency distribution in Table 5.1 could be calculated using equation (6.1)

$$\bar{x} = \frac{\sum_{i=1}^{n} x_i}{n} = \frac{(1.2 + 1.2 + 1.3 + \cdots + 1.7 + 1.8 + 1.8)\,\text{cm}}{50} = 1.50\,\text{cm}$$

Fortunately there is a simpler method that utilizes the frequency distribution and the following modification of the basic formula

$$\bar{x} = \frac{\sum_{i=1}^{k} f_i x_i}{\sum_{i=1}^{k} f_i} \tag{6.9}$$

$$= \frac{\sum_{i=1}^{k} f_i x_i}{n} \tag{6.10}$$

where now x_i represents the ith category of variable X and f_i represents the frequency of this ith category (see Example 4.3).

Similarly, this is the formula for calculating a population mean μ from a nongrouped frequency distribution

$$\mu = \frac{\sum\limits_{i=1}^{k} f_i x_i}{N} \tag{6.11}$$

EXAMPLE 6.3   Using equation (6.10), calculate a mean of the sample in Table 5.1.

Solution

To calculate this mean requires only the addition of a third column: $f_i x_i$ to Table 5.1. This column and the resulting calculation of the arithmetic mean are shown in Table 6.1.

Table 6.1

Length (cm) x_i	Frequency f_i	$f_i x_i$ (cm)
1.2	2	2.4
1.3	7	9.1
1.4	10	14.0
1.5	12	18.0
1.6	10	16.0
1.7	7	11.9
1.8	2	3.6
$\sum$	50	75.0 cm

$$\bar{x} = \frac{\sum f_i x_i}{n} = \frac{75.0 \text{ cm}}{50} = 1.50 \text{ cm}$$

Note: From the frequency histogram of this data in Fig. 5-3, you can see that 1.50 cm is the exact balance point of the center of gravity of this unimodal, symmetric distribution; that a vertical line above it would divide the histogram into equal areas on both sides. When we calculate the median for this data (see Example 6.13), you will see that it is identical to the arithmetic mean.

6.7   CALCULATING APPROXIMATE ARITHMETIC MEANS FROM GROUPED FREQUENCY DISTRIBUTIONS

An arithmetic mean calculated from a grouped frequency distribution (see Section 4.4) only approximates the exact value calculated directly from the data, and therefore it is called an *approximate arithmetic mean*. To make this calculation from the grouped data requires an assumption that all values in a class are equal to the class mark, m_i. Then, the approximate arithmetic mean is calculated with this formula for a population

$$\mu \approx \frac{\sum\limits_{i=1}^{k} f_i m_i}{N} \tag{6.12}$$

and this for a sample

$$\bar{x} \approx \frac{\sum\limits_{i=1}^{k} f_i m_i}{n} \tag{6.13}$$

where the symbol $\approx$ means approximately equal to.

Using the assumption that all values in a class are equal to the class mark is not unreasonable, as the class mark is the arithmetic mean of the class limits (see Section 4.4). We have made use of the assumption twice before: in estimating total car sales in Problem 4.24(e), and in plotting polygons from grouped data in Section 5.6.

EXAMPLE 6.4   Calculate the *exact arithmetic mean* of the 30 marathon times from the ungrouped frequency distribution in Table 4.3. Then, calculate the approximate arithmetic mean of the same data from the grouped frequency distribution in Table 4.4.

Solution

The modified version of Table 4.3 and the resulting calculation of the exact (or *true*) arithmetic mean using equation (6.11) are shown in Table 6.2. The modified version of Table 4.4 and the resulting calculation of the approximate arithmetic mean using equation (6.12) are shown in Table 6.3.

Table 6.2

Time (min) x_i	Frequency f_i	$f_i x_i$ (min)
129	1	129
130	2	260
131	0	0
132	0	0
133	1	133
134	1	134
135	1	135
136	2	272
137	0	0
138	3	414
139	0	0
140	0	0
141	3	423
142	4	568
143	5	715
144	2	288
145	5	725
$\sum$	30	4,196 min

$$\mu = \frac{\sum f_i x_i}{N} = \frac{4,196 \text{ min}}{30} = 139.9 \text{ min}$$

Note: Approximate descriptive measures are less accurate in the statistical sense (see Section 2.14) than the exact measures, and should only be calculated when it is not possible to calculate the exact measures. In this problem you can see that the approximate population arithmetic mean (139.6 min) underestimates the exact population arithmetic mean (139.9 min) by 0.3 min, which could significantly distort further calculations involving the mean. We will later calculate *exact* and *approximate medians* for this data (see Example 6.14) and, as you would expect for this negatively skewed distribution (see Fig. 5-6), the medians are to the right of the means.

Table 6.3

Time (min)	Class mark m_i	Frequency f_i	$f_i m_i$ (min)
128–130	129	3	387
131–133	132	1	132
134–136	135	4	540
137–139	138	3	414
140–142	141	7	987
143–145	144	12	1,728
Σ		30	4,188 min

$$\mu \approx \frac{\sum f_i m_i}{N} = \frac{4,188 \text{ min}}{30} = 139.6 \text{ min}$$

6.8   CALCULATING ARITHMETIC MEANS WITH CODED DATA

When a computer is not available and statistical measures such as the arithmetic mean must be calculated from data sets composed of either very large or very small numbers, then it is useful to *transform* the data into simpler numbers by using the *coding formula*

$$c_i = a + bx_i \tag{6.14}$$

where x_i is the ith measurement of the variable X, a and b are constants, and c_i is the *transformed* (or *coded*) *value* of the ith measurement.

If $a \neq 0$, and $b = 1$, then if a is a positive number the variable X is being coded by adding the same constant a to every measurement value, and if a is negative then the coding is done by subtracting a from every measurement. When a constant is either added or subtracted, the origin of the measurement scale is shifted, and this process is called a *translation of the data*.

If $a = 0$, $b > 0$, and $b \neq 1$, then the coding is being done by multiplying every measurement value by the constant b. If $b > 1$, then there is an *expansion of the measurement scale*. If $0 < b < 1$, then b is a fraction and there is a *contraction of the measurement scale*. All three forms of coding (*expansion*, *contraction*, and *translation*) are *linear transformations* of the data.

If, after coding, the arithmetic mean of the c_i values is calculated with the formula

$$\bar{c} = \frac{\sum_{i=1}^{n} c_i}{n} \tag{6.15}$$

then this *decoding formula for the sample arithmetic mean* can be used to find the mean of the original data

$$\bar{x} = \frac{1}{b}(\bar{c} - a) \tag{6.16}$$

EXAMPLE 6.5   For the following sample of length measurements (in cm), first calculate $\bar{x}$ directly from the data, and then calculate it by means of equations (6.14), (6.15), and (6.16), using $a = -490$ cm and $b = 1$ as the coding constants: 492, 493, 495, 496, 498, 500.

Solution

The direct calculation of $\bar{x}$ and the calculation using the coding and decoding formulas are shown in Table 6.4.

Table 6.4

Length (cm) x_i	$c_i = -490$ cm $+ x_i$ cm
492	2
493	3
495	5
496	6
498	8
500	10
$\sum$ 2,974 cm	34 cm

$$\bar{x} = \frac{\sum x_i}{n} = \frac{2{,}974 \text{ cm}}{6} = 495.7 \text{ cm}$$

$$\bar{c} = \frac{\sum c_i}{n} = \frac{34 \text{ cm}}{6} = 5.66667 \text{ cm}$$

$$\bar{x} = \frac{1}{b}(\bar{c} - a) = \frac{1}{1}[5.66667 \text{ cm} - (-490 \text{ cm})] = 495.7 \text{ cm}$$

6.9 WEIGHTED MEANS

The formulas for the arithmetic mean in Section 6.2 state that first all values in a sample (or population) are summed, and then the sum is divided by the number of values in the sample (or population). These formulas thus assume that all the data values have *equal importance* and therefore should be given *equal weight* in the calculation of the mean. However, when several types of data make different contributions to the mean, then each type of data should be assigned a *weight* proportional to its importance prior to calculation of the mean. When this has been done for a sample, the mean is calculated with the following *weighted mean* (or *weighted arithmetic mean*) *formula*

$$\bar{x}_w = \frac{\sum_{i=1}^{k} w_i x_i}{\sum_{i=1}^{k} w_i} \tag{6.17}$$

where $\bar{x}_w$ is the sample weighted mean, x_i is the ith measurement of the variable X, w_i is the weight assigned to the ith measurement, and k is the number of measurement categories. For a population, the weighted mean (μ_w) formula is

$$\mu_w = \frac{\sum_{i=1}^{k} w_i x_i}{\sum_{i=1}^{k} w_i} \tag{6.18}$$

EXAMPLE 6.6   A toy manufacturer has 50 employees: 15 are paid \$5.25 per hour, 25 are paid \$5.75 per hour, and 10 are paid \$6.30 per hour. Use equation (6.18) to find the average hourly wage for all 50 employees.

Solution

For this problem, x_i is the hourly wage, and the relative importance (w_i) of each wage level to the calculation is the number of employees (f_i) who are paid that wage. Therefore, the weighted mean for this population is

$$\mu_w = \frac{\sum_{i=1}^{k} w_i x_i}{\sum_{i=1}^{k} w_i} = \frac{\sum_{i=1}^{3} f_i x_i}{\sum_{i=1}^{3} f_i} = \frac{(15 \times \$5.25) + (25 \times \$5.75) + (10 \times \$6.30)}{15 + 25 + 10}$$

$$= \frac{\$285.50}{50} = \$5.71$$

Note: From this example, it can be seen that the formulas for calculating arithmetic means from nongrouped frequency distributions (see Section 6.6) are special cases of the weighted mean formula, where $w_i = f_i$. Similarly, the basic formulas for calculating the arithmetic mean (see Section 6.2) are each a special case, where $w_i = 1$, $k = n$ for a sample or $k = N$ for a population. Because $w_i = 1$, these basic arithmetic means are also called *unweighted arithmetic means* or *simple arithmetic means*. The formulas for the approximate arithmetic mean (see Section 6.7) are also special cases of the weighted mean, where $w_i = f_i$, $x_i = m_i$, and k is the number of classes in the grouped distribution.

6.10   THE OVERALL MEAN

The *overall mean* (also called the *grand mean*, *pooled mean*, or *common mean*) is the appropriate way to combine arithmetic means from several samples. The formula for the overall mean is a version of the weighted mean [equation (6.17)]

$$\text{Overall mean} = \bar{x}_w = \frac{\sum_{i=1}^{k} n_i \bar{x}_i}{\sum_{i=1}^{k} n_i} \tag{6.19}$$

where w_i is the sample size n_i, x_i is the mean of the sample $\bar{x}_i$, and k is the number of samples being considered. You can get an intuitive feeling for why it is the appropriate way to combine sample means from the fact that the numerator of the formula is $\sum_{i=1}^{k} n_i \bar{x}_i$, and from equation (6.8) we know that $n\bar{x} = \sum_{i=1}^{n} x_i$. Therefore, the numerator of the overall mean formula is the sum of all the data values in all the samples. As the denominator is the sum of the sample sizes, the overall mean is really the sum of all the data values divided by the number of values.

EXAMPLE 6.7   The effects of a new blood-pressure drug are being studied in three different hospitals. One measurement taken from groups of female patients in each hospital before and after treatment is resting heart rate in beats per minute. The results for this measurement when taken before treatment are: Hospital 1, $n_1 = 30$ patients, $\bar{x}_1 = 76.2$ beats/min; Hospital 2, $n_2 = 25$ patients, $\bar{x}_2 = 79.3$ beats/min; Hospital 3, $n_3 = 16$ patients, $\bar{x}_3 = 80.1$ beats/min. Combine these three arithmetic means to get an overall mean for this pretreatment measurement.

Solution

$$\text{Overall mean} = \frac{\sum_{i=1}^{3} n_i \bar{x}_i}{\sum_{i=1}^{3} n_i} = \frac{[(30 \times 76.2) + (25 \times 79.3) + (16 \times 80.1)] \text{ beats/min}}{30 + 25 + 16}$$

$$= \frac{5,550.1 \text{ beats/min}}{71} = 78.2 \text{ beats/min}$$

6.11   THE GEOMETRIC MEAN

For a set of positive numbers $x_1, x_2, \ldots, x_n$, the *geometric mean* is the principal nth *root* (see Section 1.7) of the product of the n numbers. In symbolic form, the formula is

$$\text{Geometric mean} = \sqrt[n]{x_1, x_2 \cdots x_n} = \sqrt[n]{\prod_{i=1}^{n} x_i} \tag{6.20}$$

where $\prod$, the capital form of the greek letter pi, means take the product of.

EXAMPLE 6.8   For the following sample, find both the arithmetic mean and the geometric mean: 1, 3, 5, 6, 8.

Solution

$$\bar{x} = \frac{\sum_{i=1}^{5} x_i}{5} = \frac{23}{5} = 4.6$$

$$\text{Geometric mean} = \sqrt[5]{\prod_{i=1}^{5} x_i} = \sqrt[5]{1 \times 3 \times 5 \times 6 \times 8} = \sqrt[5]{720} = 3.727919, \text{ or } 3.7$$

Note: For calculating the geometric mean, remember that $\sqrt[n]{b} = b^{1/n}$ [see Section 1.16(b)].

6.12   THE HARMONIC MEAN

The *harmonic mean* of a set of data $x_1, x_2, \ldots, x_n$ is the reciprocal of the arithmetic mean of the reciprocals of the data. In symbolic form

$$\text{Harmonic mean} = \frac{1}{\dfrac{1}{n}\displaystyle\sum_{i=1}^{n}\dfrac{1}{x_i}} \tag{6.21}$$

EXAMPLE 6.9   Calculate the harmonic mean for the sample in Example 6.8.

Solution

$$\text{Harmonic mean} = \frac{1}{\dfrac{1}{5}\left(\dfrac{1}{1}+\dfrac{1}{3}+\dfrac{1}{5}+\dfrac{1}{6}+\dfrac{1}{8}\right)}$$

$$= \frac{1}{0.2(1 + 0.333333 + 0.2 + 0.166667 + 0.125)} = 2.73973, \text{ or } 2.7$$

6.13   THE MEDIAN AND OTHER QUANTILES

For a set of data $x_1, x_2, \ldots, x_n$ organized into an array (see Section 4.1), the *median* of the data is the value that divides the array into two equal parts; there are as many data values below the median as above it. The *odd–even rules* can be used to find the median of such an array.

If there is an *odd number* of values in an array, then the median is the *middle value* of the array; if there is an *even number* of values in an array, then the median is the *arithmetic mean of the two middle values.*

If a frequency histogram or a frequency curve is constructed from the data, and a vertical line is drawn above the median value on the X axis, then the line will divide the histogram or curve into two equal areas (see Fig. 6-1). There are no universally accepted symbols for the median, but we will use a pair of symbols that are now fairly common in statistics books: $\tilde{x}$ (read x-tilde) for the sample median, and $\tilde{\mu}$ (read mu-tilde) for the population median.

The median is one of many possible *quantiles* that can be calculated from a data set organized into an ascending array. Each quantile in an array, designated by the symbol $Q_{j/m}$, is the *x value below which are j mths of the data*. Thus, for example, $\frac{1}{2}$ of the data are below $Q_{2/4}$ and $\frac{1}{4}$ of the data are below $Q_{1/4}$. For a frequency histogram or frequency curve, if a perpendicular line is drawn above $Q_{j/m}$, then j mths of the area will be to the left of the line. In this chapter we consider three types of quantiles: *quartiles, deciles,* and *percentiles*.

There are three quartiles: *first quartile* ($Q_{1/4}$ *or* Q_1), *second quartile* ($Q_{2/4}$ *or* Q_2), and *third quartile* ($Q_{3/4}$ *or* Q_3). Together they divide arrays, frequency histograms, and frequency curves into four equal parts. There are nine deciles: *first decile* ($Q_{1/10}$ *or* D_1), *second decile* ($Q_{2/10}$ *or* D_2), and so on to the *ninth decile* ($Q_{9/10}$ *or* D_9). Together they divide arrays, frequency histograms, and frequency curves into ten equal parts. Finally, there are 99 percentiles (also known as *centiles*): *first percentile* ($Q_{1/100}$ *or* P_1), *second percentile* ($Q_{2/100}$ *or* P_2), and so on to the *ninety-ninth percentile* ($Q_{99/100}$ *or* P_{99}). Together they divide arrays, frequency histograms, and frequency curves into 100 equal parts.

The median and other quantiles are measures of *relative location*—the location of $Q_{j/m}$ relative to the boundaries of the array or distribution. The median is a measure of central tendency or central location in that it shows the location of the exact midpoint or center of gravity of an array or distribution. Like the arithmetic mean, the median is a measure of average value in that there is typically a clustering of values near the center. While the arithmetic mean is calculated from all data values, and is thus influenced by extreme values, the median only deals with the ranking of the values and is thus unaffected by extremes. This is why the median is often recommended as a measure of average value for skewed data. As to level of measurement, the median is legitimate for ordinal-, interval-, and ratio-level measurements.

EXAMPLE 6.10   The median is equal to which quantiles?

Solution

$$\tilde{x}(\text{ or } \tilde{\mu}) = Q_2 = D_5 = P_{50}$$

6.14   THE QUANTILE-LOCATING FORMULA FOR ARRAYS

Statistics books agree on the odd–even rules for finding the median of an array, but they differ on techniques and formulas for finding other quantiles in an array. A general *quantile-locating formula for arrays* that we will use is this

$$Q_{j/m} = x_i \tag{6.22}$$

where x_i is the value in an array of x values below which are j mths of the data, and $i =$ the location in the array $= \dfrac{[j \times n(\text{or } N)]}{m} + \dfrac{1}{2}$.

EXAMPLE 6.11   Using both the odd–even rules from Section 6.13 and equation (6.22), find the median for the following samples: (*a*) 12, 13, 14,   (*b*) 12, 13, 14, 15.

Solution

In Section 6.13, it is stated that if there is an odd number of values in an array, then the median is the middle value of the array. Therefore: (*a*) $n = 3$, so $\tilde{x} = 13$. Where there is an even number of values in an array, then the median is the arithmetic mean of the two middle values. Therefore: (*b*) $n = 4$, so

$\bar{x} = \dfrac{13 + 14}{2} = 13.5$. If equation (6.22) is used to find the median, $Q_{1/2}$, of the samples, then these are the results:

(a) $i = \left(\dfrac{1 \times 3}{2}\right) + \dfrac{1}{2} = 2$; $\tilde{x} = x_2$; therefore the median is the second value in the array: 13.

(b) $i = \left(\dfrac{1 \times 4}{2}\right) + \dfrac{1}{2} = 2.5$; $\tilde{x} = x_{2.5}$; therefore the median is midway between the second and third values in the array: 13.5.

6.15   THE QUANTILE-LOCATING FORMULA FOR NONGROUPED FREQUENCY DISTRIBUTIONS

Often using the odd–even rules of Section 6.13 or equation (6.22), a quantile will be located among *tied values* (identical values). The solution for the median or any other quantile of accepting one of the tied values as the quantile is an acceptable solution found in many statistics books. However, it has obvious problems. Thus, it is likely that the definition of a quantile will be violated, and also many of the quantiles may be identical.

To avoid such problems with ungrouped data, many statistics books recommend that when there are tied values at the quantile, the quantile should be calculated using the following *quantile-locating formula for nongrouped frequency distributions*:

$$Q_{j/m} = b + \left[\dfrac{\dfrac{(j \times n)}{m} - Cf}{f}\right](w) \tag{6.23}$$

where $Q_{j/m}$ is the x value below which are j $m$ths of the data, b is the lower boundary of the implied range for the *quantile category* (the measurement category that contains the quantile), n is the sample size (or N is the population size), Cf is the cumulative frequency from all categories less than the quantile category, f is the frequency in the quantile category, and w is the width of the implied range of the quantile category.

EXAMPLE 6.12   Use the odd–even rules (see Section 6.13) and equations (6.22) and (6.23) to find the median of the following sample of weight measurements (in lb): 1.1, 1.2, 1.2, 1.3, 1.3, 1.3, 1.3, 1.3, 1.4, 1.5.

Solution

Using the odd-even rules, as $n = 10$, $\tilde{x} = \dfrac{1.3\,\text{lb} + 1.3\,\text{lb}}{2} = 1.3\,\text{lb}$.

Using equation (6.22), $i = \left(\dfrac{1 \times 10}{2}\right) + \dfrac{1}{2} = 5.5$, so again $\tilde{x}$ is located midway between x_5 (1.3 lb) and x_6 (1.3 lb): 1.3 lb.

To use equation (6.23), we have converted the sample into the frequency distribution shown in Table 6.5 and the "less than" cumulative frequency distribution shown in Table 6.6. We want to use the formula with these distributions to calculate $\tilde{x} = Q_{j/m} = Q_{1/2}$. By definition, $\frac{1}{2}$ of the data values are less than $Q_{1/2}$, so, as $n = 10$, five values must be less than $Q_{1/2}$. From the "less than" cumulative frequency distribution in Table 6.6 we see that three values are less than 1.3 lb and eight values are less than 1.4 lb, so 1.3 lb is the *quantile category*, in this case the *median category* (the category containing the median). Thus $\tilde{x}$ must be somewhere within the implied range of that category (1.25 lb to 1.35 lb). If we *assume that values in the quantile category are evenly distributed across its implied range*, then we can locate the median by defining the components of

Table 6.5

Weight (lb) x_i	Frequency f_i
1.1	1
1.2	2
1.3	5
1.4	1
1.5	1
$\sum$	10

Table 6.6

Weight (lb)	Cumulative frequency
Less than 1.1	0
Less than 1.2	1
Less than 1.3	3
Less than 1.4	8
Less than 1.5	9
Less than 1.6	10

the formula as follows:

$$Q_{j/m} = \tilde{x} = Q_{1/2}$$

b = lower boundary of the median category = 1.25 lb

n = sample size = 10

Cf = cumulative frequency from categories less than median category = 3

f = frequency in the median category = 5

w = width of the median category = 1.35 lb − 1.25 lb = 0.10 lb

Therefore

$$\tilde{x} = Q_{1/2} = 1.25 \text{ lb} + \left[\frac{\frac{(1 \times 10)}{2} - 3}{5} \right](0.10 \text{ lb})$$

$$= 1.25 \text{ lb} + (0.4 \times 0.10 \text{ lb}) = 1.29 \text{ lb}$$

Note: Putting what was done in words, we found that the median was 2/5 of the way across the implied range of the median category and so we multiplied the width of the median category by 0.4 and added this to the lower boundary of the implied range for the median category. In using this formula, it is always assumed that the values in the quantile category are evenly distributed across its implied range.

EXAMPLE 6.13   For the length data (in cm) summarized in Table 6.1, use equation (6.23) to find Q_1, Q_2, and Q_3.

Solution

In Table 6.7 we have converted the frequency distribution in Table 6.1 into a "less than" cumulative frequency distribution. As $Q_1 = Q_{1/4}$, $\frac{1}{4}$ of the 50 values, or 12.5 values, must be less than Q_1. Therefore, from

Table 6.7

Length (cm)	Cumulative frequency
Less than 1.2	0
Less than 1.3	2
Less than 1.4	9
Less than 1.5	19
Less than 1.6	31
Less than 1.7	41
Less than 1.8	48
Less than 1.9	50

the "less than" cumulative frequency distribution we can see that the Q_1 *category* is 1.4 cm. Thus

$$Q_{j/m} = b + \left[\frac{\frac{(j \times n)}{m} - Cf}{f} \right](w)$$

$$Q_1 = Q_{1/4} = 1.35 \text{ cm} + \left[\frac{\frac{(1 \times 50)}{4} - 9}{10} \right](0.10 \text{ cm})$$

$$= 1.35 \text{ cm} + (0.35 \times 0.10 \text{ cm})$$

$$= 1.385 \text{ cm, or } 1.38 \text{ cm}$$

As $Q_2 = \tilde{x} = Q_{1/2}$, $\frac{1}{2}$ of the 50 values, or 25 values, must be less than Q_2. Therefore, from the "less than" cumulative frequency distribution, 1.5 cm is the Q_2 *category* (median category). Thus

$$Q_2 = \tilde{x} = Q_{1/2} = 1.45 \text{ cm} + \left[\frac{\frac{(1 \times 50)}{2} - 19}{12} \right](0.10 \text{ cm})$$

$$= 1.45 \text{ cm} + (0.50 \times 0.10 \text{ cm})$$

$$= 1.50 \text{ cm}$$

This confirms what was said in Example 6.3, that for this unimodal, symmetric distribution the arithmetic mean (1.50 cm) is identical to the median.

Finally, as $Q_3 = Q_{3/4}$, $\frac{3}{4}$ of the 50 values, or 37.5 values, must be less than Q_3. Therefore, from the "less than" cumulative frequency distribution the Q_3 *category* is 1.6 cm. Thus

$$Q_3 = Q_{3/4} = 1.55 \text{ cm} + \left[\frac{\frac{(3 \times 50)}{4} - 31}{10} \right](0.10 \text{ cm})$$

$$= 1.55 \text{ cm} + (0.65 \times 0.10 \text{ cm})$$

$$= 1.615 \text{ cm, or } 1.62 \text{ cm}$$

6.16   THE QUANTILE-LOCATING FORMULA FOR GROUPED FREQUENCY DISTRIBUTIONS

When all the measurement values in a sample or population are known, then the median found by using either the odd–even rules or the quantile-locating formulas is called the *exact* (or *true*) *median*. As with exact and approximate means (see Section 6.7), a median calculated from a grouped frequency distribution only approximates the exact median, and is therefore called an *approximate median*. The formula used to find such approximate medians, or any other approximate quantile, is the following *quantile-locating formula for grouped frequency distributions*:

$$Q_{j/m} \approx b_c + \left[\frac{\frac{(j \times n)}{m} - Cf_c}{f_c} \right] (w_c) \tag{6.24}$$

where $Q_{j/m}$ is the quantile, b_c is the lower class boundary for the *quantile class* (the class containing the quantile), n is the sample size (or N is the population size), Cf_c is the cumulation of frequencies from all classes less than the quantile class, f_c is the frequency in the quantile class, and w_c is the class width of the quantile class.

EXAMPLE 6.14   First using equation (6.23), find the exact median for the 30 marathon times in Table 6.2. Then, using equation (6.24), find the approximate median for the grouped version of the same data in Table 6.3.

Solution

As $\frac{1}{2}$ of the 30 values (or 15 values) must be less than $\tilde{\mu} = Q_2$ ($\tilde{\mu}$ because we have treated these 30 runners as a population), the Q_2 *category* (from Table 6.2) is 142 min. Thus, the exact median is

$$Q_{j/m} = b + \left[\frac{\frac{(j \times N)}{m} - Cf}{f} \right] (w)$$

$$Q_2 = \tilde{\mu} = Q_{1/2} = 141.5\,\text{min} + \left[\frac{\frac{(1 \times 30)}{2} - 14}{4} \right] (1.0\,\text{min})$$

$$= 141.5\,\text{min} + (0.25 \times 1.0\,\text{min})$$

$$= 141.75\,\text{min, or } 141.8\,\text{min}$$

To calculate any approximate quantile with equation (6.24), we must *assume that the values in the quantile class are evenly distributed across its class width*. Then to calculate the approximate median (here, $\tilde{\mu} = Q_{1/2}$), $\frac{1}{2}$ of the 30 values (or 15 values) must be less than the median. Therefore, as 11 values are below the (140 to 142) min class (less than 139.5 min), and 18 values are below the (143 to 145) min class (less than 142.5 min), the *median class* is (140 to 142) min. The components of the formula are then defined as follows:

$$Q_{j/m} = \tilde{\mu} = Q_{1/2}$$
$$b_c = \text{lower boundary of median class} = 139.5\,\text{min}$$
$$N = \text{population size} = 30$$
$$Cf_c = \text{cumulative frequency from classes less than the median class} = 11$$
$$f_c = \text{frequency in the median class} = 7$$
$$w_c = \text{class width of the median class} = 3.0\,\text{min}$$

Therefore

$$Q_2 = \tilde{\mu} = Q_{1/2} \approx 139.5\,\text{min} + \left[\frac{\frac{(1 \times 30)}{2} - 11}{7} \right] (3.0\,\text{min})$$

$$\approx 139.5\,\text{min} + (0.571429 \times 3.0\,\text{min})$$

$$\approx 141.214287\,\text{min, or } 141.2\,\text{min}$$

Note: These results confirm what was said in Example 6.4, that the exact (141.8 min) and approximate (141.2 min) medians for this negatively skewed distribution are to the right of (larger than) the exact (139.9 min) and approximate (139.6 min) arithmetic means.

6.17   THE MIDRANGE, THE MIDQUARTILE, AND THE TRIMEAN

The *midrange* (or *range midpoint*) is the arithmetic mean of the extreme values in a data set, x_s and x_l, or stated symbolically,

$$\text{Midrange} = \frac{x_s + x_l}{2} \tag{6.25}$$

The *midquartile* is the arithmetic mean of the first and third quartiles, or, stated symbolically,

$$\text{Midquartile} = \frac{Q_1 + Q_3}{2} \tag{6.26}$$

The *trimean* is the arithmetic mean of the median Q_2 and the midquartile, or, stated symbolically,

$$\text{Trimean} = \frac{Q_2 + \frac{Q_1 + Q_3}{2}}{2} \tag{6.27}$$

Multiplying both sides of the equation by $\frac{2}{2}$ and rearranging components

$$\text{Trimean} = \frac{Q_1 + 2Q_2 + Q_3}{4} \tag{6.28}$$

EXAMPLE 6.15   For the length data summarized in Table 6.1 and the quartiles calculated in Example 6.13, determine the midrange, midquartile, and trimean.

Solution

$$\text{Midrange} = \frac{1.2\,\text{cm} + 1.8\,\text{cm}}{2} = 1.50\,\text{cm}$$

$$\text{Midquartile} = \frac{1.385\,\text{cm} + 1.615\,\text{cm}}{2} = 1.50\,\text{cm}$$

$$\text{Trimean} = \frac{1.385\,\text{cm} + 2(1.50\,\text{cm}) + 1.615\,\text{cm}}{4} = 1.50\,\text{cm}$$

Note: It can be seen for this symmetrical, unimodal distribution that

$$\bar{x} = \tilde{x} = \text{midrange} = \text{midquartile} = \text{trimean}$$

6.18   THE MODE

The basic definition of a *mode* is: The mode of a set of data is the measurement value in the set that occurs most frequently. When in an arrayed data set there are two *consecutive values* that have the same frequency, which is greater than the frequency of any other value in the set, then, generally, the mode is considered to be the arithmetic mean of the consecutive values. When there are two *nonconsecutive values* in an arrayed data set that have the same frequency, which is greater than the frequency of any other value in the set, then both values are called modes. Finally, when all values in a data set have the same frequency, the set does not have a mode.

EXAMPLE 6.16   Determine the mode for each of the following samples: (*a*) 2, 3, 3, 3, 4, 5, 6,   (*b*) 2, 3, 4, 5, 6, 6, 6,   (*c*) 2, 3, 3, 4, 4, 4, 5, 5, 6,   (*d*) 2, 3, 3, 3, 4, 4, 4, 5,   (*e*) 2, 3, 3, 3, 4, 5, 6, 6, 6, 8,   (*f*) 2, 2, 3, 3, 4, 4, 5, 5, 6, 6.

Solution

(*a*)   Mode = 3

(*b*)   Mode = 6

(*c*)   Mode = 4

(*d*)   Mode = $\dfrac{3+4}{2} = 3.5$

(*e*)   Mode = 3, and mode = 6

(*f*)   There is no mode.

6.19   MODE-LOCATING FORMULA FOR GROUPED FREQUENCY DISTRIBUTIONS

There are two accepted techniques for determining the approximate mode from a grouped frequency distribution: (1) determining the *modal class* (the class with the highest frequency) and then using its class mark as the approximate mode, and (2) using the following *mode-locating formula for grouped frequency distributions* (which can be used only when the grouped distribution has *equal class widths*):

$$\text{Mode} \approx b_c + \left(\frac{d_1}{d_1 + d_2} \right)(w_c) \tag{6.29}$$

where b_c is the lower boundary of the modal class, d_1 is the difference between the modal-class frequency and the frequency in the class preceding it in the distribution, d_2 is the difference between the modal-class frequency and the frequency in the class following it, and w_c is the class width of the modal class.

EXAMPLE 6.17   The 64 second exam scores in Table A.2 were converted to a grouped distribution in Table 4.22, and now they have been converted in Fig. 6-2 to an *ascending-array stem-and-leaf display* (see Problem 6.18). From Fig. 6-2 determine the exact mode, and then from Table 4.22 determine the approximate mode.

Solution

From Fig. 6-2 it can be seen that the score with the highest frequency is 90. Therefore: exact mode = 90. Using technique (1) on the grouped distribution in Table 4.22, the modal class is 90 to 94, and the class mark for this class is 92. Therefore: mode $\approx$ 92. Using technique (2) on the distribution in Table 4.22, where the modal class is 90 to 94: $b_c = 89.5$, $d_1 = 17 - 9 = 8$, $d_2 = 17 - 5 = 12$, and $w_c = 5$. Therefore

$$\text{Mode} \approx 89.5 + \left(\frac{8}{8 + 12} \right)(5) = 89.5 + (0.4 \times 5) = 91.5.$$

4	9	(1)
5	5799	(4)
6	44457899	(8)
7	124688999	(9)
8	00012334444566777888	(20)
9	000000011112334456789	(22)
		(64)

Fig. 6-2

Solved Problems

THE ARITHMETIC MEAN

6.1   Using equation (6.6), calculate the arithmetic mean for the following population: 5.47×10^{-4} cm, 6.831×10^{-8} cm, 2.1211×10^{-5} cm.

Solution

$$\mu = \frac{\sum x_i}{N} = \frac{5.47 \times 10^{-4}\,\text{cm} + 6.831 \times 10^{-8}\,\text{cm} + 2.1211 \times 10^{-5}\,\text{cm}}{3}$$

Before proceeding with the division, the numerator values are converted to decimal notation (see Problem 2.15).

$$\mu = \frac{0.000547\,\text{cm} + 0.00000006831\,\text{cm} + 0.000021211\,\text{cm}}{3}$$

$$= \frac{0.00056827931\,\text{cm}}{3}$$

$$= \frac{0.000568\,\text{cm}}{3}, \text{ after rounding off the numerator}$$

$$= 0.00018933333\,\text{cm, or } 0.000189\,\text{cm (or } 1.89 \times 10^{-4}\,\text{cm) after rounding off the answer.}$$

6.2   Show with the data from Example 6.1(*a*), and using six digits for the mean, that $\sum_{i=1}^{n}(x_i - \bar{x}) = 0$.

Solution

$$\sum_{i=1}^{n}(x_i - \bar{x}) = (1\,\text{g} - 3.42857\,\text{g}) + (3\,\text{g} - 3.42857\,\text{g}) + (2\,\text{g} - 3.42857\,\text{g})$$

$$+ (7\,\text{g} - 3.42857\,\text{g}) + (5\,\text{g} - 3.42857\,\text{g}) + (4\,\text{g} - 3.42857\,\text{g}) + (2\,\text{g} - 3.42857\,\text{g})$$

$$= -2.42857\,\text{g} - 0.42857\,\text{g} - 1.42857\,\text{g} + 3.57143\,\text{g} + 1.57143\,\text{g} + 0.57143\,\text{g} - 1.42857\,\text{g}$$

$$= 0.00001\,\text{g}$$

Thus, using a six-digit arithmetic mean in the calculations, the sum of the deviations is zero to the fourth decimal place. Had the calculations been done with the 12-digit mean, 3.42857142857 g, then

$$\sum_{i=1}^{n}(x_i - \bar{x}) = 0.00000000001\,\text{g}$$

6.3   A car dealership has set a goal for its 15 salespeople of selling 150 new cars in an eight-week period. In the first six weeks of this period, an average $\bar{x} = 19.5$ cars were sold each week. How many cars must be sold in the remaining two weeks to achieve the 150-car goal?

Solution

   If x_i is the total number of cars sold each week, and $\sum_{i=1}^{6} x_i$ is the total number of cars sold over six weeks, then from equation (6.8)

$$\sum_{i=1}^{6} x_i = n\bar{x} = 6 \times 19.5 \text{ cars} = 117 \text{ cars}$$

Therefore, the number of cars that must be sold in the remaining two weeks is 150 cars $-$ 117 cars $= 33$ cars

6.4   Using equation (6.10), find the arithmetic mean of the sample that is summarized in Table 5.4.

Solution

   The modified table and resulting calculation of the arithmetic mean are shown in Table 6.8.

Table 6.8

Weight (g) x_i	Frequency f_i	$f_i x_i$ (g)
14	2	28
15	2	30
16	4	64
17	18	306
18	24	432
19	35	665
20	5	100
$\sum$	90	1,625 g

$$\bar{x} = \frac{\sum f_i x_i}{n} = \frac{1,625 \text{ g}}{90} = 18.1 \text{ g}$$

   Note: The relative-frequency histogram for this data (see Fig. 5-16) shows that the distribution is negatively skewed. While it is less apparent for this distribution than it was for the perfectly symmetric distribution in Fig. 5-3, again the mean is the center of gravity. This does not indicate, however, that in Fig. 5-16 there are equal areas on both sides of the mean. It is the median that divides the distribution into two equal areas, and when we calculate the median for this distribution (see Problem 6.20) we will find it to be to the right of the mean, farther away from the skewed tail.

6.5   Using equation (6.10), find the arithmetic mean of the sample that is summarized in Table 5.5.

Solution

   The modified table and resulting calculation of the arithmetic mean are shown in Table 6.9.
   Note: For this positively skewed distribution (see Fig. 5-17) the arithmetic mean is, as always, the center of gravity, but now it is to the right of the median (see Problem 6.21). It is always true for skewed distributions that the median is farther away from the skewed tail than is the mean.

Table 6.9

Temperature (°F) x_i	Frequency f_i	$f_i x_i$ (°F)
100	10	1,000
101	45	4,545
102	25	2,550
103	10	1,030
104	5	520
105	0	0
106	3	318
107	2	214
$\sum$	100	10,177°F

$$\bar{x} = \frac{\sum f_i x_i}{n} = \frac{10,177°F}{100} = 101.8°F$$

CALCULATING APPROXIMATE ARITHMETIC MEANS FROM GROUPED FREQUENCY DISTRIBUTIONS

6.6   First, directly from Table A.2, calculate the exact arithmetic mean of the 64 second lecture exam scores (column 3). Then, using equation (6.13), calculate the approximate arithmetic mean of the same data from the grouped frequency distribution in Table 4.22.

Solution

The exact arithmetic mean is

$$\bar{x} = \frac{\sum_{i=1}^{n} x_i}{n} = \frac{5,221}{64} = 81.6$$

The modified grouped frequency distribution and the resulting calculation of the approximate arithmetic mean using equation (6.13) are shown in Table 6.10.

Note: The arithmetic-mean calculations were done by treating these discrete ratio data as if they were continuous ratio measurements presented at the units-digit level of precision. The answers were then rounded off to the tenths digit (see Section 6.3). Again, the approximate arithmetic mean (81.4) underestimates the exact arithmetic mean (81.6). It is not surprising for this negatively skewed distribution (see Fig. 5-25) that the exact and approximate medians for the data (see Problem 6.22) are to the right of the exact and approximate arithmetic means. This is also true for the graphic estimation of this median (84.5, see Problem 5.29).

6.7   For the golf winnings in Problem 4.17, first round off the amounts to the nearest $1,000, and then calculate the exact arithmetic mean of the winnings. Next, calculate the approximate arithmetic mean of the winnings from the grouped frequency distribution in Table 4.26. (In making the grouped distribution, the winnings were rounded off to the nearest $1,000.)

Solution

The data rounded off to the nearest $1,000 are presented in Table 6.11. (To save space, the zero-frequency categories are not included.) The calculation of the exact arithmetic mean μ for the population of 70 golfers, using equation (6.11), is shown at the bottom of the table. As the data was presented at the thousands digit, μ was rounded off to the hundreds digit. The modified grouped frequency distribution and the resulting calculation of the approximate arithmetic mean, using equation (6.12), are shown in Table 6.12.

Table 6.10

2d Exam	Class mark m_i	Frequency f_i	$f_i m_i$
45–49	47	1	47
50–54	52	0	0
55–59	57	4	228
60–64	62	3	186
65–69	67	5	335
70–74	72	3	216
75–79	77	6	462
80–84	82	11	902
85–89	87	9	783
90–94	92	17	1,564
95–99	97	5	485
$\sum$		64	5,208

$$\bar{x} \approx \frac{\sum f_i m_i}{n} = \frac{5,208}{64} = 81.4$$

Note: The same approximation technique was used for this grouped distribution with unequal class widths as was used for the grouped distribution with equal widths. Also note that the approximate mean does not always underestimate the exact mean; here it overestimates. Finally, note that as class marks are required for all classes to do the approximation technique, *approximate arithmetic means can not be calculated for open-ended grouped frequency distributions*. As you will see (see Problem 6.23), exact and approximate medians calculated from this data are to the left of both arithmetic means.

Table 6.11

Winnings ($) x_i	Frequency f_i	$f_i x_i$ ($)
2,000	18	36,000
3,000	7	21,000
4,000	5	20,000
6,000	7	42,000
8,000	7	56,000
10,000	3	30,000
15,000	7	105,000
22,000	5	110,000
30,000	3	90,000
36,000	2	72,000
40,000	1	40,000
45,000	1	45,000
83,000	3	249,000
200,000	1	200,000
$\sum$	70	$1,116,000

$$\mu = \frac{\sum f_i x_i}{N} = \frac{\$1,116,000}{70} = \$15,900$$

Table 6.12

Winnings ($)	Class mark m_i ($)	Frequency f_i	$f_i m_i$ ($)
2,000–4,000	3,000	30	90,000
5,000–10,000	7,500	17	127,500
11,000–20,000	15,500	7	108,500
21,000–30,000	25,500	8	204,000
31,000–40,000	35,000	3	106,500
41,000–45,000	43,000	1	43,000
46,000–81,000	63,500	0	0
82,000–83,000	82,500	3	247,500
84,000–198,000	141,000	0	0
199,000–200,000	199,500	1	199,500
Σ		70	$1,126,500

$$\mu \approx \frac{\sum f_i m_i}{N} = \frac{\$1,126,500}{70} = \$16,100$$

CALCULATING ARITHMETIC MEANS WITH CODED DATA

6.8 For the following sample of weight measurements (in grams), first calculate $\bar{x}$ directly from the data, and then calculate it by means of equations (6.14), (6.15), and (6.16), using $a = 0$ g and $b = \dfrac{1}{10,000}$ as the coding constants: 22,000.0; 30,000.0; 29,000.0; 27,500.0; 25,500.0; 24,000.0.

Solution

The direct calculation of $\bar{x}$ and the calculation using the coding and decoding formulas are shown in Table 6.13.

6.9 Two clothing factories are paying their workers an average of $\bar{x} = \$5.39$ per hour. Factory A had a good year and their management decides to reward all workers with a 5% per hour raise. Factory B had an equally good year but their management decides to give all workers a raise of $0.05 per hour. Use equation (6.14) to determine the new average hourly wage for both factories. Which factory is more generous?

Solution

For factory A, if x_i is the previous hourly wage, then each worker now receives $c_i = \$0.00 + 1.05x_i$. Therefore, the new average is

$$\bar{c} = 1.05\bar{x} = 1.05(\$5.39) = \$5.66$$

For factory B, each worker now receives per hour $c_i = \$0.05 + x_i$. Therefore, the new average is

$$\bar{c} = \$0.05 + \bar{x} = \$0.05 + \$5.39 = \$5.44$$

Clearly factory A is more generous, now paying its workers an average of $0.22 more per hour.

Table 6.13

Weight (g) x_i	$c_i = 0.0001 x_i$ g
22,000.0	2.20000
24,000.0	2.40000
25,500.0	2.55000
27,500.0	2.75000
29,000.0	2.90000
30,000.0	3.00000
$\sum$ 158,000.0 g	15.80000 g

$$\bar{x} = \frac{\sum x_i}{n} = \frac{158,000.0 \text{ g}}{6} = 26,333.33 \text{ g}$$

$$\bar{c} = \frac{\sum c_i}{n} = \frac{15.80000 \text{ g}}{6} = 2.633333 \text{ g}$$

$$\bar{x} = \frac{1}{b}(\bar{c} - a) = \frac{1}{0.0001}(2.633333 \text{ g}) = 26,333.33 \text{ g}$$

OTHER MEANS: WEIGHTED, OVERALL, GEOMETRIC, AND HARMONIC

6.10 The final grade in a biology course is determined by a score from 0 to 100, which has three components: a laboratory component of 25%, two hour-exams that together contribute 25%, and a final exam that contributes 50%. There are 100 possible points for the laboratory, 50 possible points for each hour exam, and 100 possible points for the final. A student in the course got 75 points for the laboratory, 40 and 38 points for the two hour-exams, and 85 points for the final. Use equation (6.17) to determine his overall score (from 0 to 100) for the course.

Solution

To determine the score, we let $w_i =$ the % contribution of the component, $x_i =$ the points achieved in the component, and $k = 4$. Therefore

$$\bar{x} = \frac{\sum_{i=1}^{4} w_i x_i}{\sum_{i=1}^{4} w_i} = \frac{0.25(75) + 0.25(40 + 38) + 0.50(85)}{0.25 + 0.25 + 0.50}$$

$$= \frac{80.75}{1.0} = 80.75, \text{ which would be reported as } 80.8$$

6.11 You develop a new hybrid corn and want to determine its "days to maturity": when the first ripe ears can be picked. To do this, you plant this corn in four different fields and then measure *days to maturity* for a randomly selected sample of 100 plants in each field. Calculating the arithmetic mean for each sample, the results are: $\bar{x}_1 = 70.1$ days, $\bar{x}_2 = 71.3$ days, $\bar{x}_3 = 69.5$ days, and $\bar{x}_4 = 69.2$ days. Combine these four means to get an overall mean.

Solution

Using equation (6.19) and the properties of summation notation (see Problems 1.42 and 1.43)

$$\text{Overall mean} = \frac{\sum_{i=1}^{4} n_i \bar{x}_i}{\sum_{i=1}^{4} n_i} = \frac{100 \sum_{i=1}^{4} \bar{x}_i}{400} = \frac{28{,}010 \text{ days}}{400} = 70.0 \text{ days}$$

6.12   For the following samples, find the arithmetic mean and the geometric mean: (a) 1, 1, 1, 2, 3, 8, 14,   (b) 2, 2, 2, 2, 2,   (c) 1, 3, 5, 9, 9, 9, 9.

Solution

Using equations (6.1) and (6.20)

(a) $\bar{x} = \dfrac{\sum_{i=1}^{7} x_i}{7} = \dfrac{30}{7} = 4.3$

$$\text{Geometric mean} = \sqrt[7]{\prod_{i=1}^{7} x_i} = \sqrt[7]{1 \times 1 \times 1 \times 2 \times 3 \times 8 \times 14}$$

$$= \sqrt[7]{672} = 672^{1/7} = 2.534603, \text{ or } 2.5$$

(b) $\bar{x} = \dfrac{\sum_{i=1}^{5} x_i}{5} = \dfrac{10}{5} = 2.0$

$$\text{Geometric mean} = \sqrt[5]{\prod_{1}^{5} x_i} = \sqrt[5]{2 \times 2 \times 2 \times 2 \times 2} = \sqrt[5]{32} = 32^{0.2} = 2.0$$

(c) $\bar{x} = \dfrac{\sum_{i=1}^{7} x_i}{7} = \dfrac{45}{7} = 6.4$

$$\text{Geometric mean} = \sqrt[7]{\prod_{i=1}^{7} x_i} = \sqrt[7]{1 \times 3 \times 5 \times 9 \times 9 \times 9 \times 9}$$

$$= \sqrt[7]{98{,}415} = 98{,}415^{0.142857} = 5.167658, \text{ or } 5.2$$

Note: These problems demonstrate several things about the geometric mean: (1) when the data are positively skewed, as in part (a), the geometric mean (2.5) is less affected by the skewing than the arithmetic mean (4.3); (2) when all the data have the same value, as in part (b), $\bar{x}$ = geometric mean; and (3) when the data are not all the same value, $\bar{x}$ > geometric mean.

6.13   Show that the geometric mean of a set of data is the antilogarithm of the arithmetic mean of the common logarithms of the data. (For a review of operations with logarithms, see Section 1.10.) Then use this technique to recalculate the geometric mean for the data in Example 6.8.

Solution

For a set of data $x_1, x_2, \ldots, x_n$, the common logarithms of the data are $\log_{10} x_1, \log_{10} x_2, \ldots, \log_{10} x_n$. The arithmetic mean of these logarithms ($\bar{x}_{\log}$) is

$$\bar{x}_{\log} = \frac{1}{n} \sum_{i=1}^{n} (\log_{10} x_1 + \log_{10} x_2 + \cdots + \log_{10} x_n)$$

The geometric mean of the data is

$$\text{Geometric mean} = \sqrt[n]{\prod_{i=1}^{n} x_i} = \sqrt[n]{x_1 x_2 \cdots x_n} = (x_1 x_2 \cdots x_n)^{1/n}$$

Taking the $\log_{10}$ of both sides

$$\log_{10}(\text{geometric mean}) = \log_{10}\left[(x_1 x_2 \cdots x_n)^{1/n}\right]$$
$$= \frac{1}{n}(\log_{10} x_1 + \log_{10} x_2 + \cdots + \log_{10} x_n)$$
$$= \bar{x}_{\log}$$

Therefore

$$\text{Geometric mean} = \text{antilog } \bar{x}_{\log}$$

The common logarithms of the data in Example 6.8 are: $\log_{10} 1 = 0.0000$, $\log_{10} 3 = 0.4771$, $\log_{10} 5 = 0.6990$, $\log_{10} 6 = 0.7782$, $\log_{10} 8 = 0.9031$. Therefore

$$\bar{x}_{\log} = \frac{\sum_{i=1}^{5} \log_{10} x_i}{5} = \frac{0.0000 + 0.4771 + 0.6990 + 0.7782 + 0.9031}{5}$$
$$= 0.57148$$
$$\text{Geometric mean} = \text{antilog } \bar{x}_{\log}$$
$$= 3.728039, \text{ or } 3.7, \text{ which is the value found in Example 6.8}$$

6.14 You measure (in cm) the diameter D and length L of four cylinders. As these measures are ratio level, you can calculate for each cylinder these two ratios: D/L and L/D. The results for D/L are: 2/10, 5/10, 2/10, 5/10; and for L/D are: 10/2, 10/5, 10/2, 10/5. Using these two sets of ratios, show why it is said that the geometric mean is a better way to average ratios than the arithmetic mean.

Solution

If we let $x_i = D/L$ and $y_i = L/D$, then the arithmetic means of the two sets of ratios are

$$\bar{x} = \frac{\sum_{i=1}^{4} x_i}{4} = \frac{0.2 + 0.5 + 0.2 + 0.5}{4} = 0.35$$

$$\bar{y} = \frac{\sum_{i=1}^{4} y_i}{4} = \frac{5 + 2 + 5 + 2}{4} = 3.5$$

The geometric means of the two sets are

$$\text{Geometric mean for } x_i = \sqrt[4]{\prod_{i=1}^{4} x_i} = \sqrt[4]{(0.2)(0.5)(0.2)(0.5)} = 0.316228$$

$$\text{Geometric mean for } y_i = \sqrt[4]{\prod_{i=1}^{4} y_i} = \sqrt[4]{5 \times 2 \times 5 \times 2} = 3.16228$$

The ratio D/L is the reciprocal of the ratio L/D: $D/L = \frac{1}{L/D}$. Therefore, we want the average of the D/Ls to be the reciprocal of the average of the L/Ds. This is not true for the arithmetic means, where $\bar{x} = 0.35$ is not the reciprocal of $\bar{y} = 3.5$: $0.35 \neq \dfrac{1}{3.5} = 0.285714$. It is true, however, for the geometric means where

$$\text{Geometric mean for } x_i = 0.316228 = \frac{1}{\text{geometric mean for } y_i} = \frac{1}{3.16228}$$

This is why the geometric mean is recommended as the preferred average to use for ratios.

6.15 Using equation (6.21), calculate the harmonic means for the three samples in Problem 6.12.

Solution

(a) Harmonic mean $= \dfrac{1}{\dfrac{1}{7}\left(\dfrac{1}{1}+\dfrac{1}{1}+\dfrac{1}{1}+\dfrac{1}{2}+\dfrac{1}{3}+\dfrac{1}{8}+\dfrac{1}{14}\right)}$

$= \dfrac{1}{0.142857(1+1+1+0.5+0.333333+0.125+0.071429)} = 1.73708,\ \text{or}\ 1.7$

(b) Harmonic mean $= \dfrac{1}{\dfrac{1}{5}\left(\dfrac{1}{2}+\dfrac{1}{2}+\dfrac{1}{2}+\dfrac{1}{2}+\dfrac{1}{2}\right)} = \dfrac{1}{0.2(2.5)} = 2.0$

(c) Harmonic mean $= \dfrac{1}{\dfrac{1}{7}\left(\dfrac{1}{1}+\dfrac{1}{3}+\dfrac{1}{5}+\dfrac{1}{9}+\dfrac{1}{9}+\dfrac{1}{9}+\dfrac{1}{9}\right)}$

$= \dfrac{1}{0.142857[1+0.333333+0.2+4(0.111111)]} = 3.53933,\ \text{or}\ 3.5$

Note: Comparing the results of this problem with those in Problem 6.12, we find the following to be true:

If all data values are the same, then the

$$(\text{harmonic mean}) = (\text{geometric mean}) = (\text{arithmetic mean})$$

If all data values are not the same, then the

$$(\text{harmonic mean}) < (\text{geometric mean}) < (\text{arithmetic mean})$$

The harmonic mean is rarely used, but under some conditions it is recommended for time rates, such as miles per hour.

THE MEDIAN AND OTHER QUANTILES

6.16 Using both the odd–even rules from Section 6.13 and equation (6.22), find the median for this sample: 12, 12, 12, 14, 17, 19, 21.

Solution

Using the rules, $n = 7$, so $\tilde{x} = 14$.

Using the equation, $i = \left(\dfrac{1 \times 7}{2}\right) + \dfrac{1}{2} = 4$; $\tilde{x} = x_4$; therefore the median is the fourth value in the array: 14.

6.17 For the following sample, use equation (6.22) to locate D_3 and P_{67}: 1, 6, 7, 8, 9, 11, 13, 15, 20, 21, 28.

Solution

$D_3 = \text{third decile} = Q_{3/10}$; therefore $i = \left(\dfrac{3 \times 11}{10}\right) + \dfrac{1}{2} = 3.8$; $D_3 = x_{3.8}$; therefore D_3 is 0.8 of the way between the third and fourth values in the array: $7 + 0.8\,(8 - 7) = 7.8$.

$$P_{67} = \text{sixty-seventh percentile} = Q_{67/100}; \quad \text{therefore } i = \left(\frac{67 \times 11}{100}\right) + \frac{1}{2} = 7.87;$$

$P_{67} = x_{7.87}$; therefore P_{67} is 0.87 of the way between the seventh and eighth values in the array: $13 + 0.87 (15 - 13) = 14.74$, or 14.7.

6.18 For the following sample of weight measurements (in mg), arrange the sample into a simple stem-and-leaf display that has single-digit starting parts and leaves, and a stem width of 1 mg (see Problem 5.43). Then transform the display into an ascending array, and finally use equation (6.22) to find D_4 and Q_2:

8.1, 4.9, 6.5, 6.3, 5.8, 3.7, 2.2, 1.1, 5.7, 7.4, 2.8, 3.3, 6.9, 3.9, 3.1, 2.0, 5.3, 7.0, 1.3, 1.9, 7.9, 6.2, 5.0, 5.2, 2.5, 2.1, 4.2, 3.6, 7.6, 4.5.

Solution

The simple stem-and-leaf display allows a quick organization of the data for calculating quantiles. Scanning the units digits in the data, we find they go from 1 to 8, producing the display shown in Fig. 6-3(a). In Fig. 6-3(b), the display has been transformed into an ascending array by reorganizing the leaves on each stem to go from small-to-large.

(a)			(b)		
1	139	(3)	1	139	(3)
2	28051	(5)	2	01258	(5)
3	73916	(5)	3	13679	(5)
4	925	(3)	4	259	(3)
5	87302	(5)	5	02378	(5)
6	5392	(4)	6	2359	(4)
7	4096	(4)	7	0469	(4)
8	1	(1)	8	1	(1)
		(30)			(30)

Fig. 6-3

$$D_4 = \text{fourth decile} = Q_{4/10}; \quad \text{therefore } i = \left(\frac{j \times n}{m}\right) + \frac{1}{2} = \left(\frac{4 \times 30}{10}\right) + \frac{1}{2} = 12.5;$$

$D_4 = x_{12.5}$. D_4 is midway between the 12th and 13th values in the array. These are found by counting down the check-count column in Fig. 6-3(b), where it is seen that values x_9 to x_{13} are in the third stem down, and that $x_{12} = 3.7$ mg and $x_{13} = 3.9$ mg. Therefore,

$$D_4 = \frac{3.7 \, \text{mg} + 3.9 \, \text{mg}}{2} = 3.80 \, \text{mg}$$

$Q_2 = \tilde{x} = Q_{1/2}$; therefore $i = \left(\frac{1 \times 30}{2}\right) + \frac{1}{2} = 15.5$; $Q_2 = x_{15.5}$. Q_2 is midway between the 15th and 16th values in the array. Counting down the check-count column, we find that x_{15} and x_{16} are in the fourth stem, and that $x_{15} = 4.5$ mg and $x_{16} = 4.9$ mg. Therefore,

$$Q_2 = \frac{4.5 \, \text{mg} + 4.9 \, \text{mg}}{2} = 4.70 \, \text{mg}$$

6.19 For the following sample of length measurements (in mm), first arrange the sample in a stem-and-leaf display that has single-digit starting parts, two-digit leaves, and a stem width of 0.1 mm (see Problem 5.44), then transform the display into an ascending array, and finally, use equation (6.22) to find D_7 and P_{13}:

0.948, 0.513, 0.687, 0.231, 0.299, 0.717, 0.379, 0.310, 0.785, 0.542, 0.222, 0.593, 0.827, 0.309, 0.784, 0.502, 0.272, 0.492, 0.256, 0.651, 0.329, 0.358, 0.447, 0.699, 0.589.

Solution

The requested stem-and-leaf displays are shown in Fig. 6-4.

$$D_7 = \text{seventh decile} = Q_{7/10}; \quad \text{therefore,} \quad i = \left(\frac{j \times n}{m}\right) + \frac{1}{2} = \left(\frac{7 \times 25}{10}\right) + \frac{1}{2} = 18;$$

$D_7 = x_{18}$. Therefore, D_7 is the 18th value in the array: 0.651 mm.

$$P_{13} = \text{thirteenth percentile} = Q_{13/100}; \quad \text{therefore,} \quad i = \left(\frac{13 \times 25}{100}\right) + \frac{1}{2} = 3.75;$$

$P_{13} = x_{3.75}$. Therefore, P_{13} is 0.75 of the way between x_3 (0.256 mm) and x_4 (0.272 mm): $P_{13} = 0.256$ mm + 0.75(0.272 mm − 0.256 mm) = 0.2680 mm.

Original stem-and-leaf display		Display transformed into array		
2	31, 99, 22, 72, 56	2	22, 31, 56, 72, 99	(5)
3	79, 10, 09, 29, 58	3	09, 10, 29, 58, 79	(5)
4	92, 47	4	47, 92	(2)
5	13, 42, 93, 02, 89	5	02, 13, 42, 89, 93	(5)
6	87, 51, 99	6	51, 87, 99	(3)
7	17, 85, 84	7	17, 84, 85	(3)
8	27	8	27	(1)
9	48	9	48	(1)
				(25)

Fig. 6-4

6.20 For the weight data summarized in Table 6.8, use equation (6.23) to find Q_1, Q_2, and Q_3.

Solution

In Table 6.14, we have converted the frequency distribution in Table 6.8 into a "less than" cumulative frequency distribution. As $\frac{1}{4}$ of the 90 values, or 22.5 values, must be less than Q_1, the Q_1 category is 17 g. Thus

$$Q_{j/m} = b + \left[\frac{\frac{(j \times n)}{m} - Cf}{f}\right](w)$$

$$Q_1 = Q_{1/4} = 16.5 \, \text{g} + \left[\frac{\frac{(1 \times 90)}{4} - 8}{18}\right](1.0 \, \text{g})$$

$$= 16.5 \, \text{g} + (0.805556 \times 1.0 \, \text{g})$$

$$= 17.305556 \, \text{g, or } 17.3 \, \text{g}$$

As $\frac{1}{2}$ of the 90 values, or 45 values, must be less then Q_2, the Q_2 category is 18 g. Thus

$$Q_2 = \tilde{x} = Q_{1/2} = 17.5 \, \text{g} + \left[\frac{\frac{(1 \times 90)}{2} - 26}{24}\right](1.0 \, \text{g})$$

$$= 18.291667 \, \text{g, or } 18.3 \, \text{g}$$

Table 6.14

Weight (g)	Cumulative frequency
Less than 14	0
Less than 15	2
Less than 16	4
Less than 17	8
Less than 18	26
Less than 19	50
Less than 20	85
Less than 21	90

This confirms what we said in Problem 6.4, that for this negatively skewed distribution the median will be to the right of (larger than) the arithmetic mean (18.1 g).

As $\frac{3}{4}$ of the 90 values, or 67.5 values, must be less than Q_3, the Q_3 *category* is 19 g. Thus

$$Q_3 = Q_{3/4} = 18.5\,\text{g} + \left[\frac{\frac{(3 \times 90)}{4} - 50}{35} \right](1.0\,\text{g})$$

$$= 18.5\,\text{g} + (0.5 \times 1.0\,\text{g}) = 19.0\,\text{g}$$

6.21 For the temperature data (in °F) summarized in Table 6.9, use equation (6.23) to find Q_2 and P_{87}.

Solution

In Table 6.15 we have converted the frequency distribution in Table 6.9 into a "less than" cumulative frequency distribution. As $\frac{1}{2}$ of the 100 values, or 50 values, must be less than Q_2, the Q_2 *category* is also 101°F. Thus

$$Q_{j/m} = b + \left[\frac{\frac{(j \times n)}{m} - Cf}{f} \right](w)$$

$$Q_2 = \tilde{x} = Q_{1/2} = 100.5°\text{F} + \left[\frac{\frac{(1 \times 100)}{2} - 10}{45} \right](1.0°\text{F})$$

$$= 100.5°\text{F} + (0.888889 \times 1.0°\text{F})$$

$$= 101.388889°\text{F}, \text{ or } 101.4°\text{F}$$

This confirms what we said in Problem 6.5, that for this positively skewed distribution the arithmetic mean (101.8°F) is to the right of (larger than) the median.

As 87/100 of the 100 values, or 87 values, must be less than P_{87}, the P_{87} *category* is 103°F. Thus

$$P_{87} = Q_{87/100} = 102.5°\text{F} + \left[\frac{\frac{(87 \times 100)}{100} - 80}{10} \right](1.0°\text{F})$$

$$= 102.5°\text{F} + (0.7 \times 1.0°\text{F})$$

$$= 103.2°\text{F}$$

Table 6.15

Temperature (°F)	Cumulative frequency
Less than 100	0
Less than 101	10
Less than 102	55
Less than 103	80
Less than 104	90
Less than 105	95
Less than 106	95
Less than 107	98
Less than 108	100

CALCULATING QUANTILES FROM GROUPED FREQUENCY DISTRIBUTIONS

6.22   The 64 second exam scores from Table A.2 (column 3) are shown organized into an ascending array in the simple stem-and-leaf display in Fig. 6-5. Treating this display as a nongrouped frequency distribution, use equation (6.23) to find the exact version of Q_2. Then, using equation (6.24), find the approximate version of Q_2 from the grouped distribution of this data in Table 6.10.

4	9	(1)
5	5799	(4)
6	44457899	(8)
7	124688999	(9)
8	00012334444566777888	(20)
9	0000000111123344456789	(22)
		(64)

Fig. 6-5

Solution

As $\frac{1}{2}$ of the 64 values, or 32 values, must be less than Q_2, the Q_2 *category* in the display is 84. Thus, the exact median value is

$$Q_{j/m} = b + \left[\frac{\frac{(j \times n)}{m} - Cf}{f} \right] (w)$$

$$Q_2 = \tilde{x} = Q_{1/2} = 83.5 + \left[\frac{\frac{(1 \times 64)}{2} - 29}{4} \right] (1.0)$$

$$= 83.5 + (0.75 \times 1.0)$$

$$= 84.25, \text{ or } 84.2$$

To use the quantile-locating formula for grouped frequency distributions, we have converted the grouped frequency distribution in Table 6.10 into the grouped "less than" cumulative frequency distribution in Table 6.16. To find the approximate version of Q_2, again 32 values must be less than Q_2, so the Q_2 *class* is 80 to 84.

Table 6.16

2d Exam	Cumulative frequency
Less than 45	0
Less than 50	1
Less than 55	1
Less than 60	5
Less than 65	8
Less than 70	13
Less than 75	16
Less than 80	22
Less than 85	33
Less than 90	42
Less than 95	59
Less than 100	100

Thus, the approximate Q_2 value is

$$Q_{j/m} \approx b_c + \left[\frac{\dfrac{(j \times n)}{m} - Cf_c}{f_c}\right](w_c)$$

$$Q_2 = \tilde{x} = Q_{1/2} \approx 79.5 + \left[\frac{\dfrac{(1 \times 64)}{2} - 22}{11}\right](5.0)$$

$$\approx 79.5 + (0.909091 \times 5.0)$$

$$\approx 84.045455, \text{ or } 84.0$$

Note: In Problem 6.6 we showed for this negatively skewed distribution that the median of 84.5 found by the graphic estimation technique is to the right of (larger than) either the exact (81.6) or approximate (81.4) arithmetic means. Now we have found that the exact (84.2) and approximate (84.0) medians are also to the right of these arithmetic means.

6.23 First, using equation (6.23), find the exact versions of Q_1 and Q_2 from the nongrouped distribution of golf winnings in Table 6.11. Then, using equation (6.24), find the approximate versions of Q_1 and Q_2 from the grouped distributions of this data in Table 6.12.

Solution

For the exact version of Q_1 from Table 6.11, 1/4 of the 70 values, or 17.5 values, must be less than Q_1. Therefore, the Q_1 *category* is $2,000. The exact Q_1 value is

$$Q_{j/m} = b + \left[\frac{\dfrac{(j \times N)}{m} - Cf}{f}\right](w)$$

$$Q_1 = Q_{1/4} = \$1,500 + \left[\frac{\dfrac{(1 \times 70)}{4} - 0}{18}\right](\$1,000)$$

$$= \$1,500 + (0.972222 \times \$1,000)$$

$$= \$2,472.222, \text{ or } \$2,500$$

For the exact version of Q_2, as 1/2 of the 70 values, or 35 values, must be less than Q_2, the Q_2 *category* is $6,000. Thus, the exact Q_2 value is

$$Q_2 = \tilde{\mu} = Q_{1/2} = \$5,500 + \left[\frac{\frac{(1 \times 70)}{2} - 30}{7} \right] (\$1,000)$$

$$= \$5,500 + (0.714286 \times \$1,000)$$

$$= \$6,214.286, \text{ or } \$6,200$$

The fact that the grouped frequency distribution in Table 6.12 has unequal class widths does not change the techniques used for calculating approximate quantiles. The quantile-locating formula for grouped frequency distributions is used on the grouped "less than" cumulative frequency distribution version of this distribution shown in Table 6.17. To find the approximate version of Q_1, again 17.5 values must be less than Q_1, so the Q_1 *class* is $2,000 to $4,000. Thus, the approximate Q_1 value is

$$Q_{j/m} \approx b_c + \left[\frac{\frac{(j \times n)}{m} - Cf_c}{f_c} \right] (w_c)$$

$$Q_1 = Q_{1/4} \approx \$1,500 + \left[\frac{\frac{(1 \times 70)}{4} - 0}{30} \right] (\$3,000)$$

$$\approx \$1,500 + (0.583333 \times \$3,000)$$

$$\approx \$3,249.999, \text{ or } \$3,200$$

Table 6.17

Winnings ($)	Cumulative frequency
Less than 1,500	0
Less than 4,500	30
Less than 10,500	47
Less than 20,500	54
Less than 30,500	62
Less than 40,500	65
Less than 45,500	66
Less than 81,500	66
Less than 83,500	69
Less than 198,500	69
Less than 200,500	70

To find the approximate version of Q_2, again 35 values must be less than Q_2, so the Q_2 *class* is $5,000 to $10,000. Thus, the approximate Q_2 value is

$$Q_2 = \tilde{\mu} = Q_{1/2} \approx \$4,500 + \left[\frac{\frac{(1 \times 70)}{2} - 30}{17} \right] (\$6,000)$$

$$\approx \$4,500 + (0.294118 \times \$6,000)$$

$$\approx \$6,264.708, \text{ or } \$6,300$$

Note: As you would expect for this positively skewed distribution, both versions of this median we have calculated (exact: \$6,200; approximate: \$6,300) are to the left of (smaller than) the arithmetic means from Problem 6.7 (exact: \$15,900; approximate: \$16,100).

6.24   Use the graphic estimation technique on the "less than" percentage ogive for the golf winnings (see Fig. 5-38) to find Q_1 and Q_3.

Solution

The relevant portion of Fig. 5-38 has been expanded in Fig. 6-6. To estimate Q_1, a horizontal line is drawn from the 25% point on the vertical axis to the ogive, which intersects it at E_1. A vertical line is then dropped from E_1 to the X axis, which it intersects at D_1, or roughly \$3,000, which is our first estimate of Q_1. For the more exact estimate we use the similar triangles $A_1B_1C_1$ and $A_1D_1E_1$, in which the sides have the relationship

$$\frac{A_1D_1}{A_1B_1} = \frac{D_1E_1}{B_1C_1}$$

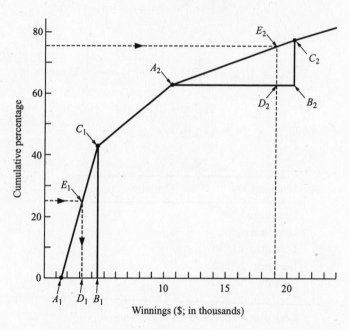

Fig. 6-6

We know that: $A_1B_1 = Q_1$ *class* width $= \$3,000$; $B_1C_1 = 42.8571 - 0.0 = 42.8571$; and $D_1E_1 = 25.0 - 0.0 = 25.0$. So now, solving for A_1D_1

$$\frac{A_1D_1}{\$3,000} = \frac{25.0}{42.8571}$$

$$A_1D_1 = \$3,000(0.583334) = \$1,750.00$$

Therefore

$$Q_1 = \$1,500 + \$1,750 = \$3,250, \text{ or } \$3,200$$

To estimate Q_3, a horizontal line is drawn from the 75% point that intersects the ogive at E_2, from which a vertical line is dropped to the X axis. The intersection with the X axis, at roughly \$19,000, is our first estimate of Q_3. For the more exact estimate, we use the similar triangles $A_2B_2C_2$ and $A_2D_2E_2$, where

$$\frac{A_2D_2}{A_2B_2} = \frac{D_2E_2}{B_2C_2}$$

We know that: $A_2B_2 = Q_3$ *class* width $= \$10,000$; $B_2C_2 = 77.1429 - 62.1429 = 15.0$; and $D_2E_2 = 75.0 - 62.1429 = 12.8571$. So now, solving for A_2D_2

$$\frac{A_2D_2}{\$10,000} = \frac{12.8571}{15.0}$$

$$A_2D_2 = \$10,000(0.857140) = \$8,571.40$$

Therefore

$$Q_3 = \$10,500 + \$8,571.40 = \$19,071.40, \text{ or } \$19,000$$

THE MIDRANGE, THE MIDQUARTILE, AND THE TRIMEAN

6.25 For the weight data summarized in Table 6.8 and the quartiles calculated in Problem 6.20, determine the midrange, midquartile, and trimean.

Solution

$$\text{Midrange} = \frac{x_s + x_l}{2} = \frac{14\,\text{g} + 20\,\text{g}}{2} = 17.0\,\text{g}$$

$$\text{Midquartile} = \frac{Q_1 + Q_3}{2} = \frac{17.305556\,\text{g} + 19.0\,\text{g}}{2} = 18.152778\,\text{g}, \text{ or } 18.2\,\text{g}$$

$$\text{Trimean} = \frac{Q_1 + 2Q_2 + Q_3}{4} = \frac{17.305556\,\text{g} + 2(18.291667\,\text{g}) + 19.0\,\text{g}}{4}$$

$$= 18.222222\,\text{g}, \text{ or } 18.2\,\text{g}$$

It can be seen for this negatively skewed distribution that

$$\text{midrange} < \bar{x} < \text{midquartile} < \text{trimean} < \tilde{x}$$

6.26 For the temperature data summarized in Table 6.9 and the quartiles calculated in Problems 6.21 and 6.51, determine the midrange, midquartile, and trimean.

Solution

$$\text{Midrange} = \frac{x_s + x_l}{2} = \frac{100°\text{F} + 107°\text{F}}{2} = 103.5°\text{F}$$

$$\text{Midquartile} = \frac{Q_1 + Q_3}{2} = \frac{100.833333°\text{F} + 102.3°\text{F}}{2} = 101.566666°\text{F}, \text{ or } 101.6°\text{F}$$

$$\text{Trimean} = \frac{Q_1 + 2Q_2 + Q_3}{4} = \frac{100.833333°\text{F} + 2(101.388889°\text{F}) + 102.3°\text{F}}{4}$$

$$= 101.477778°\text{F}, \text{ or } 101.5°\text{F}$$

It can be seen for this positively skewed distribution that

$$\tilde{x} < \text{trimean} < \text{midquartile} < \bar{x} < \text{midrange}$$

6.27 For the golf-winnings data summarized in Table 6.11 and the exact quartiles calculated in Problems 6.23 and 6.53, determine the midrange, midquartile, and trimean.

Solution

$$\text{Midrange} = \frac{x_s + x_l}{2} = \frac{\$2{,}000 + \$200{,}000}{2} = \$101{,}000$$

$$\text{Midquartile} = \frac{Q_1 + Q_3}{2} = \frac{\$2{,}472.222 + \$15{,}285.714}{2} = \$8{,}878.968, \text{ or } \$8{,}900$$

$$\text{Trimean} = \frac{Q_1 + 2Q_2 + Q_3}{4} = \frac{\$2{,}472.222 + 2(\$6{,}214.286) + \$15{,}285.714}{4}$$

$$= \$7{,}546.627, \text{ or } \$7{,}500$$

It can be seen for the exact values calculated for this positively skewed distribution that

$$\tilde{\mu} < \text{trimean} < \text{midquartile} < \mu < \text{midrange}$$

6.28 For the grouped golf-winnings data summarized in Table 6.12 and the approximate quartiles calculated from this data in Problems 6.23 and 6.53, determine the approximate versions of the midrange, midquartile, and trimean.

Solution

If only grouped data are available, such as the grouped frequency distribution in Table 6.12, then there are two accepted techniques for approximating x_s and x_l: (1) for x_s, use the class mark for the class with the smallest values; for x_l, use the class mark for the class with the largest values; (2) for x_s, use the lower boundary of the class with the smallest values; for x_l, use the upper boundary of the class with the largest values. As midrange $= \dfrac{x_s + x_l}{2}$, therefore using technique (1)

$$\text{Midrange} \approx \frac{\$3{,}000 + \$199{,}500}{2} = \$101{,}250, \text{ or } \$101{,}200$$

and using technique (2)

$$\text{Midrange} \approx \frac{\$1{,}500 + \$200{,}500}{2} = \$101{,}000$$

As midquartile $= \dfrac{Q_1 + Q_3}{2}$, therefore using the approximate values from Problems 6.23 and 6.53

$$\text{Midquartile} \approx \frac{\$3{,}249.999 + \$18{,}357.140}{2} = \$10{,}803.5695, \text{ or } \$10{,}800$$

As trimean $= \dfrac{Q_1 + 2Q_2 + Q_3}{4}$, therefore using the approximate values from Problems 6.23 and 6.53

$$\text{Trimean} \approx \frac{\$3{,}249.999 + 2(\$6{,}264.708) + \$18{,}357.140}{4} = \$8{,}534.139, \text{ or } \$8{,}500$$

It can be seen that the size ordering for these approximate versions is the same as we found for the exact versions in Problem 6.27.

$$\tilde{\mu} < \text{trimean} < \text{midquartile} < \mu < \text{midrange}$$

THE MODE

6.29 For the following ungrouped frequency distributions, first determine the mode and then the size order for $\bar{x}$, $\tilde{x}$, and mode: (a) Table 6.1, (b) Table 6.8, (c) Table 6.9.

Solution

(a) For the distribution in Table 6.1: mode $= 1.5$ cm. From Table 6.1 and Example 6.13 we know for this unimodal, symmetric distribution that $\bar{x} = 1.50$ cm and $\tilde{x} = 1.50$ cm. Therefore: $\bar{x} = \tilde{x} = $ mode.

(b) For the distribution in Table 6.8: mode $= 19$ g. From Table 6.8 and Problem 6.20 we know for this unimodal, negatively skewed distribution that $\bar{x} = 18.1$ g and $\tilde{x} = 18.3$ g. Therefore: $\bar{x} < \tilde{x} <$ mode.

(c) For the distribution in Table 6.9: mode $= 101°$F. From Table 6.9 and Problem 6.21 we know for this unimodal, positively skewed distribution that $\bar{x} = 101.8°$F and $\tilde{x} = 101.4°$F. Therefore: mode $< \tilde{x} < \bar{x}$.

Note: This problem shows that while $\bar{x} = \tilde{x} =$ mode for a unimodal symmetric distribution, if the distribution is unimodal and skewed then these statistical measures are separated, with $\bar{x}$ closest to the exteme values in the skewed tail, the mode farthest from the skewed tail, and $\tilde{x}$ between $\bar{x}$ and mode.

6.30 For the following data, determine $\bar{x}$, $\tilde{x}$, and mode, and then their size order:

$$x_1 = x_2 = x_3 < x_4 < x_5 < x_6 < x_7.$$

Solution

$$\bar{x} = \frac{\sum_{i=1}^{7} x_i}{7}, \tilde{x} = x_4, \text{ mode} = x_1 = x_2 = x_3$$

From Problem 6.29(c) we know for a positively skewed distribution that: mode $< \tilde{x} < \bar{x}$

6.31 It is stated in many statistics books that for a unimodal frequency distribution that is "moderately skewed," there is this relationship

$$\bar{x} - \tilde{x} = \frac{1}{3}(\bar{x} - \text{mode})$$

In words, this states that the distance between $\bar{x}$ and $\tilde{x}$ is a third of the distance from $\bar{x}$ to the mode. This is called an *empirical rule* because it has been found through examining patterns in sets of numbers rather than by mathematical derivation. For the following samples, find $\bar{x}$, $\tilde{x}$, and mode, and then see if this relationship is correct: (a) 1.5, 3.0, 3.0, 3.0, 4.0, 5.0, 6.0, 7.0, 8.0,   (b) 1.5, 3.0, 3.0, 3.0, 4.0, 5.0, 7.0, 9.0, 10.0.

Solution

(a) $\bar{x} = \dfrac{\sum_{i=1}^{9} x_i}{9} = \dfrac{40.5}{9} = 4.5$; by the odd-even rules, $\tilde{x} = 4.0$; and mode $= 3.0$

Therefore

$$\bar{x} - \tilde{x} = 4.5 - 4.0 = 0.5$$
$$\bar{x} - \text{mode} = 4.5 - 3.0 = 1.5$$

thus confirming for this unimodal "moderately" positively skewed distribution that

$$\bar{x} - \tilde{x} = \frac{1}{3}(\bar{x} - \text{mode})$$
$$0.5 = \frac{1}{3}(1.5)$$

(b) The mode and $\tilde{x}$ remain the same as in (a), but now the positive skewing has been increased so $\bar{x}$ has become larger (moved in the positive direction)

$$\bar{x} = \frac{\sum_{i=1}^{9} x_i}{9} = \frac{45.5}{9} = 5.055556$$

Therefore

$$\bar{x} - \tilde{x} = 5.055556 - 4.0 = 1.055556$$

$$\bar{x} - \text{mode} = 5.055556 - 3.0 = 2.055556$$

Thus, with the increase in positive skewing, the 1/3 relationship is no longer correct. Instead

$$\bar{x} - \tilde{x} = \frac{1}{1.9473678}(\bar{x} - \text{mode})$$

Note: In general, this empirical 1/3 rule is approximately correct for moderately skewed, unimodal frequency distributions, but as skewing increases and $\bar{x}$ moves farther from $\tilde{x}$ and the mode, the rule breaks down.

6.32   What is the mode of the distribution shown as a *bimodal frequency histogram* in Fig. 6-7?

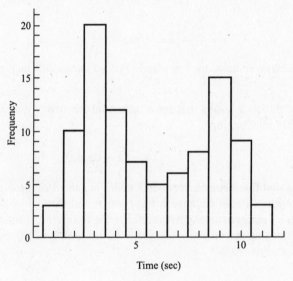

Time (sec)

Fig. 6-7

Solution

According to the definition in Section 6.18, "the mode of a set of data is the measurement value in the set that occurs most frequently," there is one mode for the distribution in Fig. 6-7: 3 sec. However, according to what was said in Example 5.2, a frequency histogram with two peaks is called bimodal. How is this seeming contradiction resolved?

If a distribution has two distinct peaks that have the same height (frequency) in a histogram, then (see Section 6.18) the distribution has two modes and there is no contradiction in calling it bimodal. The problem arises when, as in Fig. 6-7, there are two such distinct peaks but now they have different heights. By convention such distributions are also called bimodal, but many statistics books say they have only one *true mode*. Other books get around this problem by saying that the measurement value with the higher peak is the *major mode* and the value with the lower peak is the *minor mode*. The two modes are also called the *global mode* and the *local mode*.

Similarly, if a distribution has three distinct peaks of equal height then there are three modes and the distribution is called *trimodal*. If there are three such prominent peaks of unequal heights, the distribution is still called trimodal but now it is said to have one true mode, or a major and two minor modes, or a global and two local modes.

As we indicated in Problem 5.37, bimodal distributions often mean that the sample contains measurements taken from two different populations. Trimodal distributions can also be symptomatic of problems in sampling or measurement.

6.33 Frequency histograms are shown for four distributions in Fig. 6-8. For each histogram, select from the following list of terms and relationships, which have been discussed throughout the chapter, those that are true for the illustrated distributions: symmetric, positively skewed, negatively skewed, unimodal, bimodal, no mode, $\sum(x_i - \bar{x}) = 0$, $\bar{x} = \tilde{x}$, $\bar{x} = \tilde{x} =$ mode, mode $< \tilde{x} < \bar{x}$, $\bar{x} < \tilde{x} <$ mode, $\bar{x} =$ trimean, $\bar{x} <$ trimean, trimean $< \bar{x}$.

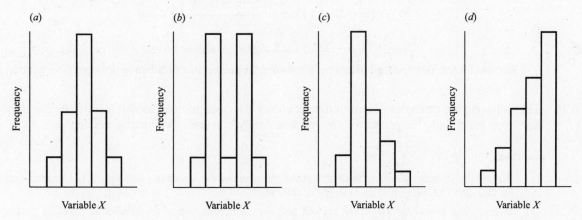

Fig. 6-8

Solution

(a) Symmetric, unimodal, $\sum(x_i - \bar{x}) = 0$, $\bar{x} = \tilde{x} =$ mode, $\bar{x} =$ trimean

(b) Symmetric, bimodal, $\sum(x_i - \bar{x}) = 0$, $\bar{x} = \tilde{x}$, $\bar{x} =$ trimean

(c) Positively skewed, unimodal, $\sum(x_i - \bar{x}) = 0$, mode $< \tilde{x} < \bar{x}$, trimean $< \bar{x}$

(d) Negatively skewed, unimodal, $\sum(x_i - \bar{x}) = 0$, $\bar{x} < \tilde{x} <$ mode, $\bar{x} <$ trimean

6.34 For the frequency histogram illustrated in Fig. 6-9, find $\bar{x}$, $\tilde{x}$, and mode.

Solution

$$\bar{x} = \frac{\sum_{i=1}^{5} f_i x_i}{\sum_{i=1}^{5} f_i} = \frac{(5 \times 2 \text{ sec}) + (5 \times 3 \text{ sec}) + (5 \times 4 \text{ sec}) + (5 \times 5 \text{ sec}) + (5 \times 6 \text{ sec})}{5 + 5 + 5 + 5 + 5}$$

$$= \frac{100 \text{ sec}}{25} = 4.0 \text{ sec}$$

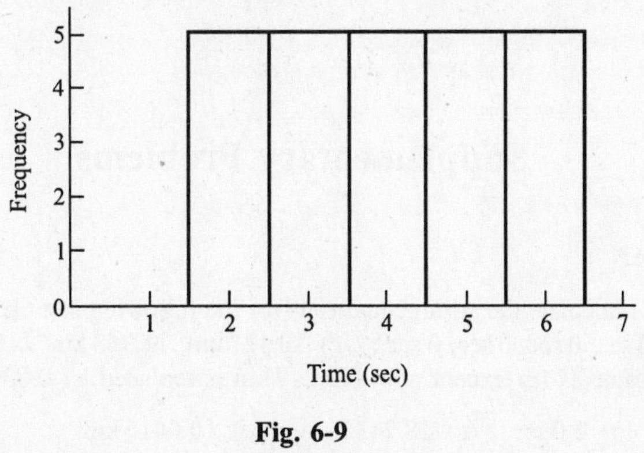

Fig. 6-9

Using equation (6.23)

$$Q_{j/m} = b + \left[\frac{\frac{(j \times n)}{m} - Cf}{f} \right](w)$$

$$Q_2 = \tilde{x} = Q_{1/2} = 3.5 \, \text{sec} + \left[\frac{\frac{(1 \times 25)}{2} - 10}{5} \right](1.0 \, \text{sec})$$

$$= 3.5 \, \text{sec} + (0.5 \times 1.0 \, \text{sec}) = 4.0 \, \text{sec}$$

Because all five measurement values have the same frequency, this distribution does not have a mode.

6.35 The following are examples of descriptive measures discussed in this chapter. Identify the measure for: (a) mean daily temperature, (b) average family income, (c) batting average.

Solution

(a) A mean daily temperature is typically the arithmetic mean of the lowest (x_s) and highest (x_l) temperatures for that day. Therefore, it is an example of midrange.

(b) Income data is usually positively skewed and for such data, while the arithmetic mean is greatly influenced by extreme values, the median is essentially unaffected. This is why average family income is typically an example of median. However, as the trimean is calculated from the median and two other quartiles, it is becoming increasingly popular as a measure of average value for skewed data.

(c) Batting average is an example of arithmetic mean.

CALCULATING MODES FROM GROUPED FREQUENCY DISTRIBUTIONS

6.36 Using techniques (1) and (2) from Section 6.19, determine the approximate mode for the grouped frequency distribution of male heights in Table 4.38.

Solution

The modal class for male heights in Table 4.38 is (68.50 to 69.49) in, with a class mark of 69.00 in. Therefore, using technique (1): mode ≈ 69.00 in.

Using technique (2)

$$\text{mode} \approx b_c + \left(\frac{d_1}{d_1 + d_2} \right)(w_c)$$

$$\approx 68.495 \, \text{in} + \left(\frac{6}{6 + 5} \right)(1.000 \, \text{in})$$

$$\approx 69.040455 \, \text{in}, \quad \text{or} \quad 69.040 \, \text{in}$$

Supplementary Problems

THE ARITHMETIC MEAN

6.37 Using equation (6.3), calculate the arithmetic means for the following samples: (a) 0 sec, 0 sec, 0 sec, 0 sec, 0 sec, (b) 10 sec, 0 sec, 0 sec, 0 sec, 0 sec, (c) 10.127 km, 11.963 km, 112.217 km, 9.777 km, 13.833 km, 14.542 km, (d) the same as (c) except now 112.217 km is replaced by 0.007 km.

Ans. (a) 0.0 sec, (b) 2.0 sec, (c) 28.7432 km, (d) 10.0415 km

6.38   Using equation (6.6), calculate the arithmetic mean for the following population: 100,000 pigs, 115,100 pigs, 152,643 pigs.

   Ans.   123,000 pigs, or 1.23×10^5 pigs, after rounding off the answer

6.39   Knowing that $n = 5$, $\bar{x} = 14.6$, and $\sum_{i=1}^{4} x_i = 58$, what is x_5?

   Ans. $\sum_{x=1}^{5} x_i = 5 \times 14.6 = 73$, and thus $x_5 = \sum_{i=1}^{5} x_i - \sum_{i=1}^{4} x_i = 73 - 58 = 15$

6.40   Using equation (6.10), find the arithmetic mean of the sample presented in Table 5.6.

   Ans.   18.0

CALCULATING APPROXIMATE ARITHMETIC MEANS FROM GROUPED FREQUENCY DISTRIBUTIONS

6.41   Using equation (6.13), calculate the approximate arithmetic mean of the grouped frequency distribution of lengths in Table 4.20.

   Ans. ≈ 1.72 mm

CALCULATING ARITHMETIC MEANS WITH CODED DATA

6.42   For the following sample of weight measurements (in grams), first calculate $\bar{x}$ directly from the data and then calculate it by means of equations (6.14), (6.15), and (6.16), using $a = -770$ g and $b = 10^8$ (see Problems 2.21 and 2.22 for calculation procedures involving numbers written in scientific notation): 7.77×10^{-6}, 7.72×10^{-6}, 7.74×10^{-6}, 7.73×10^{-6}, 7.79×10^{-6}, 7.75×10^{-6}.

   Ans.   The direct calculation of $\bar{x}$ and the calculation using the coding and decoding formulas are shown in Table 6.18.

Table 6.18

Weight (g) x_i	$c_i = -770 \text{ g} + 10^8 x_i \text{ g}$
7.72×10^{-6}	2
7.73×10^{-6}	3
7.74×10^{-6}	4
7.75×10^{-6}	5
7.77×10^{-6}	7
7.79×10^{-6}	9
$\sum$ 0.00004650 g	30 g

$$\bar{x} = \frac{\sum x_i}{n} = \frac{0.00004650 \text{ g}}{6} = 0.000007750 \text{ g} = 7.750 \times 10^{-6} \text{ g}$$

$$\bar{c} = \frac{\sum c_i}{n} = \frac{30 \text{ g}}{6} = 5.0 \text{ g}$$

$$\bar{x} = \frac{1}{b}(\bar{c} - a) = \frac{1}{10^8}[5.0 \text{ g} - (-770 \text{ g})] = \frac{775.0 \text{ g}}{10^8} = 7.750 \times 10^{-6} \text{ g}$$

OTHER MEANS: WEIGHTED, OVERALL, GEOMETRIC, AND HARMONIC

6.43   At the University of Colorado, a student's *overall grade point average* (*GPA*) is determined by the formula

$$\text{GPA} = \frac{\sum_{i=1}^{k} w_i x_i}{\sum_{i=1}^{k} w_i}$$

where w_i is credits, x_i is points per credit, and k is the number of courses being considered. The points per credit are determined by the following system: $A = 4.0$, $A^- = 3.7$, $B^+ = 3.3$, $B = 3.0$, $B^- = 2.7$, $C^+ = 2.3$, and so on. Determine the GPA for a freshman who has earned these grades in four courses: A in course 1, 4 credits; B^+ in course 2, 6 credits; C^+ in course 3, 2 credits; and B^- in course 4, 3 credits.

Ans.

$$\text{GPA} = \frac{\sum_{i=1}^{4} w_i x_i}{\sum_{i=1}^{4} w_i} = \frac{(4 \times 4.0) + (6 \times 3.3) + (2 \times 2.3) + (3 \times 2.7)}{4 + 6 + 2 + 3} = \frac{48.5}{15} = 3.23$$

6.44   You repeat an experiment three times, getting these results: $\bar{x}_1 = 19.2\,\text{cm}$, $n_1 = 10$; $\bar{x}_2 = 17.4\,\text{cm}$, $n_2 = 15$; $\bar{x}_3 = 18.5\,\text{cm}$, $n_3 = 8$. What is the overall mean of these results?

Ans.   18.2 cm

6.45   For the sample 9, 9, 11, 7, find the: (*a*) arithmetic mean,   (*b*) geometric mean,   (*c*) harmonic mean.

Ans.   (*a*) 9.0,   (*b*) 8.0,   (*c*) 8.8

THE MEDIAN AND OTHER QUANTILES

6.46   Using both the odd–even rules from Section 6.13 and equation (6.22), find the median for this sample: 12, 12, 12, 12, 13, 13, 13, 13.

Ans.   Using the rules, $\tilde{x} = 12.5$; using the formula, $\tilde{x} = 12.5$

6.47   For the sample in Problem 6.17, use equation (6.22) to locate Q_1 and $\tilde{x}$.

Ans. $Q_1 = 7.25$, or 7.2; $\tilde{x} = 11$

6.48   For the sample in Problem 6.18 and Fig. 6-3, use equation (6.22) to locate P_{82}.

Ans.   6.91 mg

6.49   For the sample in Problem 6.19 and Fig. 6-4, use equation (6.22) to locate Q_3.

Ans.   0.690 mm

6.50   For the weight data in Tables 6.8 and 6.14, use equation (6.23) to find D_4.

Ans.   17.916667 g, or 17.9 g

6.51   For the temperature data in Tables 6.9 and 6.15, use equation (6.23) to find Q_1 and Q_3.

Ans. $Q_1 = 100.833333°\text{F}$, or 100.8°F; $Q_3 = 102.3°\text{F}$

CALCULATING QUANTILES FROM GROUPED FREQUENCY DISTRIBUTIONS

6.52   For the exam-score data in Fig. 6-5, use equation (6.23) to find the exact version of P_{70}. For the version of this data in Tables 6.10 and 6.16, use equation (6.24) to find the approximate version of P_{70}.

   Ans.   Exact is 89.9, approximate is 90.3

6.53   For the golf-winnings data in Table 6.11, use equation (6.23) to find the exact version of Q_3. From the versions of this data in Tables 6-12 and 6-17, use equation (6.24) to find the approximate version of Q_3.

   Ans.   Exact is \$15,285.714, or \$15,300; approximate is \$18,357.140, or \$18,400

THE MIDRANGE, THE MIDQUARTILE, AND THE TRIMEAN

6.54   For the weight measurements given in Problem 6.18 and Fig. 6-3, we know that: $x_s = 1.1$ mg, $x_l = 8.1$ mg, and $Q_2 = 4.70$ mg. After first determining Q_1 and Q_3, find the midrange, midquartile, and trimean for this data.

   Ans. $Q_1 = 2.8$ mg, $Q_3 = 6.3$ mg, midrange $= 4.60$ mg, midquartile $= 4.55$ mg, trimean $= 4.62$ mg

THE MODE

6.55   For both the female and male hair-color samples summarized in Tables 4-17 and 4-18, determine $\bar{x}$, $\tilde{x}$, and mode.

   Ans.   While for this nominal-level data it is not possible to determine $\bar{x}$ or $\tilde{x}$, it is possible to determine the mode. Thus, for females: mode $=$ blonde; and for males: mode $=$ brown. Such modes are also called *modal categories*.

6.56   For the following data determine $\bar{x}$, $\tilde{x}$, and mode, and then the size order for these measures: $x_1 < x_2 < x_3 < x_4 = x_5$.

   Ans. $\bar{x} = \dfrac{\sum\limits_{i=1}^{5} x_i}{5}$, $\tilde{x} = x_3$, mode $= x_4 = x_5$. From Problem 6.29(b) we know for a unimodal, negatively skewed

   distribution that $\bar{x} < \tilde{x} <$ mode.

Chapter 7

Descriptive Statistics: Measures of Dispersion

7.1 WHY THE RANGE HAS LIMITED VALUE AS A MEASURE OF DISPERSION

Chapter 6 dealt with descriptive measures of central tendency, average value, and location—numerical values that summarize these characteristics of a data set typically with a single number. In this chapter we deal with descriptive measures of another defining characteristic of a data set: how dispersed or spread out the data are, typically in relation to one of the measures from Chapter 6. The measures in this chapter are called *measures of dispersion*, or *measures of variation*, or *measures of variability*.

To see the need for such measures of dispersion, consider the two frequency curves shown in Fig. 7-1. They are both unimodal and symmetric with the same means and medians, but while one rises sharply on both sides of the mean the other shows less concentration of the data near the mean with more dispersion outward.

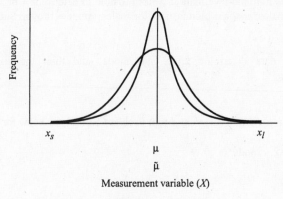

Fig. 7-1

At this point in the book we have considered two measures of dispersion: the range (see Section 4.1) and the deviation from the mean (see Section 6.4). While the deviation from the mean will play a critical role in the measures of dispersion presented in this chapter, it is clear from Fig. 7-1 why the range has limited value as a measure of dispersion. While both curves differ markedly in dispersion, they have the same range, $x_l - x_s$. Certainly the range is important in identifying the outer limits of a data distribution, but it gives no information on what is occuring between these limits. Also, the range is unreliable, being highly sensitive to extreme values that tend to vary from sample to sample.

7.2 THE MEAN DEVIATION

Of all the measures of central tendency presented in Chapter 6, the arithmetic mean is by far the most important and commonly used. Because of this it is necessary to have a measure of dispersion around the mean. (From this chapter on, unless otherwise specified, *the term mean will refer to the arithmetic mean*.) The ideal version of this measure should: (1) be calculated from all the data, (2) show with a single number the typical or average dispersion from the mean, and (3) increase, from data set to data set, with increasing dispersion.

182

For any sample the obvious measure would be the deviation from the mean $x_i - \bar{x}$ (see Section 7.1), and so it would seem that the obvious measure of typical or average dispersion around the mean would be the arithmetic mean of these deviations

$$\frac{\sum_{i=1}^{n}(x_i - \bar{x})}{n} \tag{7.1}$$

This formula certainly satisfies the first two criteria, being calculated from all the data and showing average dispersion from the mean, but it does not satisfy the third criterion. Whatever the dispersion of the data, all calculations with this formula will always result in zero for an answer. This is because the numerator of the formula is $\sum_{i=1}^{n}(x_i - \bar{x})$, and we showed in Section 6.4 that this sum will always equal zero.

The problem with equation (7.1) lies in a fundamental property of the mean: It is *the center of gravity of a distribution* (see Section 6.4). Because of this the sum of the positive deviations from the mean is always equal to the sum of the negative deviations, and thus the sum of these two sums is always zero. There are two accepted ways to solve this problem, both of which eliminate the negative signs from the calculations. The first way is shown in this formula

$$\frac{\sum_{i=1}^{n}|x_i - \bar{x}|}{n} \tag{7.2}$$

where the numerator is now the sum of the *absolute values* (see Section 1.5) of the deviations, and absolute values are always positive in sign. The second way to solve this problem, which we consider when we deal with the *variance* and the *standard deviation* (see Sections 7.5 and 7.9), is to square each deviation and use the *sum of the squared deviations* in the calculations.

The first of these solutions, equation (7.2), is called the *mean deviation* (or the *average deviation* or the *mean absolute deviation*). It shows the average size of the deviations from the mean without regard to direction of deviation. It is zero when all values in a sample are the same and increases across samples with increasing dispersion. While the mean deviation is a legitimate measure of dispersion from the mean, it is rarely used because it has limited value in theoretical statistics.

There is a comparable mean deviation formula for populations

$$\frac{\sum_{i=1}^{N}|x_i - \mu|}{N} \tag{7.3}$$

and both formulas are sometimes modified to show average deviations from the median

$$\frac{\sum_{i=1}^{n}|x_i - \tilde{x}|}{n} \tag{7.4}$$

or

$$\frac{\sum_{i=1}^{N}|x_i - \tilde{\mu}|}{N} \tag{7.5}$$

EXAMPLE 7.1 Calculate the range and the mean deviation for the samples in Example 6.1(*a*) and (*b*) (see Section 6.3 for a discussion of rounding off).

Solution

(*a*) Range $= x_l - x_s = 7 \, \text{g} - 1 \, \text{g} = 6 \, \text{g}$

$$\bar{x} = \frac{\sum_{i=1}^{7} x_i}{7} = \frac{24 \, \text{g}}{7} = 3.42857 \, \text{g}$$

$$\sum_{i=1}^{7} |x_i - \bar{x}| = |1 - 3.42857| \, \text{g} + |3 - 3.42857| \, \text{g} + |2 - 3.42857| \, \text{g} + |7 - 3.42857| \, \text{g}$$

$$+ |5 - 3.42857| \, \text{g} + |4 - 3.42857| \, \text{g} + |2 - 3.42857| \, \text{g}$$

$$= 11.42857 \, \text{g}$$

$$\text{Mean deviation} = \frac{\sum_{i=1}^{7} |x_i - \bar{x}|}{7} = \frac{11.42857 \, \text{g}}{7} = 1.63265 \, \text{g}, \text{ or } 1.6 \, \text{g}$$

(*b*) Range $= x_l - x_s = 200 \, \text{g} - 1 \, \text{g} = 199 \, \text{g}$

$$\bar{x} = \frac{\sum_{i=1}^{7} x_i}{7} = \frac{222 \, \text{g}}{7} = 31.7143 \, \text{g}$$

$$\sum_{i=1}^{7} |x_i - \bar{x}| = |1 - 31.7143| \, \text{g} + |3 - 31.7143| \, \text{g} + |2 - 31.7143| \, \text{g} + |7 - 31.7143| \, \text{g}$$

$$+ |5 - 31.7143| \, \text{g} + |4 - 31.7143| \, \text{g} + |200 - 31.7143| \, \text{g}$$

$$= 336.5715 \, \text{g}$$

$$\text{Mean deviation} = \frac{\sum_{i=1}^{7} |x_i - \bar{x}|}{7} = \frac{336.5715 \, \text{g}}{7} = 48.0816 \, \text{g}, \text{ or } 48.1 \, \text{g}$$

7.3 FREQUENCY-DISTRIBUTION FORMULA FOR MEAN DEVIATION

For calculating the mean deviation, just as there was a frequency-distribution formula for sample mean $\bar{x}$ (see Section 6.6) there is also a *frequency-distribution formula for sample mean deviation* (with a comparable formula for population mean deviation).

$$\text{Mean deviation} = \frac{\sum_{i=1}^{k} f_i |x_i - \bar{x}|}{\sum_{i=1}^{k} f_i} \tag{7.6}$$

$$= \frac{\sum_{i=1}^{k} f_i |x_i - \bar{x}|}{n} \tag{7.7}$$

EXAMPLE 7.2   Calculate the range and the mean deviation for the sample summarized in Table 6.1.

Solution

Range $= 1.8$ cm $- 1.2$ cm $= 0.6$ cm

To calculate the mean deviation for this data requires only the addition to Table 6.1 of two columns: $x_i - \bar{x}$ and $f_i|x_i - \bar{x}|$. The table with these columns and the resulting calculation of mean deviations are shown in Table 7.1.

Table 7.1

Length (cm) x_i	Frequency f_i	$f_i x_i$ (cm)	$(x_i - \bar{x})$(cm)	$f_i|x_i - \bar{x}|$ (cm)
1.2	2	2.4	-0.30	0.60
1.3	7	9.1	-0.20	1.40
1.4	10	14.0	-0.10	1.00
1.5	12	18.0	0.00	0.00
1.6	10	16.0	0.10	1.00
1.7	7	11.9	0.20	1.40
1.8	2	3.6	0.30	0.60
$\sum$	50	75.0 cm		6.00 cm

$$\bar{x} = \frac{\sum f_i x_i}{n} = \frac{75.0 \text{ cm}}{50} = 1.50 \text{ cm}$$

$$\text{Mean deviation} = \frac{\sum f_i|x_i - \bar{x}|}{n} = \frac{6.00 \text{ cm}}{50} = 0.12 \text{ cm}$$

7.4   THE APPROXIMATE MEAN DEVIATION

A mean deviation calculated from a grouped frequency distribution only approximates the exact value calculated directly from the data, so it is called an *approximate mean deviation*. To make this calculation from grouped data we must assume, as we did for the approximate arithmetic mean (see Section 6.7), that all values in a class are equal to the class mark m_i. Then, the approximate mean deviation can be calculated with this formula for a population

$$\text{Population mean deviation} \approx \frac{\sum_{i=1}^{k} f_i|m_i - (\approx \mu)|}{N} \tag{7.8}$$

or this formula for a sample

$$\text{Sample mean deviation} \approx \frac{\sum_{i=1}^{k} f_i|m_i - (\approx \bar{x})|}{n} \tag{7.9}$$

where the symbols $(\approx \mu)$ and $(\approx \bar{x})$ represent approximate means.

EXAMPLE 7.3   Calculate the approximate range and approximate mean deviation for the 30 marathon times in the grouped distribution in Table 6.3.

Solution

In Problem 6.28 we gave two accepted techniques for approximating x_s and x_l for grouped data. Therefore, using technique (1) the approximate range is

Range ≈ 144 min $- 129$ min $= 15$ min

Using technique (2) it is

$$\text{Range} \approx 145.5 \text{ min} - 127.5 \text{ min} = 18.0 \text{ min}$$

To calculate an approximate population mean deviation for the data in Table 6.3 using equation (7.8) requires the addition of two new columns to the table: $m_i - (\approx \mu)$ and $f_i|m_i - (\approx \mu)|$. The modified table with these columns and the resulting calculation of the approximate population mean deviation are shown in Table 7.2.

Table 7.2

Time (min)	Class mark m_i (min)	Frequency f_i	$f_i m_i$ (min)	$[m_i - (\approx \mu)]$ (min)	$f_i|m_i - (\approx \mu)|$ (min)
128–130	129	3	387	−10.6	31.8
131–133	132	1	132	−7.6	7.6
134–136	135	4	540	−4.6	18.4
137–139	138	3	414	−1.6	4.8
140–142	141	7	987	1.4	9.8
143–145	144	12	1,728	4.4	52.8
Σ		30	4,188 min		125.2 min

$$\mu \approx \frac{\sum f_i m_i}{N} = \frac{4,188 \text{ min}}{30} = 139.6 \text{ min}$$

$$\text{Population mean deviation} \approx \frac{\sum f_i|m_i - (\approx \mu)|}{N} = \frac{125.2 \text{ min}}{30} = 4.17333 \text{ min, or } 4.2 \text{ min}$$

7.5   THE POPULATION VARIANCE: DEFINITIONAL FORMULA

In Section 7.2 we indicated that a second technique for measuring the typical deviation from the mean of a data set involved squaring each deviation from the mean and then using the sum of the squared deviations in the calculations. This sum, called the *sum of squares* (and denoted by *SS*), is

$$\sum_{i=1}^{n}(x_i - \bar{x})^2 \tag{7.10}$$

for the sample *SS* and

$$\sum_{i=1}^{N}(x_i - \mu)^2 \tag{7.11}$$

for the population *SS*.

The *variance* (or *mean squared deviation*, or *mean sum of squares*) of a population is the arithmetic mean of its squared deviations from the population mean. It is therefore defined by this *definitional formula for the population variance*

$$\sigma^2 = \frac{\sum_{i=1}^{N}(x_i - \mu)^2}{N} \tag{7.12}$$

or

$$\sigma^2 = \frac{SS}{N} \tag{7.13}$$

where σ is the lowercase Greek letter *sigma*, and thus σ^2, the symbol for the population variance, is read "*sigma* squared." (Recall from Section 1.22 that $\sum$ is the capital letter *sigma* in Greek.)

EXAMPLE 7.4   Use equation (7.12) to calculate the variance for this population of weights (in grams): 2, 3, 4, 5, 6.

   Solution

$$\mu = \frac{\sum_{i=1}^{N} x_i}{N} = \frac{2\,g + 3\,g + 4\,g + 5\,g + 6\,g}{5} = \frac{20\,g}{5} = 4\,g$$

Therefore

$$\sigma^2 = \frac{\sum_{i=1}^{N} (x_i - \mu)^2}{N}$$

$$= \frac{(2\,g - 4\,g)^2 + (3\,g - 4\,g)^2 + (4\,g - 4\,g)^2 + (5\,g - 4\,g)^2 + (6\,g - 4\,g)^2}{5}$$

$$= \frac{10\,g^2}{5} = 2\,g^2$$

Note: The units for the variance are the original measurement units squared.

7.6 THE POPULATION VARIANCE: COMPUTATIONAL FORMULAS

The definitional formula for the population variance has the disadvantage of requiring that the mean be subtracted from each data value. To eliminate this problem, so-called *computational formulas* have been derived from the definitional formula, which are algebraically equivalent to it. These are two of the derived formulas (see Problems 7.5 and 7.6 for derivations)

$$\sigma^2 = \frac{\sum_{i=1}^{N} x_i^2 - \dfrac{\left(\sum_{i=1}^{N} x_i\right)^2}{N}}{N} \tag{7.14}$$

and

$$\sigma^2 = \frac{\sum_{i=1}^{N} x_i^2}{N} - \mu^2 \tag{7.15}$$

They are called computational formulas (or *machine formulas*, or *working formulas*) because they are simpler to use in computations.

EXAMPLE 7.5   Use equation (7.14) to calculate the variance for the population of weights in Example 7.4.

   Solution

From Example 7.4 we know that $N = 5$ and $\sum x_i = 20$ g. Therefore all we need to calculate from the data is

$$\sum_{i=1}^{N} x_i^2 = 4\,g^2 + 9\,g^2 + 16\,g^2 + 25\,g^2 + 36\,g^2 = 90\,g^2$$

Thus

$$\sigma^2 = \frac{\sum\limits_{i=1}^{N} x_i^2 - \dfrac{\left(\sum\limits_{i=1}^{N} x_i\right)^2}{N}}{N} = \frac{90\ \text{g}^2 - \dfrac{(20\ \text{g})^2}{5}}{5} = 2\ \text{g}^2$$

7.7　THE SAMPLE VARIANCE: DEFINITIONAL FORMULA

For a sample, the variance is defined by the formula

$$\text{Sample variance} = s^2 = \frac{\sum\limits_{i=1}^{n}(x_i - \bar{x})^2}{n-1} \tag{7.16}$$

The numerator is, as you would expect, the sample *SS* [see equation (7.10)]. But while the denominator of the population variance is N, population size [see equation (7.12)], here the denominator is $n-1$, sample size minus one. Why $n-1$ rather than n?

To understand this, we must go back to what was said in Section 3.14. There we indicated that there are two levels of statistics: the mathematical level, where the entire integrated system of mathematical statistics is derived and proven, and the intuitive level of general statistics, which is primarily a nonmathematical discussion of statistical concepts and techniques. We warned you that, throughout this book, concepts would be brought up from the mathematical level that would have to be accepted as true without proof. This is one of those instances.

Sample descriptive measures (called statistics) have been developed to estimate population descriptive measures (called parameters) (see Section 3.4). The estimator must have several properties, including being *unbiased* (not systematically overestimating or underestimating the parameter). While it can be proven mathematically that the sample mean $\bar{x}$ is an unbiased estimator of its population mean μ, it can also be proven that the arithmetic mean of the squared deviations, $\dfrac{\sum(x_i - \bar{x})^2}{n}$, is a biased estimator of its population variance σ^2; it underestimates σ^2. It can also be proven that the exact correction for this bias is to divide the sample *SS* by $n-1$.

EXAMPLE 7.6　Use equation (7.16) to calculate the variance for this sample of lengths (in cm): 3, 4, 5, 6, 7.

Solution

$$\bar{x} = \frac{\sum x_i}{n} = \frac{25\ \text{cm}}{5} = 5\ \text{cm}$$

Therefore

$$\begin{aligned} s^2 &= \frac{\sum(x_i - \bar{x})^2}{n-1} \\ &= \frac{(3\ \text{cm} - 5\ \text{cm})^2 + (4\ \text{cm} - 5\ \text{cm})^2 + (5\ \text{cm} - 5\ \text{cm})^2 + (6\ \text{cm} - 5\ \text{cm})^2 + (7\ \text{cm} - 5\ \text{cm})^2}{5-1} \\ &= \frac{10\ \text{cm}^2}{4} = 2.5\ \text{cm}^2 \end{aligned}$$

7.8   THE SAMPLE VARIANCE: COMPUTATIONAL FORMULAS

As was true for the population variance (see Section 7.6), there are both a definitional formula [see equation (7.16)] and algebraically equivalent derived computational formulas for the sample variance. Here are three of these derived formulas (see Problem 7.8 for derivations)

$$s^2 = \frac{\displaystyle\sum_{i=1}^{n} x_i^2 - \frac{\left(\displaystyle\sum_{i=1}^{n} x_i\right)^2}{n}}{n-1} \tag{7.17}$$

$$s^2 = \frac{n\displaystyle\sum_{i=1}^{n} x_i^2 - \left(\displaystyle\sum_{i=1}^{n} x_i\right)^2}{n(n-1)} \tag{7.18}$$

$$s^2 = \frac{\displaystyle\sum_{i=1}^{n} x_i^2 - n\bar{x}^2}{n-1} \tag{7.19}$$

EXAMPLE 7.7   Use equation (7.17) to calculate the variance for the sample of lengths in Example 7.6.

Solution

From Example 7.6 we know that $n = 5$ and $\sum x_i = 25$ cm. We need

$$\sum x_i^2 = 9 \text{ cm}^2 + 16 \text{ cm}^2 + 25 \text{ cm}^2 + 36 \text{ cm}^2 + 49 \text{ cm}^2 = 135 \text{ cm}^2$$

Therefore

$$s^2 = \frac{\sum x_i^2 - \frac{\left(\sum x_i\right)^2}{n}}{n-1} = \frac{135 \text{ cm}^2 - \frac{(25 \text{ cm})^2}{5}}{5-1} = 2.5 \text{ cm}^2$$

7.9   THE POPULATION STANDARD DEVIATION

As a measure of dispersion from the mean, the variance satisfies the three criteria we gave in Section 7.2: It is calculated from all the data, it shows with a single number the typical dispersion (in squared deviations) from the mean, and it increases with increasing dispersion. However, it has two important limitations: (1) While the units for the variance are the original measurement units squared, an ideal measure of dispersion from the mean would have the same original units as the mean, and (2) the variance is too sensitive to extreme values in the data, since it involves squared quantities. A solution to both of these limitations is to take the positive square root of the variance. This quantity, called the *standard deviation* (or *root mean square deviation*), has the original measurement units and responds less to extreme values.

The *population standard deviation*, then, is defined by these definitional formulas (see Section 7.5)

$$\sigma = \sqrt{\sigma^2} \tag{7.20}$$

$$\sigma = \sqrt{\frac{\displaystyle\sum_{i=1}^{N}(x_i - \mu)^2}{N}} \tag{7.21}$$

And it has these computational formulas (see Section 7.6)

$$\sigma = \sqrt{\dfrac{\sum\limits_{i=1}^{N} x_i^2 - \dfrac{\left(\sum\limits_{i=1}^{N} x_i\right)^2}{N}}{N}} \tag{7.22}$$

$$\sigma = \sqrt{\dfrac{\sum\limits_{i=1}^{N} x_i^2}{N} - \mu^2} \tag{7.23}$$

EXAMPLE 7.8   Using equation (7.20), calculate the standard deviation from the population variance in Example 7.4.

   Solution

$$\sigma = \sqrt{\sigma^2} = \sqrt{2\,g^2} = 1.41421 \text{ g, or } 1.4 \text{ g}$$

7.10   THE SAMPLE STANDARD DEVIATION

The *sample standard deviation* is defined by these definitional formulas (see Section 7.7)

$$s = \sqrt{s^2} \tag{7.24}$$

$$s = \sqrt{\dfrac{\sum\limits_{i=1}^{n}(x_i - \bar{x})^2}{n - 1}} \tag{7.25}$$

And it has these computational formulas (see Section 7.8)

$$s = \sqrt{\dfrac{\sum\limits_{i=1}^{n} x_i^2 - \dfrac{\left(\sum\limits_{i=1}^{n} x_i\right)^2}{n}}{n - 1}} \tag{7.26}$$

$$s = \sqrt{\dfrac{n\sum\limits_{i=1}^{n} x_i^2 - \left(\sum\limits_{i=1}^{n} x_i\right)^2}{n(n - 1)}} \tag{7.27}$$

$$s = \sqrt{\dfrac{\sum\limits_{i=1}^{n} x_i^2 - n\bar{x}^2}{n - 1}} \tag{7.28}$$

   Both the population and the sample standard deviations are *always positive values* because they are defined as the positive square roots (or principal square roots, see Section 1.7) of their variances.

   While the variance has many applications in inferential statistics, and we deal with it often throughout the book, it is rarely used as a descriptive measure because of its discussed limitations. The standard deviation, on the other hand, is the most important and commonly used measure of dispersion from the mean in both descriptive and inferential statistics.

EXAMPLE 7.9   Using equation (7.24), calculate the standard deviation from the sample variance in Example 7.6.

Solution

$$s = \sqrt{s^2} = \sqrt{2.5 \text{ cm}^2} = 1.58114 \text{ cm, or } 1.6 \text{ cm}$$

7.11   ROUNDING-OFF GUIDELINES FOR MEASURES OF DISPERSION

In Section 6.3 we stated for the arithmetic mean that in calculating this measure or while using it in further calculations, rounding off should be kept to a minimum; retaining at least six digits throughout the calculations. This advice is also true for calculations involving measures of dispersion.

We also stated in Section 6.3 that for the different problem of reporting the final answer, while there are no absolute rules, one commonly accepted guideline is

If the data are all at the same level of precision (see Section 2.15), then the mean should be reported at the next level of precision.

An extension of this guideline for the measures of dispersion that we have discussed is

If the data are all at the same level of precision, then the mean deviation and the standard deviation should also be presented at the next level of precision, but the variance should be presented at the next level beyond that.

This *level-of-precision guideline* has been used to this point in this chapter for rounding off answers.

Finally we indicated in Section 6.3 that there is another common guideline for reporting the mean that is coordinated with the standard deviation. This new guideline, the *standard-deviation guideline*, can be stated as follows:

Report the standard deviation rounded off to two significant figures, and then report the mean rounded off to the last digit position of the standard deviation. The variance can be reported with as many as twice the decimal places as the standard deviation. The mean deviation should also be rounded off to two significant figures.

As both of these guidelines are equally valid, both will be used for rounding off answers in the remainder of the book. In addition, a *standard-error guideline* will be presented in Chapter 13.

EXAMPLE 7.10   You collect this sample: 4.9, 5.2, 6.1, 5.8, 7.3, 8.2, 6.5; for which $\bar{x} = 6.28571$, mean deviation $= 0.89796$, $s^2 = 1.35149$, and $s = 1.16254$. Round off these values for reporting, using: (a) the level-of-precision guideline,   (b) the standard-deviation guideline.

Solution

(a) $\bar{x} = 6.29$, mean deviation $= 0.90$, $s^2 = 1.351$, $s = 1.16$

(b) $s = 1.2$, $\bar{x} = 6.3$, $s^2 = 1.35$, mean deviation $= 0.90$

7.12   CALCULATING STANDARD DEVIATIONS FROM NONGROUPED FREQUENCY DISTRIBUTIONS

Converting the population formulas for standard deviations (see Section 7.9) for use with frequency distributions, the *definitional frequency-distribution formula for population standard deviations* is

$$\sigma = \sqrt{\frac{\sum\limits_{i=1}^{k} f_i(x_i - \mu)^2}{N}} \tag{7.29}$$

and the *computational frequency-distribution formulas for population standard deviations* are

$$\sigma = \sqrt{\frac{\sum\limits_{i=1}^{k} f_i x_i^2 - \dfrac{\left(\sum\limits_{i=1}^{k} f_i x_i\right)^2}{N}}{N}}$$

(7.30)

$$\sigma = \sqrt{\frac{\sum\limits_{i=1}^{k} f_i x_i^2}{N} - \mu^2}$$

(7.31)

Converting the sample formulas for standard deviations (see Section 7.10) for use with frequency distributions, the *definitional frequency-distribution formula for sample standard deviation* is

$$s = \sqrt{\frac{\sum\limits_{i=1}^{k} f_i (x_i - \bar{x})^2}{n - 1}}$$

(7.32)

and the *computational frequency-distribution formulas for sample standard deviations* are

$$s = \sqrt{\frac{\sum\limits_{i=1}^{k} f_i x_i^2 - \dfrac{\left(\sum\limits_{i=1}^{k} f_i x_i\right)^2}{n}}{n - 1}}$$

(7.33)

$$s = \sqrt{\frac{n \sum\limits_{i=1}^{k} f_i x_i^2 - \left(\sum\limits_{i=1}^{k} f_i x_i\right)^2}{n(n - 1)}}$$

(7.34)

$$s = \sqrt{\frac{\sum\limits_{i=1}^{k} f_i x_i^2 - n\bar{x}^2}{n - 1}}$$

(7.35)

EXAMPLE 7.11 Using equation (7.29), calculate the standard deviation for the population of marathon times summarized in Table 6.2.

Solution

To use equation (7.29) requires the addition of these three columns to Table 6.2: $x_i - \mu$, $(x_i - \mu)^2$, and $f_i(x_i - \mu)^2$. The modified table and the resulting calculation of the standard deviation are shown in Table 7.3.

7.13 CALCULATING APPROXIMATE STANDARD DEVIATIONS FROM GROUPED FREQUENCY DISTRIBUTIONS

A standard deviation calculated from a grouped frequency distribution will only approximate the exact value calculated directly from the data, and it is therefore called an *approximate standard deviation*. To make this calculation from grouped data requires the assumption that all values in a class are equal to the

Table 7.3

Time (min) x_i	Frequency f_i	$f_i x_i$ (min)	$(x_i - \mu)$ (min)	$(x_i - \mu)^2$ (min²)	$f_i(x_i - \mu)^2$ (min²)
129	1	129	−10.8667	118.0852	118.0852
130	2	260	−9.8667	97.3518	194.7036
131	0	0	−8.8667	78.6184	0
132	0	0	−7.8667	61.8850	0
133	1	133	−6.8667	47.1516	47.1516
134	1	134	−5.8667	34.4182	34.4182
135	1	135	−4.8667	23.6848	23.6848
136	2	272	−3.8667	14.9514	29.9028
137	0	0	−2.8667	8.2180	0
138	3	414	−1.8667	3.4846	10.4538
139	0	0	−0.8667	0.7512	0
140	0	0	0.1333	0.0178	0
141	3	423	1.1333	1.2844	3.8532
142	4	568	2.1333	4.5510	18.2040
143	5	715	3.1333	9.8176	49.0880
144	2	288	4.1333	17.0842	34.1684
145	5	725	5.1333	26.3508	131.7540
$\sum$	30	4,196 min			695.4676 min²

$$\mu = \frac{\sum f_i x_i}{N} = \frac{4,196 \text{ min}}{30} = 139.8667 \text{ min, or } 139.9 \text{ min}$$

$$\sigma = \sqrt{\frac{f_i(x_i - \mu)^2}{N}} = \sqrt{\frac{695.4676 \text{ min}^2}{30}} = 4.8148 \text{ min, or } 4.8 \text{ min}$$

class mark m_i. Then, the approximate standard deviation can be calculated with this *computational formula for a population*

$$\sigma \approx \sqrt{\frac{\sum_{i=1}^{k} f_i m_i^2 - \frac{\left(\sum_{i=1}^{k} f_i m_i\right)^2}{N}}{N}} \tag{7.36}$$

or this *computational formula for a sample*

$$s \approx \sqrt{\frac{n\sum_{i=1}^{k} f_i m_i^2 - \left(\sum_{i=1}^{k} f_i m_i\right)^2}{n(n-1)}} \tag{7.37}$$

EXAMPLE 7.12   Use equation (7.36) to calculate the approximate population standard deviation from the grouped frequency distribution of the marathon times in Table 6.3.

Solution

To use equation (7.36) requires the addition of two columns to Table 6.3: m_i^2 and $f_i m_i^2$. The modified table and the resulting calculation of the approximate population standard deviation are shown in Table 7.4.

Table 7.4

Time (min)	Class mark m_i (min)	Frequency f_i	$f_i m_i$ (min)	m_i^2 (min²)	$f_i m_i^2$ (min²)
128–130	129	3	387	16,641	49,923
131–133	132	1	132	17,424	17,424
134–136	135	4	540	18,225	72,900
137–139	138	3	414	19,044	57,132
140–142	141	7	987	19,881	139,167
143–145	144	12	1,728	20,736	248,832
$\sum$		30	4,188 min		585,378 min²

$$\sigma \approx \sqrt{\frac{\sum f_i m_i^2 - \dfrac{(\sum f_i m_i)^2}{N}}{N}} = \sqrt{\frac{585,378\,\text{min}^2 - \dfrac{(4,188\,\text{min})^2}{30}}{30}}$$

$$= 4.94368\,\text{min, or } 4.9\,\text{min}$$

7.14 CALCULATING VARIANCES AND STANDARD DEVIATIONS WITH CODED DATA

In Section 6.8, we used the coding formula $c_i = a + bx_i$ [equation (6.14)] to calculate sample arithmetic means with simplified numbers. Now we use the same coding formula to calculate sample variances and standard deviations. The definitions of terms in Section 6.8 will also apply here.

If, after coding a sample, the variance of c_i is calculated with this version of equation (7.16)

$$s_c^2 = \frac{\sum_{i=1}^{n}(c_i - \bar{c})^2}{n - 1} \tag{7.38}$$

or this version of equation (7.18)

$$s_c^2 = \frac{n \sum_{i=1}^{n} c_i^2 - \left(\sum_{i=1}^{n} c_i\right)^2}{n(n-1)} \tag{7.39}$$

then the following *decoding formula for the variance* can be used to find the variance of the original data

$$s_x^2 = \frac{s_c^2}{b^2} \tag{7.40}$$

Therefore the *decoding formula for the standard deviation* is

$$\sqrt{s_x^2} = \sqrt{\frac{s_c^2}{b^2}}$$

$$s_x = \frac{s_c}{b} \tag{7.41}$$

EXAMPLE 7.13   For the sample of length measurements (in cm) summarized in Table 6.4, first calculate s_x^2 and s_x directly from the data using equations (7.18) and (7.24), and then recalculate s_x^2 and s_x using equations (6.14), (7.39), (7.40), and (7.41). Use $a = -490$ cm and $b = 1$ as the coding constants in the coding formula.

Solution

The direct calculations of s_x^2 and s_x and the calculations using the coding and decoding formulas are shown in Table 7.5.

Table 7.5

Length (cm) x_i	x_i^2 (cm^2)	$(c_i = x_i - 490)$ (cm)	c_i^2 (cm^2)
492	242,064	2	4
493	243,049	3	9
495	245,025	5	25
496	246,016	6	36
498	248,004	8	64
500	250,000	10	100
$\sum$ 2,974 cm	1,474,158 cm^2	34 cm	238 cm^2

Direct calculations:

$$s_x^2 = \frac{n\sum x_i^2 - (\sum x_i)^2}{n(n-1)} = \frac{6(1,474,158 \text{ cm}^2) - (2,974 \text{ cm})^2}{6(5)}$$

$$= 9.06667 \text{ cm}^2, \text{ or } 9.07 \text{ cm}^2$$

$$s_x = \sqrt{s_x^2} = \sqrt{9.06667 \text{ cm}^2} = 3.01109 \text{ cm}, \text{ or } 3.0 \text{ cm}$$

Calculations with coding and decoding formulas:

$$s_c^2 = \frac{n\sum c_i^2 - (\sum c_i)^2}{n(n-1)} = \frac{6(238 \text{ cm}^2) - (34 \text{ cm})^2}{6(5)} = 9.06667 \text{ cm}^2$$

$$s_c = \sqrt{s_c^2} = \sqrt{9.06667 \text{ cm}^2} = 3.01109 \text{ cm}$$

Decoding s_c^2: $s_x^2 = \dfrac{s_c^2}{b^2} = \dfrac{9.06667 \text{ cm}^2}{1^2} = 9.06667 \text{ cm}^2, \text{ or } 9.07 \text{ cm}^2$

Decoding s_c: $s_x = \dfrac{s_c}{b} = \dfrac{3.01109 \text{ cm}}{1} = 3.01109 \text{ cm}, \text{ or } 3.0 \text{ cm}$

7.15 CHEBYSHEV'S THEOREM

Chebyshev's theorem can be stated as follows:

For any number $k \geq 1$, and a set of data $x_1, x_2, \ldots, x_n$ (or $x_1, x_2, \ldots, x_N$), the *proportion* of the measurements that lies within k standard deviations of their mean will be at least

$$1 - \frac{1}{k^2}.$$

What this theorem states can be understood with the aid of the frequency curve shown in Fig. 7-2. For the population illustrated, the theorem states that the proportion of the population lying in the interval from $\mu - k\sigma$ to $\mu + k\sigma$ (the shaded area under the curve) will be at least as large as $1 - \dfrac{1}{k^2}$. This is the *lower limit* (or *lower bound*) for this proportion; typically the proportion in the interval $\mu \pm k\sigma$ (or $\bar{x} \pm ks$) will be larger than $1 - \dfrac{1}{k^2}$. The theorem is also called *Chebyshev's inequality* (see Section 1.23) because what it is really saying is that the proportion will be greater-than-or-equal-to ($\geq$) $1 - \dfrac{1}{k^2}$. It can be proven mathematically that Chebyshev's theorem is true for any set of measurements, from symmetric distributions to skewed distributions to multimodal distributions.

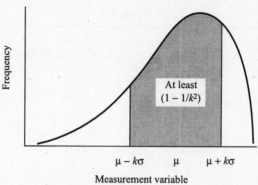

Fig. 7-2

Table 7.6

Number of σs from μ k	At least this amount in interval $\mu \pm k\sigma$	
	Proportion $1 - 1/k^2$	Percentage $[(1 - 1/k^2) \times (100)]$ (%)
1	0	00.0
$1\frac{1}{2}$		
2		
$2\frac{1}{2}$		
3		

Table 7.7

Number of σs from μ k	At least this amount in interval $\mu \pm k\sigma$	
	Proportion $1 - 1/k^2$	Percentage $[(1 - 1/k^2) \times (100)]$ (%)
1	0	00.0
$1\frac{1}{2}$	$5/9 = 0.556$	55.6
2	$3/4 = 0.750$	75.0
$2\frac{1}{2}$	$21/25 = 0.840$	84.0
3	$8/9 = 0.889$	88.9

EXAMPLE 7.14   Use Chebyshev's theorem to complete Table 7.6.

Solution

Table 7.7 is the completed table.

Note: While Chebyshev's theorem is true for any $k \geq 1$, it can be seen from the table that the theorem only gives meaningful information for $k > 1$.

7.16   THE EMPIRICAL RULE

A normal distribution is a theoretical bell-shaped curve that is defined by a specific equation called the *normal probability density function* (see Chapter 12). If for a given set of data its frequency polygon or relative frequency polygon can be precisely fitted by a normal distribution, then the data set is said to be normally distributed. If this is true, then the exact percentages of the data lying within the interval mean $\pm k$ standard deviations, can be determined by applying techniques from integral calculus to the normal probability density function. Percentages determined in this way for $\mu \pm \sigma$, $\mu \pm 2\sigma$, and $\mu \pm 3\sigma$ are shown for a normally distributed population in Fig. 7-3. These percentages hold true for any data set with polygons that are normally distributed.

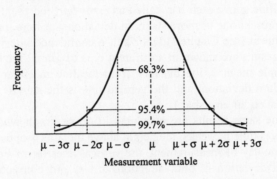

Fig. 7-3

These percentages also hold approximately true for data sets that are only approximately normally distributed (unimodal, roughly mound-shaped, essentially symmetrical). Because this generalization was determined empirically, it is called the *empirical rule* and can be stated for populations roughly as follows:

For a population that is approximately normally distributed: $\approx 68\%$ of the data lies in the interval $\mu \pm \sigma$, $\approx 95\%$ of the data lies in the interval $\mu \pm 2\sigma$, and $\approx 100\%$ of the data lies in the interval $\mu \pm 3\sigma$.

Comparable generalizations hold true for samples that are approximately normally distributed.

The empirical rule and Chebyshev's theorem are part of descriptive statistics because, given the nature of a distribution and its mean and standard deviation, they allow rapid calculations of the percentages of the data lying within specific distances from the mean. If the distribution is at least approximately normal, then the empirical rule gives the approximate percentages within one, two, or three standard deviations from the mean. If the nature of the distribution is not known or it is definitely not normal, then Chebyshev's theorem can be used to calculate the lower limit (lower bound) for the percentages lying within k standard deviations from the mean.

While the rule given here is referred to in statistics books as the empirical rule, it is not the first empirically determined rule that we have dealt with. You will recall that in Problem 6.31 we stated this "empirical rule" for moderately skewed distributions

$$\bar{x} - \tilde{x} = \frac{1}{3}(\bar{x} - \text{mode})$$

EXAMPLE 7.15   If a population is normally distributed, what percentage of the population is less than $\mu - \sigma$?

Solution

You can see in Fig. 7-3 that a normal distribution is perfectly symmetrical about the mean; that the portion of the curve to the left of the mean is a perfect reverse image of the portion to the right of the mean. Therefore, if 68.3% of the data in the normal distribution is within $\mu \pm \sigma$, then a total of $(100\% - 68.3\% = 31.7\%)$ of the data is beyond that interval on both sides. Thus, because of the symmetry of the distribution, this percentage of a normally distributed population is less than $\mu - \sigma$

$$\frac{1}{2}(31.7\%) = 15.85\%$$

7.17 GRAPHING CENTRAL TENDENCY AND DISPERSION

Three common techniques for illustrating central tendency and dispersion in a sample or population are shown in Fig. 7-4. There you can see the average height (in inches) of a sample of 10 corn plants at two and four weeks after germination. In Fig. 7-4(*a*) the heights are shown by a bar graph where the *Y* axis is the measurement scale, the height of each bar represents average plant height on the day of measurement $\bar{x}$, and the length of the vertical line above each bar, called an *error bar*, represents one standard deviation *s*. Such lines above bars can represent one or more standard deviations, as here, or they can represent one or more standard errors of the mean (see Chapter 13), or half a confidence interval (see Chapter 14).

In Fig. 7-4(*b*), the same results are shown in a different type of graph: the horizontal line through the vertical rectangle is the sample mean $\bar{x}$, the distance from the line to either the top or bottom of the rectangle is the sample standard deviation, and the vertical line is the sample range. (When such a line represents a range it is not called an error bar.)

Finally, in Fig. 7-4(*c*), the same results are presented in a line graph where the *Y* axis is again the measurement scale, the *X* axis is time in weeks, the height above the *X* axis of each black circle is the mean height $\bar{x}$ for the sample for the day of measurement, the vertical lines (error bars) above and below each circle represent one standard deviation in both directions $\bar{x} \pm s$, and consecutive circles are joined by straight-line segments to show that the phenomena are interlinked by events that are occurring over a continuous time period.

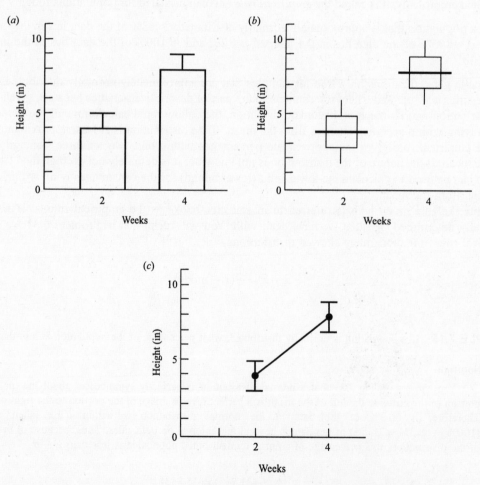

Fig. 7-4

EXAMPLE 7.16   From Fig. 7-4, what, to the nearest inch, are the mean heights, standard deviations, and ranges for weeks two and four?

Solution

Week two: $\bar{x} = 4$ in, $s = 1$ in, range $= 4$ in; week four: $\bar{x} = 8$ in, $s = 1$ in, range $= 5$ in

7.18   THE COEFFICIENT OF VARIATION

The *coefficient of variation* (also called the *coefficient of variability*, the *coefficient of dispersion*, or the *relative standard deviation*) is defined for a population by both

$$V = \frac{\sigma}{\mu} \tag{7.42}$$

and

$$V = \frac{\sigma}{\mu}(100\%) \tag{7.43}$$

Equation (7.42) expresses the standard deviation as a proportion of its mean, and equation (7.43), which is more common, expresses it as a percentage of its mean. For a sample, the formulas are

$$V = \frac{s}{\bar{x}} \tag{7.44}$$

and

$$V = \frac{s}{\bar{x}}(100\%) \tag{7.45}$$

The measures of dispersion we have dealt with previously in this chapter (range, mean deviation, variance, standard deviation) are called measures of *absolute dispersion* because they are calculated directly from the data and have the units of the original measurements or those units squared. The coefficient of variation, on the other hand, is called a measure of *relative dispersion* because it expresses a measure of absolute dispersion as a proportion (or percentage) of some measure of average value that is in the same units as the measure of dispersion. Because the numerator and denominator of the ratios in the measure have the same units, the resulting measure of relative dispersion has *no units*.

EXAMPLE 7.17   You are a biologist studying genetic variation within different species of rodents. One measure you take for each rodent is body weight in grams. For a sample of 10 males of the white-footed mouse, you get these results: $\bar{x} = 12.9$ g, $s = 1.6$ g; and for 8 males from the plains pocket gopher you get these: $\bar{x} = 545.0$ g, $s = 32.8$ g. Compare the relative dispersions of these two species using equation (7.45).

Solution

For the white-footed mouse,

$$V = \frac{s}{\bar{x}}(100\%) = \frac{1.6 \text{ g}}{12.9 \text{ g}}(100\%) = 12.4\%$$

For the plains pocket gopher,

$$V = \frac{s}{\bar{x}}(100\%) = \frac{32.8 \text{ g}}{545.0 \text{ g}}(100\%) = 6.0\%$$

These results show that there is twice as much relative dispersion of body weight among the mice as there is among the pocket gophers. This greater variation relative to the mean is not apparent from the standard deviations, which show twenty times more absolute variation among the pocket gophers.

Note: It is generally true, as here, that as means increase across samples (or populations) there is an increase in standard deviations. Therefore, when the distributions being compared have very different means, the larger forms will almost always show the larger absolute dispersions. For such distributions the coefficient of variation can be used to meaningfully compare relative dispersions. Here, the pocket gophers had a mean that was forty times larger than the mean for the mice, but this same technique could have been used to compare relative dispersions between the mice and much larger mammals, say elephants that have a mean weight of roughly 7,000,000 g ($\approx$ 540,000 times larger than the mean for mice).

7.19 THE STANDARD SCORE AND THE STANDARDIZED VARIABLE

For a population, the *standard score* (also called the *normal deviate*, or *z score*) is defined as

$$z_i = \frac{x_i - \mu}{\sigma} \tag{7.46}$$

and for a sample it is defined as

$$z_i = \frac{x_i - \bar{x}}{s} \tag{7.47}$$

For any data distribution, the standard score shows how far any given data value x_i is from the mean of the distribution in standard deviation units; how many standard deviations the value is from the mean. A positive z value indicates that x_i is larger than the mean (to its right in a histogram or polygon) and a negative z value indicates that x_i is smaller than the mean (to its left). Like the coefficient of variation, the standard score is a relative measure; while the coefficient shows absolute dispersions relative to their means, the standard score shows deviations from the mean relative to the standard deviation. Because its units are numbers of standard deviations, the standard score allows comparisons of relative positions within distributions that have very different means or different measurement units.

When for any variable X each measurement value in a sample or population is transformed into a z value, this process is known as *standardizing* (or *normalizing*) the variable, and the resulting variable Z is called a *standardized variable*.

EXAMPLE 7.18 For the following sample, first calculate $\bar{x}$ and s, and then standardize the sample: 3, 5, 7, 9, 11.

 Solution

$$\bar{x} = \frac{\sum x_i}{n} = \frac{35}{5} = 7$$

$$s = \sqrt{\frac{n \sum x_i^2 - \left(\sum x_i\right)^2}{n(n-1)}} = \sqrt{\frac{5(285) - (35)^2}{5(4)}} = 3.16228$$

To standardize the sample is to calculate a standard score z_i for each x_i. These scores are typically reported, as shown here, rounded to the nearest hundredth.

$$z_1 = \frac{x_1 - \bar{x}}{s} = \frac{3 - 7}{3.16228} = -1.26491, \text{ or } -1.26$$

$$z_2 = \frac{x_2 - \bar{x}}{s} = \frac{5 - 7}{3.16228} = -0.63246, \text{ or } -0.63$$

$$z_3 = \frac{x_3 - \bar{x}}{s} = \frac{7 - 7}{3.16228} = 0.00$$

$$z_4 = \frac{x_4 - \bar{x}}{s} = \frac{9 - 7}{3.16228} = 0.63246, \text{ or } 0.63$$

$$z_5 = \frac{x_5 - \bar{x}}{s} = \frac{11 - 7}{3.16228} = 1.26491, \text{ or } 1.26$$

7.20   THE INTERQUARTILE RANGE AND THE QUARTILE DEVIATION

The *interquartile range* is the difference between the first and third quartiles and is thus defined as

$$\text{Interquartile range} = Q_3 - Q_1 \tag{7.48}$$

This interval contains the middle 50% of the distribution.

The *quartile deviation* (or *semiinterquartile range*) is a measure of dispersion that is defined as

$$\text{Quartile deviation} = \frac{Q_3 - Q_1}{2} \tag{7.49}$$

It is thus one-half the interquartile range, and therefore one-half the interval that contains the middle 50% of the data.

EXAMPLE 7.19   For the quartiles calculated in Example 6.13 ($Q_1 = 1.385$ cm, $Q_3 = 1.615$ cm), calculate: (*a*) the interquartile range,   (*b*) the quartile deviation.

Solution

(*a*)   Interquartile range $= Q_3 - Q_1 = 1.615 \text{ cm} - 1.385 \text{ cm} = 0.230 \text{ cm}$, or 0.23 cm

(*b*)   Quartile deviation $= \dfrac{Q_3 - Q_1}{2} = \dfrac{0.230 \text{ cm}}{2} = 0.115 \text{ cm}$, or 0.12 cm

7.21   BOX PLOTS AND FIVE-NUMBER SUMMARIES

If a distribution is unimodal and perfectly symmetrical, then it would best be described statistically by its mean and standard deviation, allowing interpretation with either Chebyshev's theorem or the empirical rule. When, however, a distribution is extremely skewed or multimodal, then instead of mean and standard deviation the distribution is often described with what is called the *five-number summary*: $Q_1, Q_2, Q_3, x_s,$ x_1. The five-number summary can be shown graphically in what is called a *box plot* (or *box-and-whisker plot*).

To illustrate this graph, we show in Fig. 7-5 a box plot that was constructed from the length data in Table 6.1 and the quartiles calculated in Example 6.13. Such plots can be constructed horizontally with the measurement scale along the X axis, or, as here, vertically with the measurement scale along the Y axis. The construction of the box plot is similar to the construction of the floating-rectangle graph in Fig. 7-4(*b*), but

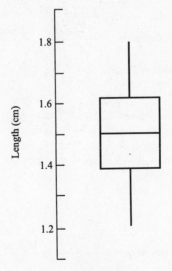

Fig. 7-5

where that graph displayed $\bar{x} \pm s$ and the range, the version of the box plot shown here displays the five-number summary. The rectangle, called the *box*, extends from Q_1 to Q_3 along the measurement scale, and thus its height is the interquartile range. The width of the box is arbitrary and thus provides no information. The horizontal line through the box is $Q_2 = \tilde{x}$. As in Fig. 7-4(b), the lines extending above and below the box show the range for the distribution. In a box plot these lines are called *whiskers*, and in the version of the box plot shown here the lower whisker extends from x_s to Q_1 and the upper whisker extends from Q_3 to x_l (where x_s is the smallest value and x_l is the largest value).

While all versions of box plots show Q_1, Q_2, and Q_3, there are different versions of the whiskers. For example, in another common version the lower whisker extends from P_{10} to Q_1, and the upper whisker extends from Q_3 to P_{90}. The difference, $P_{90} - P_{10}$, is called the *10–90 percentile range*. It is not called a five-number summary when the 10–90 percentile range is displayed instead of the range.

Solved Problems

THE MEAN DEVIATION

7.1 Calculate the range and the mean deviation for the samples in Problem 6.37(a) to (d).

Solution

(a) Range $= x_l - x_s = 0 \sec - 0 \sec = 0 \sec$

$$\bar{x} = \frac{\sum_{i=1}^{5} x_i}{5} = \frac{0 \sec}{5} = 0.0 \sec$$

$$\sum_{i=1}^{5} |x_i - \bar{x}| = |0 - 0.0| \sec + |0 - 0.0| \sec + |0 - 0.0| \sec + |0 - 0.0| \sec + |(0 - 0.0)| \sec$$

$$= 0.0 \sec$$

$$\text{Mean deviation} = \frac{\sum_{i=1}^{5} |x_i - \bar{x}|}{5} = \frac{0.0 \sec}{5} = 0.0 \sec$$

(b) Range $= x_l - x_s = 10 \sec - 0 \sec = 10 \sec$

$$\bar{x} = \frac{\sum_{i=1}^{5} x_i}{5} = \frac{10 \sec}{5} = 2.0 \sec$$

$$\sum_{i=1}^{5} |x_i - \bar{x}| = |10 - 2.0| \sec + |0 - 2.0| \sec + |0 - 2.0| \sec + |0 - 2.0| \sec + |0 - 2.0| \sec$$

$$= 16.0 \sec$$

$$\text{Mean deviation} = \frac{\sum_{i=1}^{5} |x_i - \bar{x}|}{5} = \frac{16.0 \sec}{5} = 3.2 \sec$$

(c)   Range $= x_l - x_s = 112.217\,\text{km} - 9.777\,\text{km} = 102.440\,\text{km}$

$$\bar{x} = \frac{\sum\limits_{i=1}^{6} x_i}{6} = \frac{172.459\,\text{km}}{6} = 28.7432\,\text{km}$$

$$\sum_{i=1}^{6} |x_i - \bar{x}| = |10.127 - 28.7432|\,\text{km} + |11.963 - 28.7432|\,\text{km} + |112.217 - 28.7432|\,\text{km}$$

$$+ |9.777 - 28.7432|\,\text{km} + |13.833 - 28.7432|\,\text{km} + |14.542 - 28.7432|\,\text{km}$$

$$= 166.9478\,\text{km}$$

$$\text{Mean deviation} = \frac{\sum\limits_{i=l}^{6} |x_i - \bar{x}|}{6} = \frac{166.9478\,\text{km}}{6} = 27.8246\,\text{km}$$

(d)   Range $= x_l - x_s = 14.542\,\text{km} - 0.007\,\text{km} = 14.535\,\text{km}$

$$\bar{x} = \frac{\sum\limits_{i=1}^{6} x_i}{6} = \frac{60.249\,\text{km}}{6} = 10.0415\,\text{km}$$

$$\sum_{i=1}^{6} |x_i - \bar{x}| = |10.127 - 10.0415|\,\text{km} + |11.963 - 10.0415|\,\text{km} + |0.007 - 0.007 - 10.0415|\,\text{km}$$

$$+ |9.777 - 10.0415|\,\text{km} + |13.833 - 10.0415|\,\text{km} + |14.542 - 10.0415|\,\text{km}$$

$$= 20.5980\,\text{km}$$

$$\text{Mean deviation} = \frac{\sum\limits_{i=1}^{6} |x_i - \bar{x}|}{6} = \frac{20.5980\,\text{km}}{6} = 3.4330\,\text{km}$$

7.2   For the sample summarized in Table 6.8, calculate the range and then, using equation (7.7), calculate the mean deviation.

Solution

Range $= 20\,\text{g} - 14\,\text{g} = 6\,\text{g}$

Table 7.8

Weight (g) x_i	Frequency f_i	$f_i x_i$ (g)	$(x_i - \bar{x})$(g)	$f_i|x_i - \bar{x}|$ (g)
14	2	28	−4.0556	8.1112
15	2	30	−3.0556	6.1112
16	4	64	−2.0556	8.2224
17	18	306	−1.0556	19.0008
18	24	432	−0.0556	1.3344
19	35	665	0.9444	33.0540
20	5	100	1.9444	9.7220
$\sum$	90	1,625 g		85.5560 g

$$\bar{x} = \frac{\sum f_i x_i}{n} = \frac{1,625\,\text{g}}{90} = 18.0556\,\text{g}$$

$$\text{Mean deviation} = \frac{\sum f_i |x_i - \bar{x}|}{n} = \frac{85.5560\,\text{g}}{90} = 0.950622\,\text{g}, \text{ or } 1.0\,\text{g}$$

The appropriately modified table and the resulting calculation of the mean deviation are shown in Table 7.8.

7.3   Using equation (7.9), calculate the approximate mean deviation for the grouped data in Table 6.10.

Solution

The modified table and the resulting calculations are shown in Table 7.9.

Table 7.9

2d Exam	Class mark m_i	Frequency f_i	$f_i m_i$	$[m_i - (\approx \bar{x})]$	$f_i|m_i - (\approx \bar{x})|$
45–49	47	1	47	-34.375	34.375
50–54	52	0	0	-29.375	0
55–59	57	4	228	-24.375	97.500
60–64	62	3	186	-19.375	58.125
65–69	67	5	335	-14.375	71.875
70–74	72	3	216	-9.375	28.125
75–79	77	6	462	-4.375	26.250
80–84	82	11	902	0.625	6.875
85–89	87	9	783	5.625	50.625
90–94	92	17	1,564	10.625	180.625
95–99	97	5	485	15.625	78.125
$\sum$		64	5,208		632.500

$$\bar{x} \approx \frac{\sum f_i m_i}{n} = \frac{5{,}208}{64} = 81.375, \text{ or } 81.4$$

$$\text{Sample mean deviation} \approx \frac{\sum f_i|m_i - (\approx \bar{x})|}{n} = \frac{632.500}{64} = 9.88281, \text{ or } 9.9$$

THE POPULATION VARIANCE

7.4   For the population of weights in Example 7.4, use equation (7.15) to calculate the population variance.

Solution

From Examples 7.4 and 7.5 we know that $N = 5$, $\mu = 4$ g, and $\sum x_i^2 = 90$ g^2. Therefore

$$\sigma^2 = \frac{\sum_{i=1}^{N} x_i^2}{N} - \mu^2 = \frac{90 \text{ g}^2}{5} - (4 \text{ g})^2 = 2 \text{ g}^2$$

7.5   Show that

$$\text{Population } SS = \sum_{i=1}^{N} x_i^2 - \frac{\left(\sum_{i=1}^{N} x_i\right)^2}{N} \tag{7.50}$$

Solution

In the following derivation, $\sum$ represents $\sum_{i=1}^{N}$.

$$\text{Population } SS = \sum(x_i - \mu)^2 = \sum(x_i^2 - 2x_i\mu + \mu^2)$$

$$= \sum x_i^2 - \sum 2x_i\mu + \sum \mu^2$$

As 2 and μ are constants,

$$\text{Population SS} = \sum x_i^2 - 2\mu\sum x_i + N\mu^2$$

$$= \sum x_i^2 - 2\left(\frac{\sum x_i}{N}\right)\sum x_i + N\left(\frac{\sum x_i}{N}\right)^2$$

$$= \sum x_i^2 - 2\frac{\left(\sum x_i\right)^2}{N} + \frac{\left(\sum x_i\right)^2}{N}$$

$$= \sum x_i^2 - \frac{\left(\sum x_i\right)^2}{N}$$

Note: With a similar proof, we could show

$$\text{Sample } SS = \sum_{i=1}^{n} x_i^2 - \frac{\left(\sum_{i=1}^{n} x_i\right)^2}{n} \tag{7.51}$$

7.6   Derive the computational formulas for population variance [equations (7.14) and (7.15)] from the definitional formula [equation (7.12)].

Solution

To derive equation (7.14), we substitute equation (7.50) for the numerator in equation (7.12).
To derive equation (7.15), we do the following algebraic manipulation of equation (7.14):

$$\sigma^2 = \frac{\sum x_i^2 - \frac{\left(\sum x_i\right)^2}{N}}{N}$$

$$= \frac{1}{N}\left[\sum x_i^2 - \frac{\left(\sum x_i\right)^2}{N}\right]$$

$$= \frac{\sum x_i^2}{N} - \frac{\left(\sum x_i\right)^2}{N^2}$$

$$= \frac{\sum x_i^2}{N} - \left(\frac{\sum x_i}{N}\right)^2$$

$$= \frac{\sum x_i^2}{N} - \mu^2$$

7.7   It can be proven mathematically, for any set of numbers $x_1, x_2, \ldots, x_k$, that $\sum_{i=1}^{k}(x_i - a)^2$ will be a minimum if and only if the number a is the arithmetic mean of the set of numbers. Empirically demonstrate this fact for this sample: 1, 2, 3, 4, 5.

Solution

To empirically demonstrate for this sample that $\sum\limits_{i=1}^{n}(x_i - a)^2$ is minimum if and only if

$$a = \bar{x} = \frac{\sum\limits_{i=1}^{5} x_i}{5} = \frac{15}{5} = 3$$

we calculate this summation for as that are smaller and larger than $\bar{x} = 3$. The results for five as (2.8, 2.9, 3.0, 3.1, and 3.2) are shown in Table 7.10. You can see from the totals at the bottom of each column how $\sum\limits_{i=1}^{n}(x_i - a)^2$ increases as a is made either smaller or larger than $\bar{x} = 3$.

Table 7.10

x_i	$(x_i - 2.8)^2$	$(x_i - 2.9)^2$	$(x_i - 3.0)^2$	$(x_i - 3.1)^2$	$(x_i - 3.2)^2$
1	3.24	3.61	4	4.41	4.84
2	0.64	0.81	1	1.21	1.44
3	0.04	0.01	0	0.01	0.04
4	1.44	1.21	1	0.81	0.64
5	4.84	4.41	4	3.61	3.24
$\sum$	10.20	10.05	10	10.05	10.20

THE SAMPLE VARIANCE

7.8   From the definitional formula for the sample variance [equation (7.16)], derive the three computational formulas given in Section 7.8 [equations (7.17), (7.18), and (7.19)].

Solution

To derive equation (7.17), we substitute equation (7.51) for the numerator of equation (7.16).

To derive equation (7.18), we multiply both sides of equation (7.17) by $\dfrac{n}{n}$

$$\frac{n}{n}s^2 = \frac{n\left[\sum x_i^2 - \dfrac{\left(\sum x_i\right)^2}{n}\right]}{n(n-1)}$$

$$s^2 = \frac{n\sum x_i^2 - \left(\sum x_i\right)^2}{n(n-1)}$$

Finally, to derive equation (7.19), we do the following algebraic manipulation of equation (7.16):

$$s^2 = \frac{\sum(x_i - \bar{x})^2}{n-1} = \frac{\sum(x_i^2 - 2x_i\bar{x} + \bar{x}^2)}{n-1} = \frac{\sum x_i^2 - \sum 2x_i\bar{x} + \sum \bar{x}^2}{n-1}$$

As 2 and $\bar{x}$ are constants

$$s^2 = \frac{\sum x_i^2 - 2\bar{x}\sum x_i + n\bar{x}^2}{n-1}$$

And, as $\sum x_i = n\bar{x}$

$$s^2 = \frac{\sum x_i^2 - 2\bar{x}(n\bar{x}) + n\bar{x}^2}{n-1} = \frac{\sum x_i^2 - 2n\bar{x}^2 + n\bar{x}^2}{n-1} = \frac{\sum x_i^2 - n\bar{x}^2}{n-1}$$

7.9 Use equations (7.18) and (7.19) to calculate the variance for the sample of lengths in Example 7.6.

Solution

From Examples 7.6 and 7.7, we know that $n = 5$, $\sum x_i = 25$ cm, $\sum x_i^2 = 135$ cm^2, and $\bar{x} = 5$ cm. Therefore for equation (7.18),

$$s^2 = \frac{n\sum x_i^2 - \left(\sum x_i\right)^2}{n(n-1)} = \frac{5(135 \text{ cm}^2) - (25 \text{ cm})^2}{5(5-1)} = 2.5 \text{ cm}^2$$

and for equation (7.19),

$$s^2 = \frac{\sum x_i^2 - n\bar{x}^2}{n-1} = \frac{135 \text{ cm}^2 - 5(5 \text{ cm})^2}{5-1} = 2.5 \text{ cm}^2$$

THE POPULATION STANDARD DEVIATION

7.10 For the population 9.1, 8.7, 9.0, 9.2, use equation (7.21) to calculate the population standard deviation.

Solution

$$\mu = \frac{\sum x_i}{N} = \frac{36.0}{4} = 9.0$$

$$\sigma = \sqrt{\frac{\sum(x_i - \mu)^2}{N}}$$

$$= \sqrt{\frac{(9.1 - 9.0)^2 + (8.7 - 9.0)^2 + (9.0 - 9.0)^2 + (9.2 - 9.0)^2}{4}} = 0.18708, \text{ or } 0.19$$

7.11 For the population in Problem 7.10, use equation (7.23) to calculate the population standard deviation.

Solution

From Problem 7.10 we know that $\mu = 9.0$, and so we need

$$\sum x_i^2 = (9.1)^2 + (8.7)^2 + (9.0)^2 + (9.2)^2 = 324.14$$

Therefore

$$\sigma = \sqrt{\frac{\sum x_i^2}{N} - \mu^2} = \sqrt{\frac{324.14}{4} - (9.0)^2} = 0.18708, \text{ or } 0.19$$

7.12 Use equation (7.30) to calculate the standard deviation for the population of marathon times summarized in Table 6.2.

Solution

To use equation (7.30) requires the addition of these columns to Table 6.2: x_i^2 and $f_i x_i^2$. The modified table and the resulting calculation of the standard deviation are shown in Table 7.11.

Table 7.11

Time (min) x_i	Frequency f_i	$f_i x_i$ (min)	x_i^2 (min^2)	$f_i x_i^2$ (min^2)
129	1	129	16,641	16,641
130	2	260	16,900	33,800
131	0	0	17,161	0
132	0	0	17,424	0
133	1	133	17,689	17,689
134	1	134	17,956	17,956
135	1	135	18,225	18,225
136	2	272	18,496	36,992
137	0	0	18,769	0
138	3	414	19,044	57,132
139	0	0	19,321	0
140	0	0	19,600	0
141	3	423	19,881	59,643
142	4	568	20,164	80,656
143	5	715	20,449	102,245
144	2	288	20,736	41,472
145	5	725	21,025	105,125
$\sum$	30	4,196 min		587,576 min^2

$$\sigma = \sqrt{\frac{\sum f_i x_i^2 - \frac{(\sum f_i x_i)^2}{N}}{N}}$$

$$= \sqrt{\frac{587,576\,\text{min}^2 - \frac{(4,196\,\text{min})^2}{30}}{30}} = 4.81479\,\text{min, or } 4.8\,\text{min}$$

THE SAMPLE STANDARD DEVIATION

7.13 For the sample in Example 6.1(a), use equations (7.25) and (7.27) to calculate the standard deviation.

Solution

Using equation (7.25),

$$\bar{x} = \frac{\sum x_i}{n} = \frac{24\,\text{g}}{7} = 3.42857\,\text{g}$$

$$s = \sqrt{\frac{\sum(x_i - \bar{x})^2}{n-1}} = \sqrt{\frac{25.7143\,\text{g}^2}{6}} = 2.07020\,\text{g, or } 2.1\,\text{g}$$

Using equation (7.27),

$$\sum x_i = 24\,\text{g}$$

$$\sum x_i^2 = 108\,\text{g}^2$$

$$s = \sqrt{\frac{n\sum x_i^2 - (\sum x_i)^2}{n(n-1)}} = \sqrt{\frac{7(108\,\text{g}^2) - (24\,\text{g})^2}{7(6)}}$$

$$= \sqrt{4.28571\,\text{g}^2} = 2.07020\,\text{g, or } 2.1\,\text{g}$$

7.14 For the sample in Example 6.1(b), use equations (7.25) and (7.28) to calculate the standard deviation.

Solution

Using equation (7.25),

$$\bar{x} = \frac{\sum x_i}{n} = \frac{222 \text{ g}}{7} = 31.7143 \text{ g}$$

$$s = \sqrt{\frac{\sum (x_i - \bar{x})^2}{n - 1}} = \sqrt{\frac{33,063.429 \text{ g}^2}{6}} = 74.2332 \text{ g or } 74.2 \text{ g}$$

Using equation (7.28),

$$\sum x_i^2 = 40,104 \text{ g}^2$$

$$s = \sqrt{\frac{\sum x_i^2 - n\bar{x}^2}{n - 1}} = \sqrt{\frac{40,104 \text{ g}^2 - 7,040.578 \text{ g}^2}{6}} = 74.2332 \text{ g, or } 74.2 \text{ g}$$

7.15 Use equation (7.32) to calculate the standard deviation for the sample lengths summarized in Table 6.1.

Solution

To use equation (7.32) requires the addition of three columns to Table 6.1: $x_i - \bar{x}$, $(x_i - \bar{x})^2$, and $f_i(x_i - \bar{x})^2$. The modified table and the resulting calculation of the standard deviation are shown in Table 7.12.

Table 7.12

Length (cm) x_i	Frequency f_i	$f_i x_i$ (cm)	$(x_i - \bar{x})$ (cm)	$(x_i - \bar{x})^2$ (cm^2)	$f_i (x_i - \bar{x})^2$ (cm^2)
1.2	2	2.4	−0.30	0.09	0.18
1.3	7	9.1	−0.20	0.04	0.28
1.4	10	14.0	−0.10	0.01	0.10
1.5	12	18.0	0.00	0.00	0.00
1.6	10	16.0	0.10	0.01	0.10
1.7	7	11.9	0.20	0.04	0.28
1.8	2	3.6	0.30	0.09	0.18
$\sum$	50	75.0 cm			1.12 cm^2

$$\bar{x} = \frac{\sum f_i x_i}{n} = \frac{75.0 \text{ cm}}{50} = 1.50 \text{ cm}$$

$$s = \sqrt{\frac{\sum f_i(x_i - \bar{x})^2}{n - 1}} = \sqrt{\frac{1.12 \text{ cm}^2}{49}} = 0.151186 \text{ cm, or } 0.15 \text{ cm}$$

7.16 Use equations (7.33), (7.34), and (7.35) to calculate the standard deviation for the sample of lengths summarized in Table 6.1.

Solution

To use these equations requires the addition of two columns to Table 6.1: x_i^2 and $f_i x_i^2$. The modified table and the resulting calculations of the standard deviation are shown in Table 7.13.

7.17 For what types of samples would the following be true: $s^2 = 0$, $s^2 = -1$, or $s = -1$?

Table 7.13

Length (cm) x_i	Frequency f_i	$f_i x_i$ (cm)	x_i^2 (cm^2)	$f_i x_i^2$ (cm^2)
1.2	2	2.4	1.44	2.88
1.3	7	9.1	1.69	11.83
1.4	10	14.0	1.96	19.60
1.5	12	18.0	2.25	27.00
1.6	10	16.0	2.56	25.60
1.7	7	11.9	2.89	20.23
1.8	2	3.6	3.24	6.48
$\sum$	50	75.0 cm		113.62 cm^2

$$\bar{x} = \frac{\sum f_i x_i}{n} = \frac{75.0 \text{ cm}}{50} = 1.50 \text{ cm}$$

Using equation (7.33),

$$s = \sqrt{\frac{\sum f_i x_i^2 - \frac{\left(\sum f_i x_i\right)^2}{n}}{n-1}}$$

$$= \sqrt{\frac{113.62 \text{ cm}^2 - \frac{(75.0 \text{ cm})^2}{50}}{50 - 1}} = 0.151186 \text{ cm, or } 0.15 \text{ cm}$$

Using equation (7.34),

$$s = \sqrt{\frac{n \sum f_i x_i^2 - \left(\sum f_i x_i\right)^2}{n(n-1)}}$$

$$= \sqrt{\frac{(50 \times 113.62 \text{ cm}^2) - (75.0 \text{ cm})^2}{50(50-1)}} = 0.151186 \text{ cm, or } 0.15 \text{ cm}$$

Using equation (7.35),

$$s = \sqrt{\frac{\sum f_i x_i^2 - n\bar{x}^2}{n-1}}$$

$$= \sqrt{\frac{113.62 \text{ cm}^2 - [50 \times (1.50 \text{ cm})^2]}{50-1}} = 0.151186 \text{ cm, or } 0.15 \text{ cm}$$

Solution

A sample (or a population) will have a variance and a standard deviation equal to zero if and only if all data values in the sample are the same number. A sample (or a population) will *never* have a negative variance or a negative standard deviation. The variance will always be a positive value because it is based on squared deviations, and the standard deviation will always be a positive value because it is defined to be the positive square root of the variance (see Section 7.9).

CALCULATING APPROXIMATE STANDARD DEVIATIONS FROM GROUPED FREQUENCY DISTRIBUTIONS

7.18 Use equation (7.37) to calculate the approximate sample standard deviation for the grouped frequency distribution of second lecture exam scores in Table 6.10.

Table 7.14

2d Exam	Class mark m_i	Frequency f_i	$f_i m_i$	m_i^2	$f_i m_i^2$
45–49	47	1	47	2,209	2,209
50–54	52	0	0	2,704	0
55–59	57	4	228	3,249	12,996
60–64	62	3	186	3,844	11,532
65–69	67	5	335	4,489	22,445
70–74	72	3	216	5,184	15,552
75–79	77	6	462	5,929	35,574
80–84	82	11	902	6,724	73,964
85–89	87	9	783	7,569	68,121
90–94	92	17	1,564	8,464	143,888
95–99	97	5	485	9,409	47,045
$\sum$		64	5,208		433,326

$$\bar{x} \approx \frac{\sum f_i m_i}{n} = \frac{5,208}{64} = 81.4$$

$$s \approx \sqrt{\frac{n \sum f_i m_i^2 - (\sum f_i m_i)^2}{n(n-1)}} = \sqrt{\frac{27,732,864 - 27,123,264}{64(63)}}$$

$$= 12.2690, \text{ or } 12.3$$

Solution

To use equation (7.37) requires the addition of two columns to Table 6.10: m_i^2 and $f_i m_i^2$. The modified table and the resulting calculation of the standard deviation are shown in Table 7.14.

CALCULATING VARIANCES AND STANDARD DEVIATIONS WITH CODED DATA

7.19 If $c_i = a + bx_i$, show that $s_x^2 = \frac{s_c^2}{b^2}$.

Solution

Equation (6.16) states that

$$\bar{x} = \frac{1}{b}(\bar{c} - a)$$

Therefore

$$\bar{c} = a + b\bar{x}$$

and

$$c_i - \bar{c} = (a + bx_i) - (a + b\bar{x}) = bx_i + a - a - b\bar{x} = b(x_i - \bar{x})$$

Squaring both sides of the equation,

$$(c_i - \bar{c})^2 = b^2(x_i - \bar{x})^2$$

Taking the sum of both sides of the equation,

$$\sum(c_i - \bar{c})^2 = \sum b^2(x_i - \bar{x})^2 = b^2 \sum(x_i - \bar{x})^2$$

Multiplying both sides of the equation by $\dfrac{1}{n-1}$,

$$\frac{\sum(c_i - \bar{c})^2}{n-1} = \frac{b^2 \sum(x_i - \bar{x})^2}{n-1}$$

Thus

$$s_c^2 = b^2 s_x^2$$

and

$$s_x^2 = \frac{s_c^2}{b^2}$$

7.20 For the sample of weight measurements (in grams) summarized in Table 6.13, calculate s_x^2 and s_x directly from the data using equations (7.18) and (7.24). Then, recalculate these values using equations (7.39), (7.40), and (7.41). Use $a = 0$ g and $b = 0.0001$ as the coding constants in the coding formula.

Solution

The direct calculations of s_x^2 and s_x and the calculations using the coding and decoding formulas are shown in Table 7.15.

Table 7.15

Weight (g) x_i	x_i^2 (g^2)	$(c_i = 0.0001 x_i)$ (g)	c_i^2 (g^2)
22,000.0	484,000,000	2.20	4.8400
24,000.0	576,000,000	2.40	5.7600
25,500.0	650,250,000	2.55	6.5025
27,500.0	756,250,000	2.75	7.5625
29,000.0	841,000,000	2.90	8.4100
30,000.0	900,000,000	3.00	9.0000
$\sum$ 158,000.0 g	4,207,500,000 g^2	15.80 g	42.0750 g^2

Direct calculations:

$$s_x^2 = \frac{n\sum x_i^2 - (\sum x_i)^2}{n(n-1)} = \frac{(6 \times 4{,}207{,}500{,}000 \text{ g}^2) - (158{,}000.0 \text{ g})^2}{6(5)} = 9{,}366{,}666.667 \text{ g}^2$$

$$s_x = \sqrt{s_x^2} = \sqrt{9{,}366{,}666.67 \text{ g}^2} = 3{,}060.5010 \text{ g, or } 3{,}060.50 \text{ g}$$

Calculations with coding and decoding formulas:

$$s_c^2 = \frac{n\sum c_i^2 - (\sum c_i)^2}{n(n-1)} = \frac{(6 \times 42.0750 \text{ g}^2) - (15.80 \text{ g})^2}{6(5)} = 0.0936667 \text{ g}^2$$

$$s_c = \sqrt{s_c^2} = \sqrt{0.0936667 \text{ g}^2} = 0.306050 \text{ g}$$

Decoding s_c^2: $s_x^2 = \dfrac{s_c^2}{b^2} = \dfrac{0.0936667 \text{ g}^2}{(0.0001)^2} = 9{,}366{,}670.000 \text{ g}^2$

Decoding s_c: $s_x = \dfrac{s_c}{b} = \dfrac{0.306050 \text{ g}}{0.0001} = 3{,}060.5000 \text{ g, or } 3{,}060.50 \text{ g}$

7.21 In Problem 6.9 we determined what happened to the average hourly wage in factories A and B that started with the same average ($\bar{x} = \$5.39$) but then gave different types of raises to their workers. We found that factory A, which gave a 5% per hour raise to all employees, increased their average wage to $5.66, and that factory B, which gave a $0.05 per hour raise to all employees, increased their average wage to $5.44. By a remarkable coincidence, both factories before the raises had the same standard deviations for their samples of hourly wages ($s_x = \$0.40$). Use the following formula to determine the post-raise standard deviations for both factories: $s_c = bs_x$. After the raise, which factory has the greater wage dispersion?

Solution

For factory A, if x_i is the previous hourly wage, then each worker now receives: $c_i = \$0.00 + 1.05x_i$. Thus, if s_x is the pre-raise standard deviation, s_c is the post-raise standard deviation, and the coding constants are $a = \$0.00$ and $b = 1.05$, then

$$s_c = bs_x = 1.05(\$0.40) = \$0.42$$

For factory B, each worker now receives per hour: $c_i = \$0.05 + x_i$. Thus, for B the coding constants are $a = \$0.05$ and $b = 1.00$. Therefore

$$s_c = bs_x = 1.00(\$0.40) = \$0.40$$

Thus, while the standard deviation for factory A increased from $0.40 to $0.42, for factory B the standard deviation remained the same ($0.40). Factory A, therefore, has a greater wage dispersion after the raise than does factory B.

CHEBYSHEV'S THEOREM

7.22 These descriptive measures are from a sample of time measurements: $n = 400$, $\bar{x} = 21.2$ sec, and $s = 1.7$ sec. Use Chebyshev's theorem from Section 7.15 to answer this question: At least what proportion of the data lies within $2\frac{3}{4}$ standard deviations from the arithmetic mean?

Solution

At least this proportion of the data lies within $k = 2\frac{3}{4}$ standard deviations from the mean

$$1 - \frac{1}{k^2} = 1 - \frac{1}{\left(2\frac{3}{4}\right)^2} = 1 - \frac{16}{121} = 0.867769$$

7.23 These measures describe a sample of length measurements: $n = 10,000$, $\bar{x} = 20.0$ cm, and $s^2 = 0.25$ cm^2. Using Chebyshev's theorem, determine at least *how many* of the measurements in the sample are in the interval (19.0 cm to 21.0 cm).

Solution

Finding s,

$$s = \sqrt{s^2} = \sqrt{0.25 \text{ cm}^2} = 0.5 \text{ cm}$$

To find k, we know that the interval (19.0 cm to 21.0 cm) is

$$(\bar{x} = 20.0 \text{ cm}) \pm (ks = 1.0 \text{ cm})$$

Therefore

$$k = \frac{1.0 \text{ cm}}{s} = \frac{1.0 \text{ cm}}{0.5 \text{ cm}} = 2$$

From Chebyshev's theorem we know that at least this proportion of the data lies in the interval $\bar{x} \pm 2s$

$$1 - \frac{1}{k^2} = 1 - \frac{1}{2^2} = 0.75$$

Thus, at least this many measurements in the sample lie in the interval (19.0 cm to 21.0 cm)

$$0.75(n = 10,000) = 7,500$$

7.24 A company that manufactures a type of flashlight battery wants to claim that at least 96% of these batteries last from 95 hours to 105 hours. If they test a sample of 1,000 batteries and get a mean life of $\bar{x} = 100$ hr, then what is the maximum value possible for the sample standard deviation s if they are to make the 96% claim?

Solution

From Chebyshev's theorem, we know that when at least 96% of the data lie in the interval $\bar{x} \pm ks$ then

$$1 - \frac{1}{k^2} = 0.96$$

$$\frac{1}{k^2} = 1 - 0.96 = 0.04$$

$$k = \frac{1}{0.2} = 5$$

If the interval containing at least 96% of the data is (95 hours to 105 hours), then this is equivalent to

$$(\bar{x} = 100 \text{ hr}) \pm 5 \text{ hr}$$

Therefore

$$ks = 5 \text{ hr}$$

$$s = \frac{5 \text{ hr}}{k} = \frac{5 \text{ hr}}{5} = 1 \text{ hr}$$

Thus, $s = 1$ hr is the maximum standard deviation for the sample if the company is to make the 96% claim.

7.25 A food company wants to sell a "12 ounce" bag of potato chips, and by law such a bag must contain at least 12 ounces of chips. They test their automated bag-filling machine by setting it to put 12.20 oz in each bag, and then weighing the contents of 500 bags. The results from the test are: $n = 500$, $\bar{x} = 12.20$ oz, and $s = 0.04$ oz. Can they leave the machine at this setting if they want to be at least 99% certain of obeying the law?

Solution

Putting this question in a form that can be solved with Chebyshev's theorem: Will at least 99% of the weights lie in the interval 12.20 oz $\pm k(0.04$ oz)?

From the theorem we know that for this to be true

$$1 - \frac{1}{k^2} = 0.99$$

$$k = 10$$

But for $k = 10$, the interval would extend below the mandatory 12 ounces

$$12.20 \text{ oz} \pm 10(0.04 \text{ oz})$$

$$12.20 \text{ oz} \pm 0.40 \text{ oz, or } (11.80 \text{ oz to } 12.60 \text{ oz})$$

Therefore, to be at least 99% certain of obeying the law the company must either reset the machine to increase the average per-bag weight ($\bar{x}$) or have the machine adjusted to decrease the dispersion of the per-bag weights (s).

THE EMPIRICAL RULE

7.26 If a population is normally distributed, what percentage of the population is less than $\mu - 2\sigma$?

Solution

Using the reasoning in Example 7.15, this percentage is less than $\mu - 2\sigma$

$$\frac{1}{2}(100\% - 95.4\% = 4.6\%) = 2.3\%$$

7.27 You are a biologist studying a species of snail, and as a part of your research you measure shell diameters (in mm) of a sample of 500 of these snails. Assuming, as is likely, that these diameters are essentially normally distributed, use the empirical rule (see Section 7.16) to determine the approximate number of shell diameters in this interval: $\bar{x} \pm s$.

Solution

From Section 7.16 we know that $\approx 68\%$ of the data lie in the interval $\bar{x} \pm s$. Therefore, the number of shell diameters in that interval is $\approx (0.68 \times 500 = 340)$.

7.28 Why do some statistics books define an *outlier* in a data set as a measurement that is more than three standard deviations from the mean of the set?

Solution

Sometimes there are values in a data set that are either much larger or much smaller than the rest of the data. Such extreme values are called outliers and one common definition of an outlier is any data value that is more than three standard deviations from the mean. You can see why this definition is given from the distribution percentages provided by the empirical rule. Thus, the rule states, for both symmetrical and skewed distributions, that $\approx 100\%$ of the data will be within three standard deviations from the mean. Therefore, a data value beyond three standard deviations is treated as a very unusual event—an outlier. Such outliers can be caused by procedural errors (measuring, recording, calculating), or equipment failure, or some complicating extraneous variable like taking measurements from more than one population. If there is an outlier, and good reason to believe that it is due to some technique problem, then it is sometimes legitimate to remove the outlier from the data set.

7.29 For a data set that is approximately normal, what is the relationship between its range and its standard deviation?

Solution

For such a data set, the empirical rule states that $\approx 95\%$ of the data lie in the interval (mean ± 2 standard deviations), and $\approx 100\%$ of the data lie in the interval (mean ± 3 standard deviations). Therefore, as a general rule of thumb you can assume that the range of such a data set is equal to somewhere between four and six standard deviations, and consequently the standard deviation should be somewhere between $\frac{1}{4}$ (range) and $\frac{1}{6}$ (range). This relationship can be used as a quick check of the accuracy of a standard-deviation calculation.

GRAPHING CENTRAL TENDENCY AND DISPERSION

7.30 You are a doctor at a sleep-disorder clinic, doing research on a new sleeping pill. Twenty clinic patients volunteer for the experiment, and you use a table of random numbers (see Section 3.23) to *randomly assign* 10 of them to each of two groups: *pill* and *control*. All 20 patients have their brain waves recorded during a night of sleep at the clinic. The difference between the groups is that 30 minutes before going to bed at 10 p.m., each patient in the pill group receives the new sleeping pill with a glass of milk, whereas each patient in the control group receives a sugar pill (a *placebo*) with milk. One measurement taken for each patient from their recorded brain waves is how long it takes to fall asleep: the time from 10 p.m. (0 min) to the first signs of sleep in the brain waves. The results for both groups are shown in Fig. 7-6, in the bar-graph form of Fig. 7-4(a), where the Y axis is the measurement scale, the height of each bar represents the group mean $\bar{x}$, and the length of the vertical line above a bar represents one standard deviation s. Using the information in the graph, answer these questions: (a) From Chebyshev's theorem, what time interval contains at least 68% of each group's data?   (b) From the empirical rule, what time interval contains approximately 68% of each group's data?

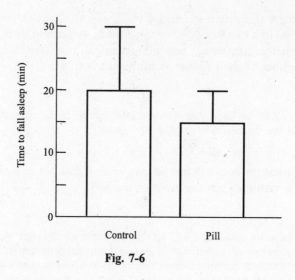

Fig. 7-6

Solution

(a)   From Chebyshev's theorem we know that when at least 68% of the data lie in the interval $\bar{x} \pm ks$, then

$$1 - \frac{1}{k^2} = 0.68$$

$$k = 1.76777$$

For each group, $\bar{x}$ and s can be determined from the bar graph by visual estimation or measuring. These values are: control group, $\bar{x} = 20$ min, s = 10 min; pill group, $\bar{x} = 15$ min, s = 5 min. Therefore, for the control group the interval containing at least 68% of the data is

$$\bar{x} \pm ks = 20 \text{ min} \pm (1.76777 \times 10 \text{ min})$$

$$= 20 \text{ min} \pm 17.6777 \text{ min, or (2.3 min to 37.7 min)}$$

And for the pill group this interval is

$$\bar{x} \pm ks = 15 \text{ min} \pm (1.76777 \times 5 \text{ min})$$

$$= 15 \text{ min} \pm 8.83885 \text{ min, or (6.2 min to 23.8 min)}$$

(b)   Assuming the group time distributions are approximately normally distributed, the empirical rule indicates $\approx 68\%$ of the data lie in the interval $\bar{x} \pm s$. For the control group

$$\bar{x} \pm s = 20 \text{ min} \pm 10 \text{ min, or (10 min to 30 min)}$$

For the pill group

$$\bar{x} \pm s = 15 \text{ min} \pm 5 \text{ min}, \text{ or } (10 \text{ min to } 20 \text{ min})$$

Note: From these results it would seem that the pill was effective: On the average, the pill group fell asleep faster and their times were less dispersed. However, these results are only *descriptions* of small samples, and what is desired is a general conclusion about all possible users and nonusers of the pill. To make such comparative population conclusions at some level of certainty (probability) requires the techniques of inferential statistics that we will begin to deal with in the next chapter.

7.31 The data from Fig. 7-6 are shown in Fig. 7-7 in the floating-rectangle-graph form of Fig. 7-4(*b*). From this graph you can answer these questions: (*a*) Are the sample distributions symmetrical? (*b*) Does either sample contain outliers?

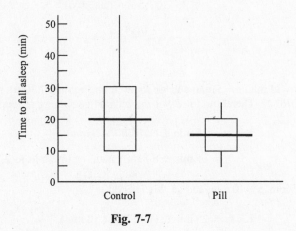

Fig. 7-7

Solution

(*a*) While the pill distribution is symmetrical about the mean, the control distribution is positively skewed (along the vertical axis).

(*b*) In Problem 7.28 we gave one common definition of an outlier: a data value that is more than three standard deviations from its mean. Here, while there are no such values in the pill group, there is at least one such outlier in the control group: $x_l = 55$ min, which is $3\frac{1}{2}$ standard deviations from its mean.

Note: While the bar-graph version in Fig. 7-6 is the more common method of presenting such information, the form of the graph in Fig. 7-7 is useful if you want to emphasize asymmetry or extreme values. Here, for example, we can see that the distribution for the control group is not "approximately normal" as was assumed in Problem 7.30(*b*), and therefore it was not really appropriate to use the empirical rule on the control group.

7.32 Because the sleeping-pill results (see Figs. 7-6 and 7-7) indicate that the pill was effective in reducing and making more predictable the time required to fall asleep, you decide to study this effect over four consecutive days. Twenty volunteers are randomly assigned (10 each) to new pill and control groups, but now the 10 people in each group repeat their versions of the experiment (described in Problem 7.30) for four consecutive days. The results for the pill group are shown in Fig. 7-8 in the line-graph form of Fig. 7-4(*c*). For days 1 and 4, use Chebyshev's theorem to determine the time interval that contains at least 68% of the data.

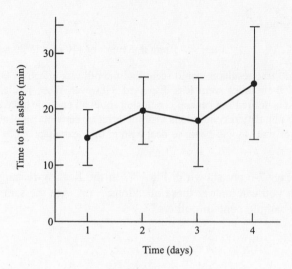

Fig. 7-8

Solution

On day 1, $\bar{x} = 15$ min, $s = 5$ min; and we know from Problem 7.30(a) that for the interval containing at least 68%, $k = 1.76777$. Therefore, for day 1 the interval containing at least 68% is

$$\bar{x} \pm ks = 15 \text{ min} \pm (1.76777 \times 5 \text{ min})$$

$$= 15 \text{ min} \pm 8.83885 \text{ min, or } (6.2 \text{ min to } 23.8 \text{ min})$$

For day 4, $\bar{x} = 25$ min, $s = 10$ min, and so this interval is

$$\bar{x} \pm ks = 25 \text{ min} \pm (1.76777 \times 10 \text{ min})$$

$$= 25 \text{ min} \pm 17.6777 \text{ min, or } (7.3 \text{ min to } 42.7 \text{ min})$$

7.33 The results for the pill group in Problem 7.32 seem to indicate that over four consecutive days the pill becomes progressively less effective: both $\bar{x}$ and s increase. To see how this compares to the control group, their results have been added to Fig. 7-8 as connected open circles in Fig. 7-9. Note how only one error bar is used with each line of circles to allow room for the two lines. Again for days 1 and 4, use Chebyshev's theorem to determine for the control group the time interval that contains at least 68% of the data.

Solution

For both days 1 and 4: $\bar{x} = 20$ min, $s = 10$ min, and so for the interval containing at least 68% of the data, $k = 1.76777$. Therefore, for both days

$$\bar{x} \pm ks = 20 \text{ min} \pm (1.76777 \times 10 \text{ min})$$

$$= 20 \text{ min} \pm 17.6777 \text{ min, or } (2.3 \text{ min to } 37.7 \text{ min})$$

Note: From Fig. 7-9 we can, at the level of descriptive statistics, compare the results for the two groups. We see that where the control group shows no particular pattern (oscillating up and down over the four nights with a constant dispersion of $s = 10$ min), the pill group shows a steady increase in $\bar{x}$ and s, until on day 4 it has a higher $\bar{x}$ and the same s as the control group. These results may indicate habituation to the pill: the same dose has progressively less effect over time. Before such a general population conclusion can be made, however, these group results must be analyzed with inferential statistics.

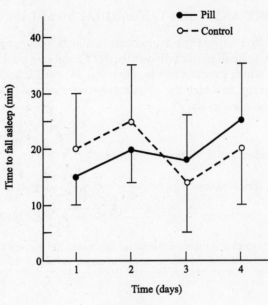

Fig. 7-9

THE COEFFICIENT OF VARIATION

7.34 An oil company takes a survey of 1,000 customers, asking these questions about a typical week: How many miles do you drive? How many gallons of gasoline do you buy? How much do you spend on gasoline? The results are: $\bar{x} = 120$ mi, $s = 9.6$ mi; $\bar{x} = 6$ gal, $s = 0.5$ gal; $\bar{x} = \$6.30$, $s = \$0.70$. Compare the relative dispersions of these three measurements using equation (7.45).

Solution

For miles per week,

$$V = \frac{s}{\bar{x}}(100\%) = \frac{9.6 \text{ mi}}{120 \text{ mi}}(100\%) = 8.0\%$$

For gallons per week,

$$V = \frac{s}{\bar{x}}(100\%) = \frac{0.5 \text{ gal}}{6 \text{ gal}}(100\%) = 8.3\%$$

For spending per week,

$$V = \frac{s}{\bar{x}}(100\%) = \frac{\$0.70}{\$6.30}(100\%) = 11.1\%$$

This problem illustrates an important use of the coefficient of variation: Because it is independent of units of measurement, the coefficient of variation can be used to compare relative dispersions across distributions that have different units. Here we see that spending per week has a greater relative dispersion than either of the other two measurements.

7.35 Why would it *not* be legitimate to calculate coefficients of variation for temperature data measured in °F or °C?

Solution

Because the coefficient of variation is a ratio, it is only legitimate to calculate it for ratio-level measurement (see Section 2.7), and temperatures in °F or °C are interval level.

THE STANDARD SCORE AND THE STANDARDIZED VARIABLE

7.36 You are employed by a university's department of athletics to monitor the academic progress of the student-athletes. A football player tells you he got 68 out of a possible 100 points on his chemistry midterm exam, in which the class results were: $\bar{x} = 74$, $s = 13$; and 74 out of a possible 100 on his history midterm exam, in which the class results were: $\bar{x} = 84$, $s = 7$. In which course did he do better relative to the class average?

Solution

Calculating standard scores for the two midterms

$$\text{for chemistry} \quad z_i = \frac{x_i - \bar{x}}{s} = \frac{68 - 74}{13} = -0.461538, \text{ or } -0.46$$

$$\text{for history} \quad z_i = \frac{x_i - \bar{x}}{s} = \frac{74 - 84}{7} = -1.42857, \text{ or } -1.43$$

From these standard scores it is clear he did better relative to the average on the chemistry midterm. While in chemistry he was 0.46 standard deviations below average, in history he was 1.43 standard deviations below average.

7.37 Show for any sample from a standardized variable Z that: (a) $\bar{z} = 0$, (b) $s_z = 1$.

Solution

(a) As

$$z_i = \frac{x_i - \bar{x}}{s}$$

then

$$\bar{z} = \text{mean of } z_i s = \frac{\sum z_i}{n} = \frac{\sum \left(\frac{x_i - \bar{x}}{s} \right)}{n} = \frac{1}{ns} \sum (x_i - \bar{x})$$

Therefore, as

$$\sum (x_i - \bar{x}) = 0$$

$$\bar{z} = \frac{1}{ns}(0) = 0$$

(b) $s_z = \text{standard deviation of } z_i s = \sqrt{\frac{\sum (z_i - \bar{z})^2}{n - 1}}$

Therefore

$$s_z = \sqrt{\frac{\sum (z_i - 0)^2}{n - 1}} = \sqrt{\frac{\sum z_i^2}{n - 1}} = \sqrt{\frac{\sum \left(\frac{x_i - \bar{x}}{s} \right)^2}{n - 1}} = \sqrt{\left(\frac{1}{s^2} \right) \left(\frac{1}{n - 1} \right) \sum (x_i - \bar{x})^2}$$

$$= \sqrt{\left(\frac{1}{s^2} \right)(s^2)}$$

$$= 1$$

Note: With a similar proof we could show for any population from a standardized variable that $\mu_z = 0$ and $\sigma_z = 1$. These properties of the mean and standard deviation of a standardized variable are important in Chapter 12, where we deal with the normal distribution of a standardized variable called the *standard normal distribution* (or the *standardized normal distribution*).

THE INTERQUARTILE RANGE AND THE QUARTILE DEVIATION

7.38 For the length data shown in Table 6.7, we calculated that: $Q_2 = \tilde{x} = 1.50$ cm (see Example 6.13), midquartile $= 1.50$ cm (see Example 6.15), and quartile deviation $= 0.12$ cm (see Example 7.19). What percent of this data lie in the intervals: (a) (midquartile) $\pm$ (quartile deviation), (b) $\tilde{x} \pm$ (quartile deviation)?

Solution

For this unimodal and perfectly symmetrical distribution, the midpoint of the interval from Q_1 to Q_3 is equal to both the midquartile and $\tilde{x}$. Therefore, the intervals defined in both (a) and (b) contain the middle 50% of the data.

7.39 For the weight data shown in Table 6.14, we calculated (see Problems 6.20 and 6.25) that: $Q_1 = 17.3056$ g, $Q_2 = \tilde{x} = 18.2917$ g, $Q_3 = 19.0$ g, and midquartile $= 18.1528$ g. From the quartiles, calculate the interquartile range and the quartile deviation. Then, determine what percent of the data lie in the intervals: (a) (midquartile) $\pm$ (quartile deviation), (b) $\tilde{x} \pm$ (quartile deviation).

Solution

Interquartile range $= Q_3 - Q_1 = 19.0$ g $- 17.3056$ g $= 1.6944$ g, or 1.7 g

Quartile deviation $= \dfrac{Q_3 - Q_1}{2} = \dfrac{1.6944 \text{ g}}{2} = 0.8472$ g, or 0.85 g

(a) For any shaped distribution the midpoint of the interval from Q_1 to Q_3 is equal to the midquartile. Therefore, for this unimodal but negatively skewed distribution the interval (midquartile) $\pm$ (quartile deviation) contains the middle 50% of the distribution.

(b) We have shown for this skewed distribution that $\tilde{x} \neq$ (midquartile). Therefore, we can *not* determine with these simple techniques what percent of the distribution lie in the interval $\tilde{x} \pm$ (quartile deviation).

Note: Because for any shaped distribution the quartile deviation is always a measure of one-half the interval containing the middle 50% of the data, it is often used in place of the standard deviation for describing dispersion in extremely skewed or multimodal distributions. When this is done, then typically the median or the midquartile is given as the measure of location instead of the mean.

7.40 For the temperature data shown in Table 6.15, we showed (see Problems 6.21, 6.26, and 6.51) that: $Q_1 = 100.8333°$F, $Q_2 = \tilde{x} = 101.3889°$F, $Q_3 = 102.3°$F, and midquartile $= 101.5667°$F. From the quartiles, calculate the interquartile range and the quartile deviation. Then, determine what percent of the data in Table 6.15 lie in the intervals: (a) (midquartile) $\pm$ (quartile deviation), (b) $\tilde{x} \pm$ (quartile deviation).

Solution

Interquartile range $= Q_3 - Q_1 = 102.3°$F $- 100.8333°$F $= 1.4667°$F, or 1.5°F

Quartile deviation $= \dfrac{Q_3 - Q_1}{2} = \dfrac{1.4667°\text{F}}{2} = 0.7333°$F, or 0.73°F

(a) For this unimodal, positively skewed distribution, the interval (midquartile) $\pm$ (quartile deviation) contains the middle 50% of the distribution.

(b) We have shown for this skewed distribution that $\tilde{x} \neq$ (midquartile), and therefore we again cannot determine with these techniques what percent of the data lie in the interval $\tilde{x} \pm$ (quartile deviation).

BOX PLOTS AND FIVE-NUMBER SUMMARIES

7.41 Using the information in Problem 6.20 construct a box plot for the weight data in Table 6.8 that shows the five-number summary.

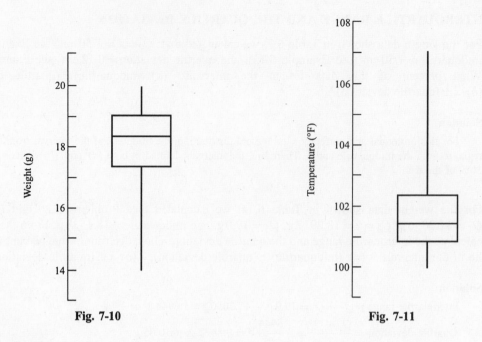

Fig. 7-10   Fig. 7-11

Solution

The box plot for this data, showing Q_1, Q_2, Q_3, x_s, x_l, is presented in Fig. 7-10.

7.42 Using the information in Problems 6.21 and 6.51, construct a box plot for the temperature data in Table 6.9 that shows the five-number summary.

Solution

The box plot for this data, showing Q_1, Q_2, Q_3, x_s, x_l, is presented in Fig. 7-11.

7.43 Using the exact quartiles from Problems 6.23 and 6.53 and the information in Table 6.11, construct a box plot for the golf winnings that shows the five-number summary.

Solution

The box plot for this data, showing Q_1, Q_2, Q_3, x_s, and x_l, is presented in Fig. 7-12.

Supplementary Problems

THE MEAN DEVIATION

7.44 Calculate the range and the mean deviation for this sample: 0.0, 15.3, 0.0, 100.3, 62.1, 0.0.

Ans. Range = 100.3, mean deviation = 34.3889, or 34.39

7.45 For the sample summarized in Table 6.9, calculate the range and then, using equation (7.6), calculate the mean deviation.

Ans. Range = 7°F, mean deviation = 1.0470°F, or 1.0°F

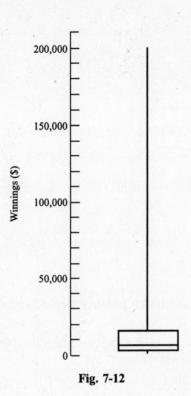

Fig. 7-12

THE POPULATION VARIANCE

7.46   Use equations (7.12) and (7.14) to calculate the variance for this population: 0, 1, 2, 11, 19.

    Ans.   53.84

7.47   Use equations (7.12) and (7.15) to calculate the variance for this population: 0.00, 0.01, 0.02, 0.11, 0.19.

    Ans.   0.005384, or 0.0054

THE SAMPLE VARIANCE

7.48   Use equations (7.16) and (7.17) to calculate the variance for this sample: 138, 129, 132.

    Ans.   21.00

7.49   Use equations (7.16) and (7.19) to calculate the variance for this sample: 152, 129, 148.

    Ans.   151.00

THE POPULATION STANDARD DEVIATION

7.50   Use equations (7.20) and (7.22) to calculate the standard deviation for the population in Problem 7.46.

    Ans.   7.3

7.51   Use equations (7.20) and (7.23) to calculate the standard deviation for the population in Problem 7.47.

    Ans.   0.073

7.52   For the golf-winnings data in Table 6.11, use equation (7.30) to calculate the standard deviation for this population.

       Ans. $28,536.14, or $28,500

THE SAMPLE STANDARD DEVIATION

7.53   Using equation (7.27), calculate the standard deviation for the samples in Problem 6.37(a) to (d).

       Ans. (a) 0 sec, (b) 4.47214 sec, or 4.5 sec, (c) 40.9383 km, (d) 5.2743 km

7.54   For the weight data in Table 6.8, use equation (7.34) to calculate the standard deviation for this sample.

       Ans. 1.24847 g, or 1.2 g

7.55   For the temperature data in Table 6.9, use equation (7.35) to calculate the standard deviation for this sample.

       Ans. 1.4485°F, or 1.4°F

CALCULATING APPROXIMATE STANDARD DEVIATIONS FROM GROUPED FREQUENCY DISTRIBUTIONS

7.56   Use equation (7.36) to calculate the approximate population standard deviation for the grouped frequency distribution of golf-winnings in Table 6.12.

       Ans. $\sigma \approx$ $28,214.58, or $28,200

CALCULATING VARIANCES AND STANDARD DEVIATIONS WITH CODED DATA

7.57   For the following sample, first calculate the standard deviation directly from the data using equation (7.27) and then calculate it by means of equations (6.14), (7.39), and (7.41): 0.0013, 0.0027, 0.0039, 0.0022. Use $a = 0$ and $b = 1,000$ as the coding constants in the coding formula.

       Ans. 0.0010844, or 0.00108

CHEBYSHEV'S THEOREM

7.58   For the sample described in Problem 7.22, use Chebyshev's theorem (see Section 7.15) to answer this question: At least what percentage of the data lie in the interval (19.3 sec to 23.1 sec)?

       Ans. At least 19.9% of the data lie in the interval.

7.59   For the sample described in Problem 7.22, use Chebyshev's theorem to answer this question: At least what percentage of the data lie in the interval $(\bar{x} = 21.2 \text{ sec}) \pm 4.1 \text{ sec}$?

       Ans. At least 82.8% of the data lie in the interval.

THE EMPIRICAL RULE

7.60   For the sample of shell diameters in Problem 7.27, use the empirical rule (see Section 7.16) to determine the approximate number of shell diameters in the interval $\bar{x} \pm 2s$.

       Ans. $\approx (0.95 \times 500 = 475)$

7.61   For the sample of shell diameters in Problem 7.27, use the empirical rule to determine the approximate number of shell diameters in the interval $\bar{x} \pm 3s$.

       Ans. $\approx (100\%$ of the 500)

7.62   If a population is normally distributed, what percentage of the population is less than $\mu \pm 2\sigma$?

   Ans.   97.7%

GRAPHING CENTRAL TENDENCY AND DISPERSION

7.63   From Fig. 7-7, which sample has the smallest and largest values?

   Ans. x_s for both samples is 5 min, but x_l is much larger for the control group (55 min vs. 25 min).

7.64   In Problem 7.29 we indicated that, as a general rule of thumb, the standard deviation should be somewhere between $\frac{1}{4}$(range) and $\frac{1}{6}$(range). Is this true for the control and pill groups in Fig. 7-7?

   Ans.   For control: $s \approx \frac{1}{5}$(range); for pill: $s \approx \frac{1}{4}$(range)

THE COEFFICIENT OF VARIATION

7.65   Calculated for the past year (250 days of market trading), company A's stock had an average daily price per share of $\mu = \$140$, $\sigma = \$8$, and company B's stock had an average daily price per share of $\mu = \$5$, $\sigma = \$0.8$. Using equation (7.43), which stock had the greater relative dispersion?

   Ans.   For company A: $V = 5.7\%$; for company B: $V = 16.0\%$

THE STANDARD SCORE AND THE STANDARDIZED VARIABLE

7.66   For a sample $x_1, x_2, \ldots, x_n$ that is approximately normally distributed: (a) use Chebyshev's theorem (see Section 7.15) to determine at least what proportion of the sample has standardized values between -2 and 2, and (b) use the empirical rule (see Section 7.16) to determine approximately what percentage of the sample has standardized values between -1 and 1.

   Ans.   (a) 0.75,   (b) $\approx 68\%$

THE INTERQUARTILE RANGE AND THE QUARTILE DEVIATION

7.67   For the distributions shown in Tables 6.7, 6.14, and 6.15, what percent of the data lie outside (above and below) the interval (midquartile) $\pm$ (quartile deviation)?

   Ans.   50%

BOX PLOTS AND FIVE-NUMBER SUMMARIES

7.68   The box plot in Fig. 7-13 shows a distribution summarized by a five-number summary: Q_1, Q_2, Q_3, x_s, x_l. Determine from the plot: interquartile range, quartile deviation, $\tilde{x}$, range, and whether the distribution is symmetrical or skewed.

   Ans.   Interquartile range $= 4$ sec,   quartile deviation $= 2$ sec, $\tilde{x} = 15$ sec,   range $= 10$ sec, distribution is symmetrical

7.69   The box plot in Fig. 7-14 shows a distribution summarized by a five-number summary: Q_1, Q_2, Q_3, x_s, x_l. Determine from the plot: interquartile range, quartile deviation, $\tilde{x}$, range, and whether the distribution is symmetrical or skewed.

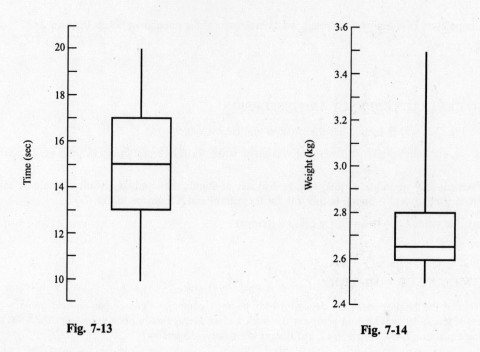

Fig. 7-13   **Fig. 7-14**

Ans.   Interquartile range $= 0.2$ kg, quartile deviation $= 0.1$ kg, $\tilde{x} = 2.65$ kg, range $= 1.0$ kg, distribution is positively skewed

Chapter 8

Probability: The Classical, Relative Frequency, Set Theory, and Subjective Interpretations

8.1 THE CLASSICAL INTERPRETATION OF PROBABILITY

The topic of probability was introduced in Chapter 3 (see Section 3.16) as part of an introductory overview of the science of statistics. Probability is an essential component of inferential statistics, where it is used to quantify the degree of uncertainty that can be assigned to any sample-to-population inference (see Section 3.5). Therefore, as the primary emphasis of this book now shifts from descriptive to inferential statistics, it is necessary to first, in this chapter and the next, develop an understanding of the basic concepts and techniques of probability.

As we indicated in Section 3.16, the probability of an event (a number between 0 and 1) indicates the likelihood or chance that the event will occur, with 0 indicating that the event cannot possibly occur and 1 indicating that the event is certain to occur. Depending on how these probability numbers are calculated, however, they can have four different interpretations: the *classical interpretation* (presented in this section), the *relative frequency interpretation* (see Section 8.2), the *set theory interpretation* (see Sections 8.3 to 8.6), and the *subjective interpretation* (see Section 8.7).

In discussing these four interpretations, we use a standard terminology that includes these concepts: *experiments*, *trials*, *outcomes*, and *events*. While in science an experiment involves the manipulation of an independent variable to see the effect on a dependent variable (see Section 1.19), in statistics an experiment is any process that yields a measurement. An example of such an experiment would be to roll a six-sided die and to observe, when the die stops rolling, the number of dots on the upward face. Each identical repetition of an experiment is called a trial (or *subexperiment*) of the experiment, and each result of a trial, the measurement, is called an outcome. Thus, each roll of the six-sided die is a trial of the experiment of rolling the die, and each trial yields one of six possible outcomes: 1, 2, 3, 4, 5, or 6 dots, which represent categories on a discrete ratio measurement scale (see Section 2.8). Any specific outcome of an experiment can be classified in different categories called events. Thus, for example, in the die-rolling experiment, rolling a specific number (e.g., a 3) is one possible event for this experiment, and rolling an even number (2, 4, or 6) is another.

The classical interpretation of probability, which is the oldest of the four interpretations, was developed in the seventeenth century from studies of the games of chance used in gambling. It deals with *idealized games* in which every trial of an experiment is done under uniform and perfect conditions. Such perfect games are always "fair" in that all possible outcomes are *equally likely to occur*. One such idealized game would be the rolling of a uniformly dense, perfectly symmetrical die onto a flawless surface, using identical hand motions with each trial. Under these conditions all six faces of the die are equally likely to be upward at the end of the roll. This equal likelihood of outcomes is a distinguishing characteristic of the classical interpretation.

For each experiment in such an idealized game, event A occurs if the experiment results in outcomes *favorable to A*. Thus, for the die-rolling experiment, if A is rolling a 3, then A occurs if the single outcome favorable to this (rolling a 3) occurs. Similarly, if A is rolling an even number, then A occurs if one of three favorable outcomes (2, 4, or 6) occurs. For such experiments, the *probability of event A* [denoted by $P(A)$] is the ratio of the number of possible outcomes favorable to A (denoted by N_A) to the total number of possible outcomes for the experiment (denoted by N), with the assumption that all possible outcomes are equally likely. Thus

$$P(A) = \frac{N_A}{N} \tag{8.1}$$

which is the symbolic version of equation (3.3).

For the die-rolling experiment, if A is rolling a 3, then there is one favorable outcome out of a total of six, and

$$P(A) = \frac{N_A}{N} = \frac{1}{6} = 0.17, \text{ or } 17\%$$

If A is rolling an even number, then there are three possible favorable outcomes out of a total of six, and

$$P(A) = \frac{N_A}{N} = \frac{3}{6} = 0.50, \text{ or } 50\%$$

These calculations illustrate that probability can be written as simple fractions, decimals, or percentages. If fractions are converted to decimals or percentages, then probabilities are typically reported, as here, to two significant figures.

The classical interpretation is appropriate for any experiment where a definite set of possible outcomes is known in advance, one of the outcomes must occur, and all outcomes are equally likely. Because in this interpretation the probability is determined before the experiment is attempted, classical probabilities are also called *a priori* (before the fact) *probabilities*.

EXAMPLE 8.1   You have a coin that is uniformly dense, perfectly flat and symmetrical, with two faces: a head and a tail. The experiment is to flip the coin into the air and observe which face lands upward. Determine the following probabilities: (*a*) P(head),   (*b*) P(tail),   (*c*) P(head or tail).

   Solution

   (*a*)   There are two equally likely outcomes for each trial of this experiment—a head or a tail—and so $N = 2$. $N_A = 1$ because only one of these outcomes (getting a head) is favorable to the event of interest. Therefore, using equation (8.1)

$$P(\text{head}) = \frac{N_A}{N} = \frac{1}{2} = 0.50, \text{ or } 50\%$$

   (*b*)   This problem has a solution identical to (*a*)

$$P(\text{tail}) = \frac{N_A}{N} = \frac{1}{2} = 0.50, \text{ or } 50\%$$

   (*c*)   Here, N is the same but now $N_A = 2$ because there are two possible outcomes favorable to the event of interest

$$P(\text{head or tail}) = \frac{N_A}{N} = \frac{2}{2} = 1.0 \text{ or } 100\%$$

   This solution indicates that for an idealized flip of a coin, the probability is 100% that the outcome will be either a head or a tail.

8.2   THE RELATIVE FREQUENCY INTERPRETATION OF PROBABILITY

The classical interpretation of probability has serious limitations: it is restricted to idealized experiments where all possible outcomes are known in advance and all are equally likely. There are many instances, however, where it is necessary to determine the probability of an event and it cannot be assumed that these classical requirements are being met. Thus in the following examples, while each experiment has two known outcomes (yes or no), these outcomes cannot be assumed to be equally likely: (1) an insurance company statistician determining the probability that a 45-year-old man will live to be 46;

(2) a geologist determining the probability that an earthquake will occur sometime within the next five years in a city located along a fault line; and (3) a manufacturer of automobile headlights determining the probability that none of the headlights in a shipment of 100 will be defective. While classical probabilities cannot be calculated for these examples, the required probabilities can be calculated using the *relative frequency interpretation* (also called the *frequentistic interpretation*, or *empirical interpretation*).

In the classical interpretation, probabilities are determined before any experiments are done. In the relative frequency interpretation, probabilities are determined from the results of previous experiments. The data from many replications of the same or "similar" experiments are analyzed to see the *relative frequency* (i.e., the proportion of times) that the event of interest occurred. This relative frequency of the event, taken from previous data, is then considered to be an estimate of the probability of future occurrences of the event. Thus, in the insurance example, the relative frequency of 45-year-old men dying before their 46th birthday would be calculated from previous data. Similarly, for the earthquake example, the geologist would first find previous situations similar to the city in question and then would calculate the relative frequency of earthquakes following those conditions within a five-year period. And the headlight manufacturer would test several 100-light samples and determine the relative frequency of defective lights in the samples.

To develop the basic probability formula for the relative frequency interpretion, consider this simpler example. You have a coin that you think may be "unfair"—in some way fixed so that the two possible outcomes of a flip (head or tail) are *not* equally likely. To investigate this possibility you flip the coin 100 times and get 70 heads and 30 tails. The relative frequency of heads is $70/100 = 7/10 = 0.70$, or 70%, which is then your estimate of the probability of getting a head in future flips of this coin. It is only an estimate, an approximation, because if you went on to repeat the experiment 1,000 times or 1,000,000 times you might get a somewhat different result.

The relative frequency interpretation of probability can be stated as follows:

The probability of event A [again denoted by $P(A)$] is approximately equal to the ratio of the number of times A occurred in a long series of trials (denoted by n_A) to the total number of trials in the series (denoted by n). Thus

$$P(A) \approx \frac{n_A}{n} \qquad (8.2)$$

Another way that this is stated is the *Law of Large Numbers* (also known as *Bernoulli's theorem*). This law states that for n trials of an experiment, if n_A is the number of times A occurred in those trials and $P(A)$ is the probability of A occurring on any one trial, then the relative frequency (n_A/n) will get closer and closer to $P(A)$ as n increases. In other words, the greater the number of trials the better the relative-frequency estimate of $P(A)$. For the coin-flipping example described above, 100 trials is sufficient to confirm your suspicion that this coin is unfair—the coin seems "fixed" to give more heads than tails.

EXAMPLE 8.2 A *random number table* (see Section 3.23) consists of thousands of digits—each of which is any one of the ten numbers from 0 to 9, and each of these ten numbers has an equal probability of appearing at any digit position in the table. Therefore, if we perform, the experiment of randomly picking a single-digit number from such a table, we can then use the classical interpretation to calculate the probabilities of events resulting from this pick (a specific number or group of numbers). Thus, for example, for P(a number less than 5) there are five numbers less than 5 (0 to 4) so $N_A = 5$, and there is a total of 10 possible numbers (0 to 9) so $N = 10$. Thus, using equation (8.1),

$$P(\# < 5) = \frac{N_A}{N} = \frac{5}{10} = 0.50, \text{ or } 50\%$$

Test this probability for Table A.1 in the Appendix by using any technique to enter the table, and then collecting 240 consecutive single-digit numbers. Consider the taking of each number to be a trial. Calculate relative frequencies by accumulating frequencies and trials after every 10th trial (e.g., the relative frequency of numbers less than 5 in the first 10 trials, in the first 20 trials, in the first 30 trials, etc.) and then construct a graph of these relative frequencies.

Solution

We entered Table A.1 at row 1 and column 1 and proceeded to the right along the row, accumulating ten digits from each consecutive five-column unit. Thus, the first ten digits collected from row 1 and columns 1 through 5 were: 1, 0, 0, 9, 7, 3, 2, 5, 3, 3; the next ten digits collected from row 1 and columns 6 through 10 were: 7, 6, 5, 2, 0, 1, 3, 5, 8, 6; and so on. To get 240 such digits we went from column 1 to column 75 along row 1 and then from column 1 to column 45 along row 2, ending with these ten digits from row 2 and columns 41 through 45: 8, 8, 6, 9, 5, 4, 1, 9, 9, 4. The results are shown in Fig. 8-1, where the vertical axis is relative frequency (n_A/n), the horizontal axis is numbers of trials (n), the dashed horizontal line represents the classical frequency for this event [$P(\# < 5) = N_A/N = 0.50$], and the connected filled circles represent the relative frequency of numbers less than 5, with the frequency and trials accumulated after every 10th trial.

It can be seen in Fig. 8-1, as predicted by the Law of Large Numbers, that as the number of trials increases, the relative frequency estimate of the probability stabilizes near the classical probability. This agreement between the classical probability and the relative frequency estimate is evidence that Table A.1 is indeed a random number table where each of the ten possible numbers is equally likely to be found at any digit position. This is not surprising because, if you read the introduction to the original source for this table (The RAND Coporation, *A Million Random Digits*, Free Press, Glencoe, Ill., 1995), you will see that the table was generated by an "electronic roulette wheel" that had been carefully tested and adjusted to produce a random table.

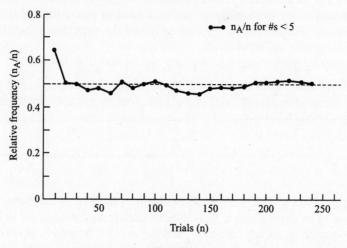

Fig. 8-1

8.3 SETS, SUBSETS, AND SAMPLE SPACES

As indicated in Section 1.17, a *set* is a specific collection of things (e.g., objects, symbols, numbers, words). Set theory is the branch of mathematics that deals with sets, their characteristics and relationships. Set theory is discussed here because it is the basis for the mathematical theory of probability. However, we will deal only with those limited apects of the theory that are essential for an intuitive-level understanding of inferential statistics (see Section 3.14).

The things that *belong to* or are *contained in* a set are called the *elements* or *members* of the set. These elements are indicated either by *listing* them or by specifying their *defining property*. For example, say the set is the integers from 1 to 5. By the listing technique this set would be shown as

$$S = \{1, 2, 3, 4, 5\}$$

where the elements of the set are listed within braces, and the braced list is made equal to the symbol S. By the defining-property technique, this set would be formally stated as

$$S = \{x | x \text{ is an integer and } 1 \leq x \leq 5\}$$

where the symbol $x|x$ (also written as $x{:}x$) means "x such that x." The entire equation is read: "S is the set of all elements x such that x is an integer and $1 \leq x \leq 5$." While this is the formal statement of a defining property, in this book we use a simpler phrase to specify the property. Thus, for this example

$$S = \{\text{all integers from 1 to 5}\}$$

If the set is specified by the listing technique, then the order of listing has no significance. It is easier to see this for sets that have no necessary ordering: the students in a history class or the introductory economics books in a college library.

A *subset* is any part of a set. Thus, if $S = \{1, 2, 3, 4, 5\}$, then the following are some of the subsets of S: $A = \{1, 2\}$; $B = \{1, 2, 3\}$; $C = \{1, 4, 5\}$. Strangely, the entire set can be considered to be a subset, as can a set that contains no elements at all, which is called the *empty set* or *null set* and denoted by $\varnothing$.

We said in Section 8.1 that in statistics an experiment is any process that yields a measurement. Each statistical experiment has a *sample space*—the set whose elements are all the possible outcomes of the experiment. Thus, for the die-rolling experiment described in Section 8.1 the sample space is $S = \{1, 2, 3, 4, 5, 6\}$, and for the coin-flipping experiment in Example 8.1 the sample space is $S = \{\text{head, tail}\}$. Note that the symbols used to denote sample spaces are the same as those used to denote sets.

The die-rolling and coin-flipping experiments each have a finite number of possible outcomes, and so the sample spaces for these experiments are called *finite sample spaces*. By contrast, experiments at higher levels of measurement can have an infinite number of possible outcomes, and so can have *infinite sample spaces*. For example, if the experiment is to measure the air temperature in °F (interval level) or the heights of children in centimeters (ratio level), then these are continuous measurements that can theoretically yield an infinite number of outcomes. This chapter deals only with finite sample spaces, but later chapters deal with both types.

8.4 EVENTS

In Section 8.1 we said that any particular outcome or group of outcomes from an experiment is called an event. Now, restating this in set-theory language, we can say that any particular subset of the sample space is an event. A *simple event* (or *elementary event*) is a subset that contains only one outcome (element) that cannot be broken down (decomposed) into a simpler, more basic outcome. A typical notation for simple events is the lowercase of the letter e with a subscript representing position in the list of outcomes for the sample space. Thus, for the die-rolling experiment where $S = \{1, 2, 3, 4, 5, 6\}$, there are six simple events in the sample space: $e_1 = \{1\}$, $e_2 = \{2\}$, $e_3 = \{3\}$, $e_4 = \{4\}$, $e_5 = \{5\}$, $e_6 = \{6\}$.

A *compound event* (or *composite event*) is defined as a subset of the sample space that contains more than one simple event. Such compound events are typically denoted by capital letters. In the die-rolling experiment, for example, compound event A could be: rolling an even number. Then, A would include three simple events: $e_2 = \{2\}$, $e_4 = \{4\}$, $e_6 = \{6\}$, and be denoted by any of the following: $A = \{e_2, e_4, e_6\}$, $A = \{2, 4, 6\}$, or $A = \{\text{rolling an even number}\}$. A compound event has occurred if any one of its component simple events has occurred.

From set theory it is known that if there are n elements in a set then there are 2^n subsets of that set. Thus, for the sample space of the die-rolling experiment, $S = \{1, 2, 3, 4, 5, 6\}$, $n = 6$ and there are ($2^n = 2^6 = 64$) subsets of the set. Each of these subsets is an event that can result from the experiment, depending upon how the outcomes are classified.

EXAMPLE 8.3 Three identical balls, numbered 1, 2, and 3, are placed in a jar. The experiment is to reach into the jar and, without looking, pull out one of the three balls. How many events (i.e., outcomes or *groups of outcomes*) can result from this experiment?

Solution

The sample space for this experiment is $S = \{1, 2, 3\}$. Since $n = 3$, there are ($2^n = 2^3 = 8$) subsets of the set. One of the outcomes could be pulling out the #2 ball (a simple event), and another could be pulling out either the #2 ball or the #3 ball (a compound event, symbolized as $\{2, 3\}$). All eight possible events are: the three simple events [{1}, {2}, {3}]; the three compound events that involve picking either one of two designated balls [{1, 2}, {1, 3}, {2, 3}]; the compound event that involves picking any one of the three balls [{1, 2, 3}, the set as a subset of itself]; and finally not picking any ball (∅, the empty set). (While not picking any ball is not a real outcome of this experiment, it is nevertheless a subset with a probability of zero.)

8.5   VENN DIAGRAMS

A *Venn diagram* shows a sample space and events within the space. Three typical ways of constructing Venn diagrams are shown in Fig. 8-2. All three diagrams represent the sample space $S = \{1, 2, 3, 4, 5, 6\}$ for the die-rolling experiment. The sample space is represented by a closed figure, such as an ellipse [Fig. 8-2(a)] or a rectangle [Fig. 8-2(b) and (c)]. Simple events within the sample space can be shown as dots, called *sample points* [Fig. 8-2(a) and (b)], or they can be assumed present but not shown [Fig. 8-2(c)]. Compound events are indicated by closed figures within the space. Compound event $A = \{$rolling an even number$\}$ is shown as a shaded ellipse in Fig. 8-2(a) and (b) and as a shaded circle in Fig. 8-2(c). Where the sample points are shown, either identified by symbols or not, the Venn diagram is also called a *Euler diagram*. Where the sample points are not shown, the area of the closed figure within the sample space has no significance; the area is not proportional to the number of sample points that the figure encloses.

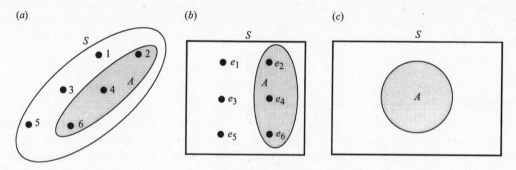

Fig. 8-2

The *complement* of any event A in a sample space S is the subset of S that contains all elements of S that are not in A. This subset is given the symbol A' (or $\bar{A}$). While A' is shown in Fig. 8-2(a), (b), and (c) as the space within S that is outside A, the most common version of the Venn diagram for A and A' is shown in Fig. 8-3. Here the circle containing A is not shaded, and A' is shown as the shaded area that is inside the rectangle but outside the circle.

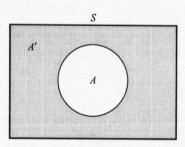

Fig. 8-3

Two events are said to be *mutually exclusive events* (or *disjoint events*) if when one occurs the other cannot occur. In set theory language, the events are mutually exclusive if they do not share any elements; there are no elements in one that are also in the other. By definition, all pairs of simple events in a sample space are mutually exclusive. Thus, for example, it is not possible to flip a coin and simultaneously observe both a head and a tail, or to roll a die and simultaneously observe both a one and a six. Compound events are mutually exclusive if they have no simple events in common.

EXAMPLE 8.4   Use a Venn diagram to depict the sample space for a die-rolling experiment, showing the two mutually exclusive events: $A = \{e_1, e_3\}$ and $B = \{e_4, e_6\}$.

Solution

A Venn diagram of the mutually exclusive events A and B in this example in which the sample points are shown, is seen in Fig. 8-4(a). A more typical Venn diagram for mutually exclusive events is shown in Fig. 8-4(b). In both diagrams, the interior figures are separated and not shaded.

The *union* of two events A and B in a sample space S is the subset of S whose elements belong to A or to B, or both A and B. In other words, the experiment can result in A or B or in both A and B. Such a union of two events is denoted in several ways: $A \cup B$, $A + B$, A or B. For the die-rolling experiment, the union of events $A = \{3, 4\}$ and $B = \{2, 4, 6\}$ is shown, in Venn diagrams, in Fig. 8-5. In part (a) of the figure, the sample points are shown and both elliptical figures are shaded. The more typical version, seen in part (b) of the figure, shows the union of A and B with the two event circles overlapping and both circles shaded.

The *intersection* of two events A and B in a sample space S is the subset of S whose elements belong to both A and B. In other words, the experiment results in the occurrence of both A and B. Such an intersection is denoted in various ways: $A \cap B$; A, B; AB; A and B. For the die-rolling experiment, a Venn diagram of the intersection of events $A = \{3, 4\}$ and $B = \{2, 4, 6\}$, showing the sample points, is seen in Fig. 8-6(a). The more typical version of a Venn diagram for the intersection of events A and B, without the sample points, is seen in Fig. 8-6(b). Notice in both figures that only the area of intersection (or overlap) is shaded.

EXAMPLE 8.5   For the die-rolling experiment where $S = \{1, 2, 3, 4, 5, 6\}$, if $A = \{1, 2\}$ and $B = \{2, 3, 4\}$ what are: (a) $A \cup B$,   (b) $A \cap B$,   (c) A or B,   (d) A and B?

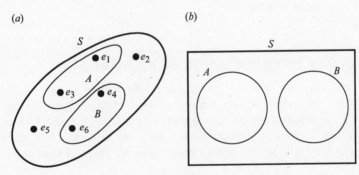

Fig. 8-4

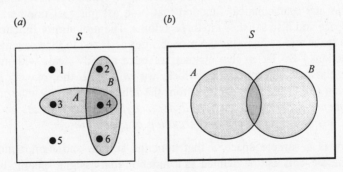

Fig. 8-5

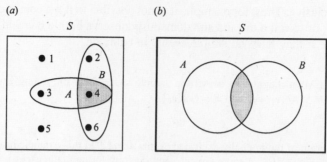

Fig. 8-6

Solution

(a) $A \cup B$ is the union of A and B and therefore is the subset of S that contains all elements of S that are in A, in B, or in both A and B. Thus

$$A \cup B = \{1, 2, 3, 4\}$$

(b) $A \cap B$ is the intersection of A and B and therefore is the subset of S that contains all elements in S that are in both A and B. Thus

$$A \cap B = \{2\}$$

(c) A or B is the same as $A \cup B$.

(d) A and B is the same as $A \cap B$.

8.6   THE SET THEORY INTERPRETATION OF PROBABILITY

The set theory concepts and operations that we have dealt with in Sections 8.3 to 8.5 form the basis of the mathematical theory of probability. As with other mathematical theories, the theory of probability begins with a given set of assumptions called *axioms* (or *postulates*) that must be accepted as true. All the rest of the theory is then derived by deductive logic from the axioms. There are three such axioms for probability theory, which we will present shortly, from which all the properties and rules of probability can be derived.

In the mathematical theory, probability is not related to specific real-world phenomena like gambling but is instead an abstract concept wholly defined and developed within the context of the theory, just as lines and points are abstract concepts in theoretical geometry. Within this theoretical context, a *probability function* is defined as any mathematical function that both assigns real numbers called probabilities to events in a sample space and also satisfies the three axioms. The probability function has as its *domain* the events in the sample space and as its *range* the probabilities assigned to these events. (See Section 1.17 for a discussion of functions.) So far in this chapter we have already dealt with two such functions: the function from the classical interpretation, which we will call the *classical probability function*, $P(A) = N_A/N$, equation (8.1); and the function from the relative frequency interpretation, which we will call the *relative frequency probability function*, $P(A) \approx n_A/n$, equation (8.2).

Stated formally, this is the *set theory interpretation of probability*:

Given that there is a sample space S that contains simple and compound events, then the probability of these events can be defined as a function that assigns specific real numbers called probabilities to each event in the sample space, with the provision that the function satisfies these axioms:

Axiom I   For event A in S

$$P(A) \geq 0$$

Axiom II   For sample space S

$$P(S) = 1$$

Axiom III   If events A and B in S are mutually exclusive, then

$$P(A \cup B) = P(A) + P(B)$$

Axiom I states that the probability of event A is always a nonnegative real number; that $P(A)$ is always greater than or equal to zero. Axiom II states that one of the events in S must occur for every trial of the experiment; that the probability of one event in S occurring is 100%. It is because of Axiom II that S is called the *certain event*. Axiom III, also called the *special addition rule*, states that the probability of the union of two mutually exclusive events A and B is equal to the sum of their separate probabilities.

While these three axioms are part of the abstract world of mathematics, they were selected as the axioms for probability theory because the known properties of classical and relative frequency probability can be derived from them. In this book, we will not do these formal mathematical derivations but instead will simply present properties and rules that can be so derived. The following are seven such properties:

Property 1.   For the empty event $\varnothing$ in S

$$P(\varnothing) = 0$$

Property 2.   For event A in S

$$0 \leq P(A) \leq 1$$

Property 3.   For event A and its complement A'

$$P(A) + P(A') = 1$$

Property 4.   If events $A_1, A_2, \ldots, A_k$ in S are all mutually exclusive, then

$$P(A_1 \cup A_2 \cup \cdots \cup A_k) = P(A_1) + P(A_2) + \cdots + P(A_k)$$

Property 5.   If S contains n simple events e_i that each have a probability $P(e_i)$, then

$$\sum_{i=1}^{n} P(e_i) = 1$$

Property 6.   If event A in S contains k simple events e_i, then

$$P(A) = \sum_{i=1}^{k} P(e_i)$$

Property 7.   If S contains N equally likely simple events e_i, then

$$P(e_i) = \frac{1}{N} \quad \text{and} \quad P(A) = N_A\left(\frac{1}{N}\right)$$

$\emptyset$ is the empty or null subset of S, also called the *empty event* or *null event*, and therefore Property 1 states that the probability of the empty event is 0%. For this reason, the empty event is also called the *impossible event*; it can never occur.

We know from Axiom I that $P(A)$ is always greater than or equal to zero and from Axiom II that the probability of the certain event S is 1. Property 2 simply combines these two axioms to state that all $P(A)$ values are nonnegative real numbers that are in the range from 0 to 1.

A and A' are mutually exclusive events. Therefore from Axiom III,

$$P(A \cup A') = P(A) + P(A')$$

We also know that A' includes all elements of the set that are not in A (see Section 8.5), so that

$$A \cup A' = S$$

Therefore

$$P(A \cup A') = P(S) = P(A) + P(A') = 1$$

Thus, Property 3 states that the probability that an event will or will not occur is 100%. Property 3 can also be written

$$P(A) = 1 - P(A')$$

or

$$P(A') = 1 - P(A)$$

Property 4 simply states that Axiom III can be generalized to apply to any number of mutually exclusive events in S.

S contains n mutually exclusive events e_i. Therefore

$$S = e_1 \cup e_2 \cup \cdots \cup e_n$$

and

$$P(e_1 \cup e_2 \cup \cdots \cup e_n) = P(S) = 1$$

We know from Property 4 that

$$P(e_1 \cup e_2 \cup \cdots \cup e_n) = P(e_1) + P(e_2) + \cdots + P(e_n) = \sum_{i=1}^{n} P(e_i)$$

Thus

$$\sum_{i=1}^{n} P(e_i) = 1$$

Therefore, Property 5 states that the sum of the probabilities of the simple events in S is 100%.

Property 6 states that the probability of event A is the sum of the probabilities of the simple events contained in A.

Finally, Property 7 states that for the special case where all simple events in S are equally likely (the conditions for the classical interpretation) then if there are N simple events e_i in S each of these will have the probability of $1/N$, and the probability of any compound event A will be the product of the number of simple events in A, N_A, times $1/N$.

EXAMPLE 8.6 In this section we indicated that any mathematical function is a probability function if it both assigns specific real-number probabilities to events in S and also satisfies the three axioms. Using examples from the die-rolling experiment, demonstrate that the function $P(A) = N_A/N$, equation (8.1), which we have called the classical probability function, is indeed a probability function.

Solution

As indicated in Section 8.1, $P(A) = N_A/N$ can be used for any experiment where the classical interpretation is appropriate. Under such conditions

$$P(A) = \frac{N_A}{N} = \frac{\text{number of outcomes favorable to } A}{\text{total number of possible outcomes}}$$

This can be stated for the set theory interpretation as

$$P(A) = \frac{N_A}{N} = \frac{\text{number of simple events in } A}{\text{total number of simple events in } S}$$

It is clear that this function assigns specific real-number probabilities to events in S and thus satisfies the first part of the definition of a probability function. To demonstrate that it also satisfies the three axioms, we must show that the axioms remain true if this function replaces $P(A)$ in the axioms.

Axiom I, $P(A) \geq 0$, states, in essence, that each event in S must have a probability greater than or equal to zero. This is clearly true for the die-rolling experiment where $S = \{1, 2, 3, 4, 5, 6\}$. $P(A) = 0$ when $N_A = 0$, which is true for an empty subset of S such as $A = \{\text{rolling a number larger than 6}\}$. For all other subsets of S, $N_A > 0$ and thus $P(A) > 0$.

Axiom II, $P(S) = 1$, states, in essence, that one simple event in S must occur for every trial of the experiment. Clearly for the die-rolling experiment, for every roll of the die one of the faces of the die must land upward and therefore one of the simple events in S must occur.

Axiom III states for mutually exclusive events A and B that $P(A \cup B) = P(A) + P(B)$. In general you can see that this is true from the following:

$$P(A \cup B) = \frac{N_{A \cup B}}{N} = \frac{N_A + N_B}{N} = \frac{N_A}{N} + \frac{N_B}{N} = P(A) + P(B)$$

We can demonstrate that it holds true for the die-rolling experiment, using the two mutually exclusive events $A = \{1, 2\}$ and $B = \{3, 4, 5\}$. For these,

$$N_A = 2, \qquad N_B = 3, \qquad N_{A \cup B} = 5, \qquad N = 6$$

and

$$P(A \cup B) = \frac{N_{A \cup B}}{N} = \frac{5}{6}$$

$$= \frac{N_A + N_B}{N} = \frac{2 + 3}{6}$$

$$= \frac{N_A}{N} + \frac{N_B}{N} = \frac{2}{6} + \frac{3}{6}$$

$$= P(A) + P(B)$$

8.7 THE SUBJECTIVE INTERPRETATION OF PROBABILITY

Probabilities determined with classical or relative frequency probability functions are called *objective probabilities*. They are determined from purely objective information: clear factual information about the likelihood of an event, that has not been distorted by personal feelings or prejudices. Thus, classical probabilities are determined by knowing in advance all possible, equally likely outcomes of an experiment, and relative frequency probabilities are determined by knowing the proportion of times an event has occurred in a long series of trials.

There are many instances, however, when it is necessary to determine probabilities and it is not possible to do this from purely objective information. These instances require "personal judgments" or "educated guesses"—the personal and unique assessment of available information to determine the probability that an event will occur. In such instances, a numerical value is assigned to a personal *degree of belief* (or *degree of certainty*) in the likelihood of the event. Such measures of degree of belief are *subjective*, and thus this version of probability is called the *subjective* (or *personalistic* or *personal*) *interpretation of probability.*

The subjective interpretation is appropriate, for example, when an experiment that has never been done before will be attempted only once. Say you are a business manager who is about to introduce a new product into a market. To determine the probability of success for the product, you evaluate available information (e.g., successes and failures of similar products, your previous experiences when introducing new products, the general economic climate in this market), consult your feelings and intuitions, and then, somehow, put it all together into this subjective probability value: there is a 0.80 probability of success. By this mental integration you have decided that the product is four times more likely to succeed than to fail.

Another example of a subjective probability for a one-time experiment is a high-school senior's judgment that there is a 50% chance she will be accepted into the college of her choice. From her assessment of available information and beliefs she has decided that she is as likely to be accepted as rejected.

Typically the probability function for such subjective probabilities cannot be externalized and formally written out. Whatever skill, experience, and integrative mental processes went into such a calculation, however, the resulting probability values must conform to the axioms and properties of Section 8.6 and other rules and properties to be discussed in Chapter 9.

8.8 THE CONCEPT OF ODDS

The *concept of odds* is an alternative method for expressing any form of probability, objective or subjective. If $P(A)$ is the probability that event A will occur and $P(A')$ is the probability that it will not occur (the probability of its complement), then the *odds in favor of the event occurring* are defined as the ratio of $P(A)$ to $P(A')$. This ratio, by convention, is expressed as the ratio of two positive integers, c and d, that have no factors in common. Thus, the odds in favor of A are

$$\frac{P(A)}{P(A')} = \frac{c}{d} \tag{8.3}$$

which is typically stated as: The odds in favor of A are c to d. The odds unfavorable to A are

$$\frac{P(A')}{P(A)} = \frac{d}{c} \tag{8.4}$$

which is typically stated as: The odds against A are d to c.

Odds are conventionally stated as odds in favor of A if $P(A) > P(A')$, and as odds against A if $P(A) < P(A')$.

EXAMPLE 8.7 A television weatherman says that there is a 70% probability that it will rain tomorrow. According to him, what are the odds that it will rain tomorrow?

Solution

In this problem, the event A is the occurrence of rain tomorrow and the weatherman has stated that $P(A) = 0.70$. From Property 3 in Section 8.6 we know that $P(A') = 1 - P(A)$. Therefore

$$P(A') = 1 - 0.70 = 0.30$$

Thus, the odds in favor of rain tomorrow are

$$\frac{P(A)}{P(A')} = \frac{0.70}{0.30} = \frac{7}{3}, \text{ or 7 to 3}$$

Betting odds, as given by commercial gambling houses or at the race track, indicate the amount that a gambler can win or lose in betting on an event. Thus, betting odds of 5 to 1 indicate that the gambler can win $5 or lose $1 on the bet. Such betting odds are set to make a profit for the gambling establishment;

they are not necessarily the same as the odds the event will occur. If the two types of odds are the same for a given bet, then the bet is called a *fair bet*.

EXAMPLE 8.8 You know that two tennis players, R and S, have played 36 matches and that R has won 24 of them. You offer to bet a friend your \$15 against his \$10 that R will win the next match. Is this a fair bet?

> **Solution**
>
> $A = \{R \text{ wins}\}$, $A' = \{S \text{ wins}\}$. With the relative frequency interpretation
>
> $$P(A) \approx \frac{24}{36} = \frac{2}{3} \qquad P(A') \approx \frac{12}{36} = \frac{1}{3}$$
>
> Therefore, the odds in favor of A are
>
> $$\frac{P(A)}{P(A')} \approx \frac{2/3}{1/3} = \frac{2}{1}, \quad \text{or approximately 2 to 1}$$
>
> Thus, for this bet to be a fair bet it should be your \$20 against his \$10. As is, it is not a fair bet.

8.9 DETERMINING PROBABILITIES FROM ODDS

In Section 8.8 we determined odds from probabilities. In this section, we determine probabilities from odds, by solving the following equation for both $P(A)$ and $P(A')$.

$$\frac{P(A)}{P(A')} = \frac{c}{d}$$

We know from Property 3 in Section 8.6 that

$$P(A') = 1 - P(A)$$

Substituting this relation in the odds equation

$$\frac{P(A)}{1 - P(A)} = \frac{c}{d}$$

and thus

$$d \times P(A) = c[1 - P(A)]$$
$$= c - c \times P(A)$$

Therefore

$$[c \times P(A)] + [d \times P(A)] = c$$
$$P(A)(c + d) = c$$

and thus

$$P(A) = \frac{c}{c + d} \tag{8.5}$$

To solve for $P(A')$ we again start with Property 3 in Section 8.6

$$P(A') = 1 - P(A)$$

Therefore

$$P(A') = 1 - \frac{c}{c+d}$$

$$= \frac{c+d}{c+d} - \frac{c}{c+d}$$

$$= \frac{c+d-c}{c+d}$$

and thus

$$P(A') = \frac{d}{c+d} \tag{8.6}$$

EXAMPLE 8.9   If the odds in favor of event A occurring are 4 to 3, what are: (a) $P(A)$,   (b) $P(A')$?

Solution

As $c = 4$ and $d = 3$:

(a) $P(A) = \dfrac{c}{c+d} = \dfrac{4}{4+3} = 0.57$, or 57%

(b) $P(A') = \dfrac{d}{c+d} = \dfrac{3}{4+3} = 0.43$, or 43%

Solved Problems

THE CLASSICAL INTERPRETATION OF PROBABILITY

8.1   In a standard 52-card deck of playing cards there are four suits (clubs, diamonds, hearts, and spades) of 13 cards each (ace, 2, 3, 4, 5, 6, 7, 8, 9, 10, jack, queen, and king). The diamonds and hearts are called red cards (marked with red symbols) and the clubs and spades are called black cards (marked with black symbols). The experiment consists of picking one card from a well-shuffled standard deck. Determine the following probabilities: (a) $P(10$ of hearts), (b) P(heart),   (c) $P(10)$,   (d) P(red card).

Solution

(a) $N = 52$ because each of the 52 cards is equally likely to be picked. $N_A = 1$ because only one of the 52 cards is favorable to this event. Therefore, using equation (8.1),

$$P(10 \text{ of hearts}) = \frac{N_A}{N} = \frac{1}{52} = 0.019, \text{ or } 1.9\%$$

(b)   Again $N = 52$, but now there are 13 possible favorable outcomes (heart cards), so $N_A = 13$. Therefore

$$P(\text{heart}) = \frac{N_A}{N} = \frac{13}{52} = 0.25, \text{ or } 25\%$$

(c)   Again $N = 52$, but now there are 4 possible favorable outcomes (cards identified as 10s), so $N_A = 4$. Therefore

$$P(10) = \frac{N_A}{N} = \frac{4}{52} = 0.077, \text{ or } 7.7\%$$

(d) Again $N = 52$, but now there are 26 possible favorable outcomes (diamond and heart cards), so $N_A = 26$. Therefore

$$P(\text{red card}) = \frac{N_A}{N} = \frac{26}{52} = 0.50, \text{ or } 50\%$$

8.2 Table 8.1 shows the age and sex of 85 students in a college history class (e.g., 15 of the students are males who are 20 years old or younger). The experiment is to use a table of random numbers (see Section 3.23) to select one of the students from the class and find his or her age and sex. Determine the following probabilities: (a) $P(\text{male 20 or younger})$, (b) $P(\text{male})$, (c) $P(\text{student 20 or younger})$, (d) $P(\text{male or female})$.

Table 8.1

Sex	Age 20 or younger	Over 20	Total
Male	15	30	45
Female	20	20	40
Total	35	50	85

Solution

(a) $N = 85$ because the equally likely outcomes of this experiment are the 85 students. They are all equally likely because the selection is random. $N_A = 15$ because there are 15 males that are of age 20 or younger. Therefore, using equation (8.1),

$$P(\text{male 20 or younger}) = \frac{N_A}{N} = \frac{15}{85} = 0.18, \text{ or } 18\%$$

(b) Again $N = 85$, but now there is a total of 45 males so $N_A = 45$. Therefore

$$P(\text{male}) = \frac{N_A}{N} = \frac{45}{85} = 0.53, \text{ or } 53\%$$

(c) Again $N = 85$, but now there is a total of 35 students 20 or younger so $N_A = 35$. Therefore

$$P(\text{student 20 or younger }) = \frac{N_A}{N} = \frac{35}{85} = 0.41, \text{ or } 41\%$$

(d) Again $N = 85$, but now all the students are either male or female so $N_A = 85$. Therefore

$$P(\text{male or female}) = \frac{N_A}{N} = \frac{85}{85} = 1.0, \text{ or } 100\%$$

8.3 Fifty marbles of different colors are placed in a jar and thoroughly mixed. Twenty-five of the marbles are blue, 20 are green, and 5 are red. If one marble is then blindly selected from the jar, determine the following probabilities: (a) $P(\text{red marble})$, (b) $P(\text{blue or red marble})$.

Solution

(a) $N = 50$ and $N_A = 5$. Using equation (8.1),

$$P(\text{red marble}) = \frac{N_A}{N} = \frac{5}{50} = 0.10, \text{ or } 10\%$$

(b) $N = 50$ and $N_A = 25 + 5 = 30$. Using equation (8.1),

$$P(\text{blue or red marble}) = \frac{N_A}{N} = \frac{30}{50} = 0.60, \text{ or } 60\%$$

8.4 Three families get together for a holiday dinner. Mr. and Mrs. Brown have three daughters and one son. Mr. and Mrs. Cruz have one daughter and two sons. Mr. and Mrs. Hansen have three sons. As a way of deciding who will carve the turkey, everyone writes his or her name on a piece of paper and places it in a hat. The pieces of paper are thoroughly mixed within the hat and, without looking, Mr. Hansen draws one of the names from the hat. Determine the probability that the name drawn is: (a) Mr. Hansen, (b) a member of the Cruz family, (c) a male, (d) not a member of the Hansen family.

Solution

(a) $N = 6 + 5 + 5 = 16$ and $N_A = 1$. Using equation (8.1),

$$P(\text{Mr. Hansen}) = \frac{N_A}{N} = \frac{1}{16} = 0.06 = 6\%$$

(b) $N = 6 + 5 + 5 = 16$ and $N_A = 5$

$$P(\text{member of Cruz family}) = \frac{N_A}{N} = \frac{5}{16} = 0.31\%, \text{ or } 31\%$$

(c) $N = 6 + 5 + 5 = 16$ and $N_A = 2 + 3 + 4 = 9$

$$P(\text{a male}) = \frac{N_A}{N} = \frac{9}{16} = 0.56, \text{ or } 56\%$$

(d) $N = 6 + 5 + 5 = 16$ and $N_A = 6 + 5 = 11$

$$P(\text{not a member of the Hansen family}) = \frac{N_A}{N} = \frac{11}{16} = 0.69, \text{ or } 69\%$$

THE RELATIVE FREQUENCY INTERPRETATION OF PROBABILITY

8.5 In Section 8.1 we used the classical interpretation to determine the probabilities for two possible events resulting from an idealized die-rolling experiment: $P(\text{rolling a 3}) = 0.17$, and $P(\text{rolling an even number}) = 0.50$. Test these probabilities for an actual die by rolling it 240 consecutive times. Try to make all trials as similar as possible. From the results, calculate relative frequencies for each event by accumulating frequencies and trials after every 12th trial (e.g., relative frequency of 3s in the first 12 trials, in the first 24 trials, in the first 36 trials, etc.) and then construct a graph of these relative frequencies.

Solution

We rolled a standard plastic die 240 times with a "uniform motion": the die was rolled into a wall some four feet away and then allowed to roll back along the floor until it stopped. The relative frequencies of getting even numbers and of getting 3s are shown in Fig. 8-7, where the vertical axis is relative frequency (n_A/n); the horizontal axis is trials (n); the two dashed horizontal lines represent the classical probabilities (N_A/N) for the two events; the connected filled circles represent the relative frequencies of 3s, accumulated after every 12th trial; and the connected open circles represent the relative frequencies of even numbers, accumulated after every 12th trial.

It can be seen, as the Law of Large Numbers predicts (see Section 8.2), that as the number of trials increases the accumulated relative frequencies become increasingly more stable, getting closer and closer to the classical probabilities for these events. If the trials had continued, the relative frequencies might have settled exactly on the classical probability lines or, due to imperfections in the technique or the die, they might have remained slightly off the lines.

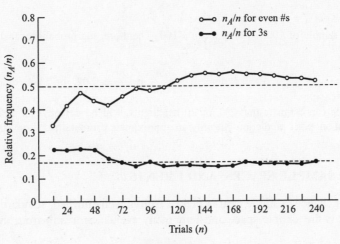

Fig. 8-7

8.6   During the past year in the maternity ward of a hospital in a large city, 1,060 males were born and 1,000 females. Assuming this data to be representative for all births, what is the probability that the next baby born in this hospital will be a boy? What is the probability it will be a girl?

Solution

For a boy, $n_A = 1,060$, and $n = (1,060 + 1,000) = 2,060$. Thus, using equation (8.2),

$$P(\text{boy}) \approx \frac{n_A}{n} = \frac{1,060}{2,060} = 0.51, \text{ or } 51\%$$

For a girl, $n_A = 1,000$, and $n = 2,060$. Thus

$$P(\text{girl}) \approx \frac{n_A}{n} = \frac{1,000}{2,060} = 0.49, \text{ or } 49\%$$

8.7   Early in the season (in late May), a baseball player had 5 hits in 26 times at bat. Later in that same season (early August), he had achieved 117 hits in 352 times at bat. What is the probability that the next time he comes to bat he will get a hit in: (a) late May,   (b) early August?

Solution

(a)   In late May the relative frequency of hits (his *batting average*) was $5/26 = 0.19$. Thus the relative frequency estimate of the probability of his getting a hit the next time at bat was

$$P(\text{getting a hit}) \approx \frac{n_A}{n} = 0.19$$

(b)   By early August the relative frequency estimate was

$$P(\text{getting a hit}) \approx \frac{n_A}{n} = \frac{117}{352} = 0.33$$

From the Law of Large Numbers we expect that the August estimate, based on almost 14 times the number of experiments (times at bat), is a much better estimate of the probability that he will get a hit.

8.8   A manufacturer of automobile headlights wants to determine the probability that the next headlight produced by his company will be defective. He tests a sample of 100 and finds 2 defectives. What is the probability the next will be defective? How many defectives can he expect to find in a shipment of 1,000?

Solution

For the sample of 100, $n_A = 2$, and $n = 100$. Therefore, the probability that the next headlight will be defective is

$$P(\text{defective}) \approx \frac{n_A}{n} = \frac{2}{100} = 0.02, \text{ or } 2\%$$

From this one can estimate that 2%, or 20 headlights, will be defective in a shipment of 1,000. This, of course, is not an exact prediction but only an approximate estimate based on 100 trials of an experiment.

SETS, SUBSETS, SAMPLE SPACES, AND EVENTS

8.9   In a game of chance, a single card is picked from a well-shuffled, standard 52-card deck of playing cards. What is the sample space and how many events can result from the experiment?

Solution

There are 52 possible outcomes but instead of listing them we will use the defining property: $S = \{$the 52 cards in a standard deck$\}$. Thus, from Section 8.4, $2^n = 2^{52}$ possible events can result from this experiment.

8.10   The experiment is to randomly select one student from the 85-student class shown in Table 8.1. What is the sample space and how many events can result from the experiment?

Solution

Using the defining property $S = \{$the 85 students in the history class$\}$, 2^{85} possible events can result from this experiment.

8.11   The experiment is to flip a coin two times and to observe, for each flip, whether the head or the tail lands face upward. What is the sample space and how many events can result from the experiment?

Solution

For two consecutive trials of the coin-flipping experiment, each simple event has two components. Using H for heads and T for tails, the sample space is $S = \{HH, HT, TH, TT\}$. Thus, there are $(2^4 = 16)$ possible events that can result from this experiment. These events are:

$\{HH\}, \{HT\}, \{TH\}, \{TT\}, \{HH, HT\}, \{HH, TH\}, \{HH, TT\}, \{HT, TH\}, \{HT, TT\}, \{TH, TT\}, \{HH, HT, TH\}, \{HH, HT, TT\}, \{HH, TH, TT\}, \{HT, TH, TT\}, \{HH, HT, TH, TT\}$, and $\varnothing$

(Recall from Section 8.4 that a compound event, such as $\{HH, HT\}$, involves getting just one of its simple events, such as HH.)

8.12   For the maternity ward in Problem 8.6, the experiment is to determine the sex of the next baby born in the ward. For this experiment, what is the sample space and how many events can result?

Solution

$S = \{$boy, girl$\}$; the $(2^2 = 4)$ possible events are: $\{$boy$\}$, $\{$girl$\}$, $\{$boy, girl$\}$, and $\varnothing$

8.13   For the baseball player in Problem 8.7, the experiment is to determine whether he gets a hit the next time at bat. For this experiment, what is the sample space and how many events can result?

Solution

$S = \{$hit, no hit$\}$; the $(2^2 = 4)$ possible events are: $\{$hit$\}$, $\{$no hit$\}$, $\{$hit, no hit$\}$, and $\varnothing$

8.14 For the headlight manufacturer in Problem 8.8, the experiment is to determine whether the next headlight is defective. For this experiment, what is the sample space and how many events can result?

Solution

$S = \{$defective, not defective$\}$; the $(2^2 = 4)$ possible events are: $\{$defective$\}$, $\{$not defective$\}$, $\{$defective, not defective$\}$, and $\varnothing$

VENN DIAGRAMS

8.15 A single card is drawn from a well-shuffled, standard deck of playing cards. For this experiment, indicate whether the following pairs of events are mutually exclusive and/or complementary: (a) $A = \{$a red card$\}$, $B = \{$a black card$\}$; (b) $A = \{$a card numbered 2, 3, 4, 6, or 10$\}$, $B = \{$a card from the diamond suit$\}$; (c) $A = \{$a card numbered 2, 3, 4, 5, 6, 7, 8, 9, or 10$\}$, $B = \{$a card that is an ace, jack, queen, or king$\}$; (d) $A = \{$a diamond card that is an ace, 2, 3, 4, 5, 6, or 7$\}$, $B = \{$a diamond card that is an 8, 9, 10, jack, queen, or king$\}$.

Solution

(a) Events A and B are mutually exclusive because no card can be both red and black. A and B are also complementary ($B = A'$) because all simple events in S that are not in A are in B.

(b) Events A and B are not mutually exclusive because they share simple events; the diamond suit has cards numbered 2, 3, 4, 6, and 10. As they are not mutually exclusive, events A and B are automatically also not complementary. Recall that A' contains all simple events in the sample space that are *not* in A.

(c) Events A and B are mutually exclusive; they do not share any simple events. Events A and B are also complementary ($B = A'$) because all simple events in S that are not in A are in B.

(d) Events A and B are mutually exclusive; no diamond card can be simultaneously both an ace, 2, 3, 4, 5, 6, or 7, and an 8, 9, 10, jack, queen, or king. Events A and B are not complementary, however, because the other three suits in the sample space are not included in either A or B.

8.16 A single student is chosen, at random, from the history class shown in Table 8.1. For this experiment, indicate whether the following pairs of events are mutually exclusive and/or complementary: (a) $A = \{$a male 20 or younger$\}$, $B = \{$a male over 20$\}$; (b) $A = \{$a male$\}$, $B = \{$a female$\}$; (c) $A = \{$a student 20 or younger$\}$, $B = \{$a student over 20$\}$; (d) $A = \{$a male 20 or younger$\}$, $B = \{$a female 20 or younger$\}$.

Solution

(a) A and B are mutually exclusive but not complementary.

(b) A and B are both mutually exclusive and complementary.

(c) A and B are both mutually exclusive and complementary.

(d) A and B are mutually exclusive but not complementary.

8.17 For the sample space shown in the Venn diagram in Fig. 8-8, which includes events A, B, and C, what are: (a) A', (b) $A \cup A'$, (c) A and A', (d) $A \cap B$, (e) A or B, (f) B and C, (g) $B \cup C$, (h) $A' \cap B$, (i) $A' \cup B$, (j) A' or B', (k) $B' \cup C'$, (l) B' and C'?

Solution

(a) A' is the complement of A and therefore contains all elements of S that are not in A. Thus

$$A' = \{e_1, e_4, e_6, e_7, e_8, e_9, e_{10}, e_{11}, e_{12}\}$$

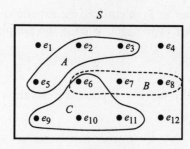

Fig. 8-8

(b) $A \cup A'$, the union of A and its complement, is the subset of S containing all elements in A, and A', or in both A and A'. Thus

$$A \cup A' = S$$

(c) A and A', the intersection of A and its complement, is the subset of S containing all elements shared by A and A'. As, by definition, there are no shared elements

$$A \text{ and } A' = \varnothing$$

(d) Events A and B do not share any elements because they are mutually exclusive. Thus

$$A \cap B = \varnothing$$

(e) A or B is the union of A and B. Thus

$$A \text{ or } B = \{e_2, e_3, e_5, e_6, e_7, e_8\}$$

(f) Events B and C share e_6. Thus

$$B \text{ and } C = \{e_6\}$$

(g) $B \cup C = \{e_6, e_7, e_8, e_9, e_{10}, e_{11}\}$

Note that while e_6 is in both B and C it is only included once in $B \cup C$.

(h) $A' \cap B = \{e_6, e_7, e_8\}$

(i) $A' \cup B = A' = \{e_1, e_4, e_6, e_7, e_8, e_9, e_{10}, e_{11}, e_{12}\}$

(j) A' or $B' = S$

(k) $B' \cup C' = \{$all elements in S except $e_6\}$

(l) B' and $C' = \{e_1, e_2, e_3, e_4, e_5, e_{12}\}$

8.18 A single card is drawn from a well-shuffled, standard deck of playing cards, in which $S = \{$the 52 cards in a standard deck$\}$. For this experiment, consider the following events: $A = \{$ace$\}$, $B = \{$diamond$\}$, $C = \{2, 3, 4, 5,$ or $6\}$, $D = \{$black card$\}$. What are: (a) B', (b) $A \cup B$, (c) A and B, (d) $A \cap B'$, (e) $B \cap D'$, (f) B or C, (g) $A \cap C$, (h) C or D, (i) A and C', (j) $A' \cap B'$, (k) $B' \cup D'$?

Solution

(a) $B' = \{$the 39 cards in the club, heart, and spade suits$\}$

(b) $A \cup B = \{$the 13 cards in the diamond suit and the ace of clubs, the ace of hearts, and the ace of spades$\}$

(c) A and $B = \{$ace of diamonds$\}$

(d) $A \cap B' = \{$ace of clubs, ace of hearts, ace of spades$\}$

(e) $B \cap D' = \{$the 13 cards in the diamond suit$\}$

(f) B or $C = \{$the 13 cards in the diamond suit, and the 2, 3, 4, 5, and 6 cards from the other three suits$\}$

(g) $A \cap C = \emptyset$

(h) C or $D = \{$the 26 black cards (clubs and spades) and the 2, 3, 4, 5, and 6 red cards (diamonds and hearts)$\}$

(i) A and $C' = \{$the 4 aces$\}$

(j) $A' \cap B' = \{$all cards in the club, heart, and spade suits except the aces$\}$

(k) $B' \cup D' = S$

THE SET THEORY INTERPRETATION OF PROBABILITY

8.19 A ticket for tonight's concert costs \$4, \$8, \$12, \$15, or \$20. The first 1,000 people to purchase tickets bought the following: 200 bought \$4 tickets, 500 bought \$8 tickets, 150 bought \$12 tickets, 100 bought \$15 tickets, and 50 bought \$20 tickets. The experiment is to predict the price of the next ticket purchased. Using this experiment as an example, demonstrate that equation (8.2), $P(A) \approx n_A/n$, which we have called the relative frequency probability function, is indeed a probability function. Use Example 8.6 as a model for this demonstration.

Solution

In the relative frequency interpretation, the probabilities for events in the sample space of an experiment are determined by the relative frequency of these events in a series of previous trials of the experiment. Thus, as was stated in Section 8.2:

$P(A)$ is approximately equal to the ratio of the number of times A occurred in a long series of trials (n_A) to the total number of trials in the series (n).

Or, stated symbolically

$$P(A) \approx \frac{n_A}{n}$$

This function assigns specific real-number probabilities to events in S and thus satisfies the first half of the definition of a probability function. It also clearly satisfies Axiom I, as n_A can never be less than zero and thus $P(A)$ can never be less than zero. $P(A)$ can be zero for an empty subset of S (e.g., purchasing a \$25 ticket) or if n_A is zero (e.g., no \$4 tickets were purchased).

Axiom II is satisfied because it is 100% certain that one of the simple events in S must occur on every trial of the experiment; that each time a ticket is purchased it must cost \$4, \$8, \$12, \$15, or \$20.

To show that Axiom III is satisfied we use the same logic as in Example 8.6 and let $A = \{$ticket $\leq$ \$8$\}$ and $B = \{$ticket $\geq$ \$15$\}$. Then

$$n_A = 700, \qquad n_B = 150, \qquad n_{A \cup B} = 850, \qquad n = 1,000$$

and

$$P(A \cup B) \approx \frac{n_{A \cup B}}{n} = \frac{850}{1,000}$$

$$\approx \frac{n_A + n_B}{n} = \frac{700 + 150}{1,000}$$

$$\approx \frac{n_A}{n} + \frac{n_B}{n} = \frac{700}{1,000} + \frac{150}{1,000}$$

$$\approx P(A) + P(B)$$

8.20 In the sample space shown in Fig. 8-9, which includes events A, B, C, and D, each sample point represents an equally likely (equally probable) simple event. Use the axioms and properties of Section 8.6 to determine the following probabilities: (a) $P(A)$, (b) $P(B)$, (c) $P(C)$, (d) $P(D)$, (e) $P(A \cup C)$, (f) $P(A')$, (g) $P(A \cup B \cup C \cup D)$, (h) $P(A \cap B)$, (i) $P(A \cup C')$.

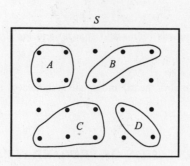

Fig. 8-9

Solution

(a) Property 7 states that if S contains N equally likely simple events e_i then

$$P(e_i) = \frac{1}{N}$$

In this sample space $N = 20$, and therefore

$$P(e_i) = \frac{1}{20}$$

Property 7 further states that

$$P(A) = N_A \left(\frac{1}{N} \right)$$

Thus, as there are four simple events in A,

$$P(A) = N_A \left(\frac{1}{N} \right) = 4 \left(\frac{1}{20} \right) = \frac{4}{20} = 0.20$$

(b) $P(B) = N_B \left(\dfrac{1}{N} \right) = 3 \left(\dfrac{1}{20} \right) = \dfrac{3}{20} = 0.15$

(c) $P(C) = N_C \left(\dfrac{1}{N} \right) = 5 \left(\dfrac{1}{20} \right) = \dfrac{5}{20} = 0.25$

(d) $P(D) = N_D \left(\dfrac{1}{N} \right) = 2 \left(\dfrac{1}{20} \right) = \dfrac{2}{20} = 0.10$

(e) A and C are mutually exclusive and therefore from Axiom III

$$P(A \cup C) = P(A) + P(C) = 0.20 + 0.25 = 0.45$$

(f) From Property 3

$$P(A') = 1 - P(A)$$
$$= 1 - 0.20 = 0.80$$

(g) A, B, C, and D are all mutually exclusive, and therefore from Property 4

$$P(A \cup B \cup C \cup D) = P(A) + P(B) + P(C) + P(D)$$
$$= 0.20 + 0.15 + 0.25 + 0.10 = 0.70$$

(h) As A and B are mutually exclusive, $A \cap B = \varnothing$, and therefore from Property 1

$$P(A \cap B) = P(\varnothing) = 0$$

(i)	A and C' are not mutually exclusive so we cannot use Axiom III. Instead we will use Property 7:

$$A \cup C' = \{\text{all simple events in } S \text{ that are outside } C\}$$

Thus

$$N_{A \cup C'} = 15$$

and

$$P(A \cup C') = N_{A \cup C'} \left(\frac{1}{N}\right) = 15\left(\frac{1}{20}\right) = \frac{15}{20} = 0.75$$

8.21	In the sample space shown in Fig. 8-10, which includes events A, B, and C, the number above each sample point is the relative frequency approximation of the probability of the simple event represented by the sample point $P(e_i) \approx \dfrac{n_{e_i}}{n}$. Use the axioms and properties of Section 8.6 to determine the following probabilities:	(a) $P(A)$,	(b) $P(B)$,	(c) $P(C)$,	(d) $P(A \cup B)$,	(e) $P(B')$,	(f) $P(A \cup B \cup C)$,	(g) $P(B \cap C)$,	(h) $P(A \cup B')$.

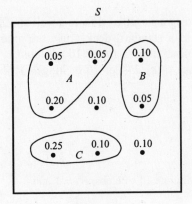

Fig. 8-10

Solution

(a)	Property 6 states

$$P(A) = \sum_{i=1}^{k} P(e_i)$$

Therefore here

$$P(A) \approx 0.05 + 0.05 + 0.20 = 0.30$$

(b)	$P(B) \approx 0.10 + 0.05 = 0.15$

(c)	$P(C) \approx 0.25 + 0.10 = 0.35$

(d)	A and B are mutually exclusive, and therefore from Axiom III

$$P(A \cup B) = P(A) + P(B) \approx 0.30 + 0.15 = 0.45$$

(e)	From Property 3

$$P(B') = 1 - P(B)$$

Therefore

$$P(B') \approx 1 - 0.15 = 0.85$$

(f)   *A, B*, and *C* are all mutually exclusive, and therefore from Property 4

$$P(A \cup B \cup C) = P(A) + P(B) + P(C)$$

and

$$P(A \cup B \cup C) \approx 0.30 + 0.15 + 0.35 = 0.80$$

(g)   As *B* and *C* are mutually exclusive, $B \cap C = \varnothing$, and therefore from Property 1:

$$P(B \cap C) = P(\varnothing) = 0$$

(h)   *A* and *B'* are not mutually exclusive, so we cannot use Axiom III. Instead we will use Property 6:

$$A \cup B' = \{\text{all simple events in } S \text{ that are outside } B\}$$

Thus

$$P(A \cup B') \approx 0.05 + 0.05 + 0.20 + 0.10 + 0.25 + 0.10 + 0.10 = 0.85$$

8.22   For the experiment of drawing a single card from a well-shuffled, standard deck of playing cards, in which $S = \{\text{the 52 cards in a standard deck}\}$, use the axioms and properties of Section 8.6 to determine the following probabilities: (a) $P(10 \text{ of clubs})$,   (b) $P(\text{king})$,   (c) $P(\text{card that is not a king})$,   (d) $P(\text{heart or king})$,   (e) $P(4 \text{ or } 10 \text{ or king})$.

Solution

(a)   Using Property 7,

$$P(e_i) = \frac{1}{N} = \frac{1}{52}$$

Therefore, as there is only one 10 of clubs, $N_A = 1$ and

$$P(10 \text{ of clubs}) = N_A \left(\frac{1}{N}\right) = 1\left(\frac{1}{52}\right) = \frac{1}{52} = 0.019$$

(b)   Now for Property 7, $N_A = 4$ and

$$P(\text{king}) = N_A \left(\frac{1}{N}\right) = 4\left(\frac{1}{52}\right) = \frac{4}{52} = 0.077$$

(c)   Using Property 3,

$$P(\text{card that is not a king}) = 1 - P(\text{king}) = 1 - 0.077 = 0.923$$

(d)   {Heart or king} = {13 heart cards and the kings of clubs, diamonds, and spades}. Using Property 7,

$$P(\text{heart or king}) = N_A \left(\frac{1}{N}\right) = 16\left(\frac{1}{52}\right) = \frac{16}{52} = 0.308$$

(e)   Since 4, 10, and king are mutually exclusive, we can use Property 4:

$$P(4 \text{ or } 10 \text{ or king}) = P(4) + P(10) + P(\text{king})$$

As

$$P(4) = P(10) = P(\text{king}) = N_A \left(\frac{1}{N}\right) = 4\left(\frac{1}{52}\right) = \frac{4}{52} = 0.077$$

then

$$P(4 \text{ or } 10 \text{ or king}) = 0.077 + 0.077 + 0.077 = 0.231$$

8.23 In Problem 8.11 we determined that 16 possible events (outcomes and groups of outcomes) can result from the experiment of flipping a coin twice. If this experiment is actually done and the result is *HH*, how many events occurred and what was the probability of each?

Solution

Prior to doing the experiment, $(2^n = 2^4 = 16)$ events were possible. Actually doing the experiment resulted in 2^{n-1} events occurring, as there are this many subsets of S that contain the outcome of the experiment. Thus, here, there are $(2^{4-1} = 2^3 = 8)$ events that contain *HH*: {*HH*}, {*HH, HT*}, {*HH, TH*}, {*HH, TT*}, {*HH, HT, TH*}, {*HH, HT, TT*}, {*HH, TH, TT*} and {*HH, HT, TH, TT*}.

We can use Property 7 from Section 8.6 to determine all these probabilities. Thus, as $S = \{HH, HT, TH, TT\}$, $N = 4$, and therefore

$$P(HH) = N_A\left(\frac{1}{N}\right) = 1\left(\frac{1}{4}\right) = \frac{1}{4} = 0.25$$

$$P(HH, HT) = P(HH, TH) = P(HH, TT) = N_A\left(\frac{1}{N}\right) = 2\left(\frac{1}{4}\right) = \frac{2}{4} = \frac{1}{2} = 0.50$$

$$P(HH, HT, TH) = P(HH, HT, TT) = P(HH, TH, TT) = N_A\left(\frac{1}{N}\right) = 3\left(\frac{1}{4}\right) = \frac{3}{4} = 0.75$$

$$P(HH, HT, TH, TT) = N_A\left(\frac{1}{N}\right) = 4\left(\frac{1}{4}\right) = 1$$

SUBJECTIVE INTERPRETATION OF PROBABILITY AND THE CONCEPT OF ODDS

8.24 For the following probabilities, first indicate whether they are classical, relative frequency, or subjective, and then, using the appropriate formula from Section 8.8, convert the probabilities into odds: (*a*) you estimate that the probability you will get the job you have applied for is 0.10, (*b*) the probability of not getting a 3 in one roll of a die is 5/6, (*c*) based on the number of boys and girls born within a maternity ward over the past year, the probability that the next baby born in the hospital will be a boy is approximately 0.51.

Solution

(*a*) Subjective probability:

$$A = \{\text{get the job}\}, \qquad A' = \{\text{do not get the job}\}, \qquad P(A) = 0.10, \qquad P(A') = 0.90$$

Using equation (8.4),

$$\frac{P(A')}{P(A)} = \frac{0.90}{0.10} = \frac{9}{1}$$

And thus, because $P(A) < P(A')$, it is stated conventionally that the odds against your getting the job are 9 to 1.

(*b*) Classical probability:

$$A = \{3\}, \qquad A' = \{1, 2, 4, 5, 6\}, \qquad P(A) = 1/6, \quad P(A') = 5/6$$

Using equation (8.4),

$$\frac{P(A')}{P(A)} = \frac{5/6}{1/6} = \frac{5}{1}$$

And thus the odds against getting a 3 are 5 to 1.

(c)   Relative frequency probability:

$$A = \{boy\}, \quad A' = \{girl\}, \quad P(A) \approx 0.51, \quad P(A') \approx 0.49$$

Using equation (8.3),

$$\frac{P(A)}{P(A')} \approx \frac{0.51}{0.49} = \frac{51}{49}$$

And thus, the odds in favor of the next baby being a boy are approximately 51 to 49.

8.25   For the following odds, first use the appropriate formula from Section 8.9 to convert the odds to probabilities, and then state whether the resulting probabilities are classical, relative frequency, or subjective: (a) odds against your six-number ticket matching the selected six numbers and winning the state lottery are 5.2 million to 1;   (b) your estimate of the odds that the last prospective buyer will actually buy your house is 3 to 2;   (c) from past data it is estimated that the odds against a high school athlete becoming a professional athlete are 500,000 to 1.

Solution

(a) $A = \{$ticket wins$\}$, $A' = \{$ticket does not win$\}$, $c = 1$, $d = 5{,}200{,}000$. Using equation (8.6),

$$P(A') = \frac{d}{c+d} = \frac{5{,}200{,}000}{1 + 5{,}200{,}000} = 0.99999981, \text{ or essentially } 100\%$$

Classical probability

(b) $A = \{$will buy$\}$, $A' = \{$will not buy$\}$, $c = 3$, $d = 2$. Using equation (8.5),

$$P(A) = \frac{c}{c+d} = \frac{3}{3+2} = \frac{3}{5} = 0.60$$

Subjective probability

(c) $A = \{$becoming a professional$\}$, $A' = \{$not becoming a professional$\}$, $c \approx 1$, $d \approx 500{,}000$. Using equation (8.6),

$$P(A') = \frac{d}{d+c} \approx \frac{500{,}000}{1 + 500{,}000} = 0.9999980 \text{ or again essentially } 100\%$$

Relative frequency probability

8.26   A real-estate developer estimates for a commercial building that: the odds against its value increasing over two years are 3 to 2; the odds against its value staying the same over two years are 7 to 3; and the odds against its value decreasing over two years are 3 to 3. Do you see anything wrong with these odds?

Solution

Converting the odds to probabilities, we use equation (8.5) to find these probabilities:

$A_1 = \{$value increases$\}$, $A_1' = \{$value does not increase$\}$, $c = 2$, $d = 3$

$$P(A_1) = \frac{c}{c+d} = \frac{2}{2+3} = \frac{2}{5} = 0.40$$

$A_2 = \{$value remains the same$\}$, $A_2' = \{$value does not remain the same$\}$, $c = 3$, $d = 7$

$$P(A_2) = \frac{c}{c+d} = \frac{3}{3+7} = \frac{3}{10} = 0.30$$

$A_3 = \{$value decreases$\}$, $A_3' = \{$value does not decrease$\}$, $c = 2$, $d = 3$

$$P(A_3) = \frac{c}{c+d} = \frac{2}{2+3} = \frac{2}{5} = 0.40$$

A_1, A_2, and A_3 are mutually exclusive. Therefore, from Property 4 in Section 8.6,

$$P(A_1 \cup A_2 \cup A_3) = P(A_1) + P(A_2) + P(A_3)$$
$$= 0.40 + 0.30 + 0.40 = 1.10$$

But this result violates Property 2, $0 \leq P(A) \leq 1$, from Section 8.6, and subjective probabilities like these must be consistent with all the axioms and properties of probability. Therefore, the developer should redo these probabilities until they are consistent.

8.27 So far, a student has taken 16 college courses and has passed 12 of them. She makes a bet with her mother about passing the statistics course that she plans to take next semester. The student bets her $25 against her mother's $10 that she will pass the course. Is this a fair bet?

Solution

$A = \{$student wins$\}$, $A' = \{$mother wins$\}$. With the relative frequency interpretation described by equation (8.2),

$$P(A) \approx \frac{n_a}{n}$$

$$P(A) \approx \frac{12}{16} = \frac{3}{4} \qquad P(A') \approx \frac{4}{16} = \frac{1}{4}$$

Therefore, using equation (8.3), the odds in favor of A are

$$\frac{P(A)}{P(A')} \approx \frac{3/4}{1/4} = \frac{3}{1}, \text{ or approximately 3 to 1}$$

For this bet to be a fair bet (see Example 8.8), it should be the student's $30 against her mother's $10. It is not a fair bet.

8.28 Two basketball teams play each other 10 times a year. Over the past eight years, team A has won 48 times and team B has won 32 times. A fan of team A offers to bet a fan of team B his $30 against her $20 that team A will win the next game. Is this a fair bet?

Solution

$A = \{$team A wins$\}$, $A' = \{$team B wins$\}$. With the relative frequency interpretation described by equation (8.2),

$$P(A) \approx \frac{n_a}{n}$$

$$P(A) \approx \frac{48}{80} = \frac{3}{5} \qquad P(A') \approx \frac{32}{80} = \frac{2}{5}$$

The odds in favor of A are

$$\frac{P(A)}{P(A')} \approx \frac{3/5}{2/5} = \frac{3}{2}, \text{ or approximately 3 to 2}$$

This is a fair bet, as $30 to $20 is the same as 3 to 2.

Supplementary Problems

THE CLASSICAL INTERPRETATION OF PROBABILITY

8.29 In a game of chance, a six-sided die is rolled once. Determine the following probabilities: (a) $P(3$ or $5)$, (b) P (odd number).

Ans. (a) 0.33, (b) 0.50

8.30   In a game of chance, a single card is drawn from a standard deck of cards. Determine the following probabilities: (*a*) *P*(queen),   (*b*) *P*(diamond or heart),   (*c*) *P*(jack, queen, or king).

   Ans.   (*a*) 0.08,   (*b*) 0.50,   (*c*) 0.23

8.31   Assume in this problem that a birthday is equally likely to occur on any one of the 365 days of the year. You have just found a new friend. What is the probability that she was born on: (*a*) the same day of the year as you were born,   (*b*) the same month of the year that you were born?

   Ans.   (*a*) 0.003,   (*b*) 0.08

8.32   Twenty names, including your name and your date's name, are written on small pieces of paper and placed in a hat. The pieces of paper are thoroughly mixed and one is drawn out. What is the probability that the name drawn from the hat is: (*a*) your name,   (*b*) either your name or your date's name?

   Ans.   (*a*) 0.05,   (*b*) 0.10

8.33   Fifty people, including five grandfathers and two great grandmothers, attend a graduation party. Each person's name is written on a small piece of paper and placed in a box, where they are thoroughly mixed. One name is drawn, blindly, from the box. What is the probability that the name is of: (*a*) a grandfather,   (*b*) a great grandmother?

   Ans.   (*a*) 0.10,   (*b*) 0.04

THE RELATIVE FREQUENCY INTERPRETATION OF PROBABILITY

8.34   At age 64, you learn that 40 of your 620 college classmates have died. What is the approximate probability that a senior graduating this year from your college will live to age 64?

   Ans.   0.94

8.35   An insurance company wants to know the probability that a car will develop a cracked windshield that will need replacement. It collects information on 20,000 cars, from July 1 of one year to July 1 of the next year, and finds that 600 of the cars developed cracked windshields. What is the approximate probability that a car will develop a cracked windshield within a one year time period?

   Ans.   0.03

8.36   A university wants to know how many students it can accept without over-enrolling the following fall semester. It examines its records for the past five years and finds that during that time it accepted 30,000 students but only 12,000 actually registered the following fall semester. What is the approximate probability that a student accepted into the university will register for classes the following fall semester?

   Ans.   0.40

8.37   A vegetable farmer normally plants his tomatoes on June 1, after which there has never been a frost in his fields. This year, he plans to plant his tomatoes on May 15 and he wants to know: What is the chance that there will be a frost between May 15 and June 1? He examines weather data, from a nearby weather station, collected over the past 30 years and finds that in 6 of the years there was a frost between May 15 and June 1. What is the approximate probability that a frost will occur this year, between May 15 and June 1?

   Ans.   0.20

8.38   A manufacturing company that makes floppy disks for computers wants to know the probability that one of its disks will be defective. It examines 2,000 disks, and finds 100 are defective. What is the approximate probability that a given disk will be defective?

   Ans.   0.05

SETS, SUBSETS, SAMPLE SPACES, AND EVENTS

8.39   Sixty marbles are placed in a jar and thoroughly mixed. Twenty of the marbles are red, 30 are blue, and 10 are green. The experiment is to draw, without looking, a single marble. What is the sample space of the experiment?

   Ans. $S = \{$red, blue, green$\}$

8.40   A yogurt company promotes its product by having a lottery, in which numbers hidden under the container tops indicate how much a consumer will win. Some container tops have no numbers, some have $1,000, some have $2,000, and others have $5,000. If the experiment is picking a yogurt container in a supermarket, what is the sample space?

   Ans. $S = \{$no number, $1,000, $2,000, $5,000$\}$

8.41   In the coin-flipping experiment, $S = \{$head, tail$\}$. How many possible events can result from a single flip of the coin?

   Ans. $2^n = 2^2 = 4$

8.42   In an experiment, a die is rolled twice and the outcome is the sum of the two rolls.   (a) Describe the subset in which event $A = 12$. Is the event simple or compound?   (b) Describe the subset in which event $A = 8$. Is the event simple or compound?

   Ans.   (a) $A = \{6, 6\}$, simple event,   (b) $A = \{(1, 7), (2, 6), (3, 5), (4, 4)\}$, compound event

8.43   A university has five categories of students: freshmen, sophomores, juniors, seniors, and graduate students. The experiment is to select three students at random, with the outcome of the experiment being the category of student. What is the sample space and how many events can result from each experiment?

   Ans. $S = \{$freshman, sophomore, junior, senior, graduate$\}$; number of possible events $= 2^5 = 32$

8.44   A winery has produced the same brand of wine in four years: 1986, 1989, 1991, and 1993, and wants to know whether a consumer can distinguish among the years. To find out, it has a wine-tasting party and arranges a table in which the four different wines are offered within 200 identical glasses (50 glasses of each kind of wine) and the glasses randomly placed on the table. The experiment is to blindfold a guest and to have him/her take four glasses and then taste the wine in each glass. What is the sample space and how many events can result from each experiment?

   Ans. $S = \{$1986, 1989, 1991, 1993$\}$; number of possible events $= 2^4 = 16$

VENN DIAGRAMS

8.45   A Venn diagram can show a sample space, an event in the sample space, the complement of an event, the union of two events, and the intersection of two events. Define each of these aspects of a set.

   Ans.   A sample space is the set of all possible outcomes of an experiment. An event is a particular outcome of an experiment. The complement of an event is the subset of the sample space that contains all elements of the set that are not part of the event. The union of event A and event B is the subset of the set whose outcomes belong to A or to B or to both A and B. The intersection of event A and event B is the subset of the set whose outcomes belong both to event A and to event B.

8.46   A university offers 30 courses in history. The set is a listing of the history courses, from 1 to 30. The experiment is to choose two students who graduated with a major in history and find out which history classes they took. Event A is the series of classes taken by student A and event B is the series taken by student B. Write out, in set theory notation, the following outcomes of the experiment: (a) the list of classes taken by student A but not student B,   (b) the list of classes taken by either student A or student B or both students,   (c) the list of classes taken by neither student A nor student B,   (d) the list of classes taken by both student A and student B.

   Ans.   (a) $A \cup B'$,   (b) $A \cup B$,   (c) $A' \cup B'$, (d) $A \cap B$

8.47 Two friends read as many new novels each year as they can, and both of them buy their books from the same book store. The set is a list of all the new novels published in the past year and offered for sale in their book store. Event A is the list of new novels read by friend A, event B is the list of new novels read by friend B. Write down, in set theory notation, the following outcomes of the experiment: (*a*) the new novels read by both friends, (*b*) the new novels read by friend A but not friend B, (*c*) the new novels read by neither friend.

Ans. (*a*) $A \cap B$, (*b*) $A \cup B'$, (*c*) $A' \cup B'$

SET THEORY INTERPRETATION OF PROBABILITY

8.48 For the experiment of drawing a single card from a well-shuffled, standard deck of playing cards, in which $S = \{$the 52 cards in a standard deck$\}$, use the axioms and properties of Section 8.6 to determine the probability of drawing a red card.

Ans. 0.50

8.49 For the card-drawing experiment in Problem 8.48, determine the probability of drawing either a heart or a diamond.

Ans. $0.25 + 0.25 = 0.50$

8.50 For the card-drawing experiment in Problem 8.48, determine the probability of drawing a heart and a diamond.

Ans. $P(\varnothing) = 0$

8.51 For the card-drawing experiment in Problem 8.48, determine the probability of drawing a heart and a king.

Ans. 0.019

SUBJECTIVE INTERPRETATION OF PROBABILITY AND THE CONCEPT OF ODDS

8.52 An insurance company estimates that the probability of a fire in your house next year is aproximately 1/125. Is this a classical, relative frequency, or subjective probability? Use the appropriate formula from Section 8.8 to convert the probability into odds.

Ans. Relative frequency probability; the odds against a fire in your house next year are approximately 124 to 1.

8.53 A coin is flipped twice and the probability of getting at least one head is 0.75. Is this a classical, relative frequency, or subjective probability? Use the appropriate formula from Section 8.8 to convert the probability into odds.

Ans. Classical probability; the odds of getting at least one head are 3 to 1.

8.54 You guess that there is a 40% chance that the piece of land you are interested in buying will increase in value by at least 60% in the next five years. Is this a classical, relative frequency, or subjective probability? Use the appropriate formula from Section 8.8 to convert the probability into odds.

Ans. Subjective probability; the odds against the land value increasing by at least 60% in the next five years are 3 to 2.

8.55 A single card is drawn from a standard deck of playing cards, and the odds of selecting a card numbered from 2 to 10 are 9 to 4. Use the appropriate formula from Section 8.9 to convert the odds to a probability, and then state whether the resulting probability is classical, relative frequency, or subjective.

Ans. 0.69; classical probability

8.56   It has been estimated that the odds against hitting oil on any given drill for oil are 100 to 1. Use the appropriate formula from Section 8.9 to convert the odds to a probability, and then state whether the resulting probability is classical, relative frequency, or subjective.

    Ans.   0.99; relative frequency probability

8.57   You estimate that the odds of the stock market being higher one year from today than it is now are 1 to 1. Use the appropriate formula from Section 8.9 to convert the odds to a probability, and then state whether the resulting probability is classical, relative frequency, or subjective.

    Ans.   0.50; subjective probability

Chapter 9

Probability: Calculating Rules and Counting Rules

9.1 CALCULATING PROBABILITIES FOR COMBINATIONS OF EVENTS

Chapter 8 was a brief overview of the four interpretations of probability: classical, relative frequency, set theory, and subjective. In that overview, we also began an examination of how probabilities are calculated for combinations of events: events that are subsets of the same sample space. Thus, in Section 8.6 we presented formulas for calculating probabilities for unions of events when the events are mutually exclusive. For two such events A and B, Axiom III, also known as the special addition rule, gives the formula

$$P(A \cup B) = P(A) + P(B)$$

and for more than two mutually exclusive events $A_1, A_2, \ldots, A_k$, Property 4 gives the formula

$$P(A_1 \cup A_2 \cup \cdots \cup A_k) = P(A_1) + P(A_2) + \cdots + P(A_k)$$

For the mutually exclusive events A and its complement A', Property 3 gives the formula

$$P(A) + P(A') = 1$$

We also examined another way that events A and B can be combined, intersections ($A \cap B$), and did the most elementary probability calculation for when A and B are mutually exclusive [see Problem 8.20(h)]:

$$P(A \cap B) = P(\varnothing) = 0$$

In this chapter we go on to show how probabilities are calculated for the unions and intersections of events that are not mutually exclusive. To do this we must first discuss new types of relationships between events, whether they are *dependent* or *independent*. We will develop mathematical definitions for these relationships (see Section 9.4), but we can now say that events are independent when the occurrence of one does not affect the probability of the occurrence of the other; and they are dependent when they are not independent. To develop these mathematical definitions and to do the rest of the work of this chapter, we must first deal with *conditional probabilities*.

9.2 CONDITIONAL PROBABILITIES

Given two events A and B, if we want to determine the probability of the intersection of the two events, $P(A \cap B)$, we answer this question: What is the probability that events A and B will both occur? If, on the other hand, we want to determine a *conditional probability* for these events, we answer a related but different question: What is the probability of A occurring given that B is known to have occurred? Or the reverse question: What is the probability of B given that A is known to have occurred.

To understand how conditional probabilities differ from intersection probabilities, we will calculate examples of both types for the experiment of a single roll of a six-sided die where $S = \{1, 2, 3, 4, 5, 6\}$, event $A = \{$number $\geq 3\}$, and event $B = \{$even number$\}$. The set and the events are diagrammed in Fig. 9-1.

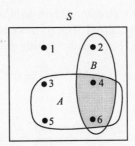

Fig. 9-5

We know from Property 7 in Section 8.6 that for any sample space S containing N equally likely simple events e_i, $P(e_i) = \dfrac{1}{N}$ and $P(A) = N_A\left(\dfrac{1}{N}\right)$. Thus, for this sample space

$$P(e_i) = \frac{1}{N} = \frac{1}{6}$$

$$P(A) = N_A\left(\frac{1}{N}\right) = 4\left(\frac{1}{6}\right) = \frac{4}{6} = 0.67$$

$$P(B) = N_B\left(\frac{1}{N}\right) = 3\left(\frac{1}{6}\right) = \frac{3}{6} = 0.50$$

We also know that the intersection of A and B, $A \cap B$, is the subset of S that contains all the elements of S that are in both A and B [see Example 8.5(b)], which we will denote by $N_{A \cap B}$. Therefore

$$P(A \cap B) = P(A \text{ and } B) = N_{A \cap B}\left(\frac{1}{N}\right) = 2\left(\frac{1}{6}\right) = \frac{2}{6} = \frac{1}{3} = 0.33$$

All three probabilities [$P(A)$, $P(B)$, and $P(A \cap B)$] were calculated for the entire sample space S; they represent the proportion of the sample points in S that are in the event of interest.

Now consider the same experiment as a conditional probability—the probability of event A occurring given the condition that event B has occurred. In this particular die-rolling experiment, the conditional probability is the probability that the number on the upward face of the die will be greater than or equal to three (event A) given that we know an even number (event B) has occurred. The conditional probability is denoted by $P(A|B)$, where the vertical bar in this denotation is read "given" (or "such that", see Section 8.3) and the whole symbol is read "the probability of A given B."

The conditional probability $P(A|B)$ is the ratio of the number of sample points in $A \cap B$ to the number of sample points in B. Unlike our calculation of the probability that both A and B occur, which considered the entire sample space S, the calculation of A given B considers only a portion of the sample space—the sample points in event B. Stated symbolically,

$$P(A|B) = \frac{N_{A \cap B}}{N_B} \tag{9.1}$$

which can also be written

$$P(A|B) = \frac{\dfrac{N_{A \cap B}}{N}}{\dfrac{N_B}{N}}$$

or

$$P(A|B) = \frac{P(A \cap B)}{P(B)}, \quad \text{provided that } P(B) \neq 0 \tag{9.2}$$

This last equation is the *general formula for conditional probabilities*. It is used to calculate the likelihood of A occurring given that we know B has occurred. The formula is valid only when $P(B) \neq 0$, because it is not permitted to divide a number by zero. For this die-rolling experiment

$$P(A|B) = \frac{N_{A \cap B}}{N_B} = \frac{2}{3} = 0.67$$

and also

$$P(A|B) = \frac{P(A \cap B)}{P(B)} = \frac{1/3}{1/2} = \frac{2}{3} = 0.67$$

There is a 0.67 probability that the upward face of the die will be a 3, 4, 5, or 6 given that it is an even number. In other words, two out of the three even numbers (2, 4, and 6) are greater than or equal to three. The conditional probability of A given B (0.67) is higher than the probability of both A and B (0.33), because the conditional probability involves only a portion of the sample space.

The probability of A given that B has occurred, or $P(A|B)$, is a conditional probability. The probability that B has occurred, or $P(B)$, is calculated independently of its occurrence with A and is an *unconditional probability*, which is also called a *simple probability*.

If we want $P(B|A)$ instead of $P(A|B)$, this probability would be written

$$P(B|A) = \frac{N_{A \cap B}}{N_A}$$

$$= \frac{P(A \cap B)}{P(A)}, \quad \text{provided } P(A) \neq 0$$

$$= \frac{1/3}{2/3} = \frac{1}{2} = 0.50$$

There is a 0.50 probability that an even number will occur given that a number greater than or equal to three has occurred. Two of the four numbers ≥ 3 are even numbers. Again, note that the formula is valid only when $P(A) \neq 0$. Here, $P(B|A)$ is a conditional probability and $P(A)$ is an unconditional probability.

EXAMPLE 9.1   A medical team has developed a possible vaccine for the common cold. They test it on a group of 160 volunteers divided into an 80-person experimental group and an 80-person control group. The members of the experimental group are vaccinated, while the members of the control group are not. After 12 months all 160 people are asked if they got a cold during the past year. The results are summarized in Table 9.1 (e.g., 48 vaccinated people got a cold). The probability experiment is to randomly select one of the 160 people. If for this experiment $S = \{$the 160 people$\}$, $A = \{$got a cold$\}$, and $B = \{$vaccinated$\}$, then find the probability $P(A|B)$.

Solution

Using equation (9.2),

$$P(A|B) = \frac{P(A \cap B)}{P(B)}$$

Table 9.1

Got vaccinated	Got cold		Total
	Yes	No	
Yes	48	32	80
No	52	28	80
Total	100	60	160

where

$$N_B = 80, \qquad N_{A \cap B} = 48, \qquad \frac{1}{N} = \frac{1}{160}$$

We find that

$$P(A \cap B) = N_{A \cap B}\left(\frac{1}{N}\right) = 48\left(\frac{1}{160}\right) = \frac{48}{160} = \frac{3}{10}$$

$$P(B) = N_B\left(\frac{1}{N}\right) = 80\left(\frac{1}{160}\right) = \frac{80}{160} = \frac{1}{2}$$

and

$$P(A|B) = \frac{P(A \cap B)}{P(B)} = \frac{3/10}{1/2} = \frac{3}{5} = 0.60$$

This states that the probability of selecting a person who did get a cold from the vaccinated group is 0.60.

EXAMPLE 9.2   For the experiment in Example 9.1: $A' = \{$did not get a cold$\}$ and $B' = \{$not vaccinated$\}$. Use the information in Example 9.1 to determine $P(A|B) + P(A'|B)$.

Solution
We know that

$$N = 60, \qquad P(B) = 1/2, \qquad P(A|B) = 0.60$$

and from Table 9.1, $N_{A' \cap B} = 32$. Therefore

$$P(A' \cap B) = N_{A' \cap B}\left(\frac{1}{N}\right) = 32\left(\frac{1}{160}\right) = \frac{1}{5}$$

Then, using equation (9.2),

$$P(A'|B) = \frac{P(A' \cap B)}{P(B)} = \frac{1/5}{1/2} = \frac{0.2}{0.5} = 0.40$$

Therefore

$$P(A|B) + P(A'|B) = 0.60 + 0.40 = 1.00$$

This states that for a vaccinated person there is a 100% probability that either they had a cold or they did not.
   Note: This example illustrates what is true in general when an event and its complement are conditioned on the same event:

$$P(Y|X) + P(Y'|X) = 1$$

There is a 100% probability that either the event or its complement will occur regardless of the conditional event.

9.3   THE GENERAL MULTIPLICATION RULE

The multiplication rules of probability deal with calculating the probability that two events both occur. The probability that both event A and event B occur is the probability of the intersection of the two events, $P(A \cap B)$.

The *general multiplication rule* applies to the intersection of dependent events (see Section 9.1) and so the concept of conditional probability applies. The expression $P(A|B)$, which indicates the probability of event A occurring given that B has occurred, can be viewed as the number of times that two events, A and B, occur together relative to the total number of times that B occurs. In the notation of probability,

$$P(A|B) = \frac{P(A \cap B)}{P(B)}$$

or, for the probability that B occurs given that A has occurred

$$P(B|A) = \frac{P(A \cap B)}{P(A)}$$

Rearranging to solve for $P(A \cap B)$

$$P(A \cap B) = P(B)P(A|B) \tag{9.3}$$

and

$$P(A \cap B) = P(A)P(B|A) \tag{9.4}$$

These two mathematical relationships form the general multiplication rule. In words, the rule states that the probability of the intersection of two events is calculated by multiplying the unconditional probability of one of the events times the conditional probability of the other event given that the first event has occurred.

EXAMPLE 9.3   If in Example 9.1 two people were randomly selected, one after the other, from the 160, then what is the probability that both have been vaccinated?

Solution

The probability of the second person being vaccinated depends on whether the first person is vaccinated, as this person is removed from the sample space before the second person is selected. Thus, the probability of the second event depends on the outcome of the first event. If we let B_1 be the event that the first person was vaccinated and B_2 be the event that the second person was vaccinated, then using the general multiplication rule,

$$P(B_1 \cap B_2) = P(B_1)P(B_2|B_1)$$

From Example 9.1,

$$P(B_1) = P(B) = \frac{80}{160} = \frac{1}{2}$$

Having removed one person from the vaccinated group (i.e., sampling without replacement; see Section 3.16),

$$P(B_2|B_1) = \frac{79}{159}$$

and

$$P(B_1 \cap B_2) = P(B_1)P(B_2|B_1)$$

$$= \frac{1}{2} \times \frac{79}{159} = \frac{79}{318} = 0.25$$

The general multiplication rule can be expanded to include more than two events. This expansion is known as the *generalization of the general multiplication rule*.

EXAMPLE 9.4   In three trials of the card-selection experiment, if the cards are not replaced between selections, then what is the probability that they will be, in any order, a jack of spades, a queen of hearts, and a king of diamonds?

Solution

For determining the intersection of k events $A_1, A_2, A_3, \ldots, A_k$ the general multiplication rule generalizes to the following:

$$P(A_1 \cap A_2 \cap A_3 \cap \cdots \cap A_k) = P(A_1)P(A_2|A_1)P(A_3|A_2 \cap A_1) \cdots P(A_k|A_{k-1} \cap \cdots \cap A_1) \tag{9.5}$$

If we let C_1 be the first of these three cards, C_2 be the second, and C_3 be the third, then

$$P(C_1) = \frac{3}{52}$$

$$P(C_2|C_1) = \frac{2}{51}$$

$$P(C_3|C_2 \cap C_1) = \frac{1}{50}$$

and

$$P(C_1 \cap C_2 \cap C_3) = P(C_1)P(C_2|C_1)P(C_3|C_2 \cap C_1)$$

$$= \frac{3}{52} \times \frac{2}{51} \times \frac{1}{50} = \frac{6}{132,600} = 0.000045$$

9.4 INDEPENDENT AND DEPENDENT EVENTS

We are now ready to mathematically define independent and dependent events. Recall that in Section 9.1 we indicated that two events are said to be *independent* if the occurrence of one does not affect the *probability* of whether or not the second will occur. This statement can be written as the following conditional probabilities for independent events A and B:

$$P(A|B) = P(A) \tag{9.6}$$

and

$$P(B|A) = P(B) \tag{9.7}$$

In essence, the conditional probability of A given B is equal to the unconditional probability of A, and if this is true it is also true that the conditional probability of B given A is equal to the unconditional probability of B. If A is independent of B, then B is automatically independent of A and the events are said to be *independent events*. If A and B are not independent, they are said to be *dependent events*.

9.5 THE SPECIAL MULTIPLICATION RULE

The *special multiplication rule* deals with the probability that two independent events will both occur. From Section 9.4 we know that whenever two events A and B are independent, we can substitute $P(A)$ for $P(A|B)$ in equation (9.3)

$$P(A \cap B) = P(B)P(A|B) = P(B)P(A)$$

and $P(B)$ for $P(B|A)$ in equation (9.4)

$$P(A \cap B) = P(A)P(B|A) = P(A)P(B)$$

These equivalent formulas are called the *special multiplication rule*, which is typically stated as

$$P(A \cap B) = P(A)P(B) \tag{9.8}$$

EXAMPLE 9.5 Use the special multiplication rule to determine the following probabilities: (a) rolling a 4 twice in a row with the die-rolling experiment,   (b) selecting a queen both times in two repetitions of the card-selection experiment, if the card is replaced and the deck reshuffled between selections.

Solution

(a) $A = \{$rolling a 4 on the first roll$\}$, $P(A) = \dfrac{1}{6}$

$B = \{$rolling a 4 on the second roll$\}$, $P(B) = \dfrac{1}{6}$

$$P(A \cap B) = P(A)P(B) = \frac{1}{6} \times \frac{1}{6} = \frac{1}{36} = 0.028$$

(b) $A = \{$selecting a queen on the first trial$\}$, $P(A) = \dfrac{4}{52} = \dfrac{1}{13}$

$B = \{$selecting a queen on the second trial after replacement and reshuffling$\}$, $P(B) = \dfrac{4}{52} = \dfrac{1}{13}$

$$P(A \cap B) = P(A)P(B) = \dfrac{1}{13} \times \dfrac{1}{13} = \dfrac{1}{169} = 0.0059$$

When determining the intersection of k *independent events* $A_1, A_2, A_3, \ldots, A_k$, the special multiplication rule [equation (9.8)] generalizes to

$$P(A_1 \cap A_2 \cap A_3 \cap \cdots \cap A_k) = P(A_1)P(A_2)P(A_3) \cdots P(A_k) \tag{9.9}$$

which is known as the *generalization of the special multiplication rule.*

9.6   THE GENERAL ADDITION RULE

In Section 8.6 in Axiom III, we gave the special addition rule for the union of mutually exclusive events A and B

$$P(A \cup B) = P(A) + P(B)$$

We are now going to develop a general rule for the union of events A and B that applies whether or not they are mutually exclusive.

From Property 7 in Section 8.6, we know that the special addition rule can be written as

$$P(A \cup B) = N_A\left(\dfrac{1}{N}\right) + N_B\left(\dfrac{1}{N}\right)$$

This equation applies only to the union of events that are mutually exclusive. If it were used for two events that are not mutually exclusive, then the sample points that lie within the intersection of the two events would be counted twice. For events that are not mutually exclusive—that have sample points in common—we modify the equation as follows

$$P(A \cup B) = N_A\left(\dfrac{1}{N}\right) + N_B\left(\dfrac{1}{N}\right) - N_{A \cap B}\left(\dfrac{1}{N}\right)$$

$$P(A \cup B) = N_{A \cup B}\left(\dfrac{1}{N}\right) = \dfrac{N_A + N_B - N_{A \cap B}}{N}$$

The numerator of this final fraction is the number of simple events in A plus the number in B minus the number in the intersection of A and B. It follows that

$$P(A \cup B) = \dfrac{N_A}{N} + \dfrac{N_B}{N} - \dfrac{N_{A \cap B}}{N}$$

and thus that

$$P(A \cup B) = P(A) + P(B) - P(A \cap B) \tag{9.10}$$

This formula is the *general addition rule.*

EXAMPLE 9.6   For the die-rolling experiment shown in Fig. 9-1, what is the probability that a toss of the die will result in event $A = \{3, 4, 5, 6\}$ or event $B = \{2, 4, 6\}$?

Solution

The two events have an intersection (4 and 6), and so are not mutually exclusive

$$P(A) = N_A\left(\frac{1}{N}\right) = \frac{4}{6}$$

$$P(B) = N_B\left(\frac{1}{N}\right) = \frac{3}{6}$$

$$P(A \cap B) = N_{A \cap B}\left(\frac{1}{N}\right) = \frac{2}{6}$$

Using the general addition rule

$$P(A \cup B) = P(A) + P(B) - P(A \cap B)$$

$$= \frac{4}{6} + \frac{3}{6} - \frac{2}{6} = \frac{4 + 3 - 2}{6} = \frac{5}{6} = 0.83$$

The general addition rule can be extended to the case of more than two events. This more general rule, which evaluates the probability of the union of two or more events, is known as the *generalization of the general addition rule.*

EXAMPLE 9.7   A single card is drawn from a well-shuffled deck of standard playing cards, in which $S = \{52$ cards$\}$, $A = \{$face card$\}$, $B = \{$black card$\}$, and $C = \{$club$\}$. Determine $P(A \cup B \cup C)$.

Solution

A, B, and C are not mutually exclusive and, for a union of the three events, the general addition rule generalizes to

$$P(A \cup B \cup C) = P(A) + P(B) + P(C) - P(A \cap B) - P(A \cap C) - P(B \cap C) + P(A \cap B \cap C) \qquad (9.11)$$

For this experiment:

$$A = \{\text{face card}\}, \qquad P(A) = \frac{12}{52}$$

$$B = \{\text{black card}\}, \qquad P(B) = \frac{26}{52}$$

$$C = \{\text{club}\}, \qquad P(C) = \frac{13}{52}$$

$$A \cap B = \{\text{black face card}\}, \qquad P(A \cap B) = \frac{6}{52}$$

$$A \cap C = \{\text{club face card}\}, \qquad P(A \cap C) = \frac{3}{52}$$

$$B \cap C = \{\text{black club card}\}, \qquad P(B \cap C) = \frac{13}{52}$$

$$A \cap B \cap C = \{\text{black club face card}\}, \qquad P(A \cap B \cap C) = \frac{3}{52}$$

Therefore

$$P(A \cup B \cup C) = P(A) + P(B) + P(C) - P(A \cap B) - P(A \cap C) - P(B \cap C) + P(A \cap B \cap C)$$

$$= \frac{12}{52} + \frac{26}{52} + \frac{13}{52} - \frac{6}{52} - \frac{3}{52} - \frac{13}{52} + \frac{3}{52}$$

$$= \frac{12 + 26 + 13 - 6 - 3 - 13 + 3}{52} = \frac{32}{52} = 0.62$$

9.7 DERIVING THE SPECIAL ADDITION RULE FROM THE GENERAL ADDITION RULE

The *special addition rule* for probabilities applies to events that are mutually exclusive, with no elements that belong to both A and B. Thus

$$N_{A \cap B} = 0$$

and

$$P(A \cap B) = 0$$

Applying the general addition rule

$$P(A \cup B) = P(A) + P(B) - P(A \cap B)$$

$$P(A \cup B) = P(A) + P(B) - 0$$

$$P(A \cup B) = P(A) + P(B) \tag{9.12}$$

Thus, the special addition rule is a special case of the general addition rule that is used when the events are mutually exclusive.

EXAMPLE 9.8 A single card is drawn from a well-shuffled deck of standard playing cards. What is the probability of drawing either a king or a 2, 3, or 4?

Solution

Let event $A = \{\text{king}\}$ and event $B = \{2, 3, 4\}$.

$$A \cap B = \{\text{king and } 2, 3, 4\} \qquad A \cup B = \{\text{king or } 2, 3, 4\}$$

$$P(A) = \frac{4}{52}$$

$$P(B) = \frac{3(4)}{52} = \frac{12}{52}$$

As event A and event B are mutually exclusive,

$$P(A \cap B) = 0$$

Therefore

$$P(A \cup B) = P(A) + P(B) - P(A \cap B)$$

$$= \frac{4}{52} + \frac{12}{52} - 0 = \frac{16}{52} = 0.31$$

9.8 CONTINGENCY TABLES, JOINT PROBABILITY TABLES, AND MARGINAL PROBABILITIES

In this section we deal with experiments in which all possible outcomes can be placed into the mutually exclusive and *exhaustive* (also called totally inclusive, see Section 2.4) categories of two variables. For example, the experiment might be a potential customer entering a car dealership, and the outcome could be classified according to gender (male or female) and purchaser (purchases car or does not purchase car). In Table 9.2 we show the outcomes, so classified, for 100 trials of the experiment: 100 potential customers entering the dealership. At a glance you can see, for example, that 70 of the potential customers are male (M), and of these 70 males, 40 purchased cars (P) and 30 did not (P'). Such a summary table is called a *contingency table* if, as here, it is possible that the two variables have an independent–dependent relationship (see Section 1.17), which is also described as one variable being *contingent* (dependent) on the other. Such tables will be used in Chapter 20 to study contingency relationships. In Table 9.2, for example, the relationship to be studied would be: Is purchase of a car contingent on gender?

Table 9.2

Gender	Purchaser Purchases (P)	Does not purchase (P')	Total
Male (M)	40	30	70
Female (F)	10	20	30
Total	50	50	100

From such a contingency table, it is a simple matter to convert these frequency values into probabilities for the experiment. This conversion has been done in Table 9.3. Each frequency value in Table 9.2 was converted to relative frequency by dividing by 100 (the total number of potential customers), which we then considered to be the probability of future events (see Section 8.2). While these probabilities are approximate, for this example we will consider them to be exact. Thus there is a 0.40 probability of the next potential customer being male and a purchaser, and a 0.30 probability of the next potential customer being female (purchaser and nonpurchaser). The probabilities in the cells of the table are probabilities of the intersection of two events. Thus $P(P \cap M) = 40/100 = 0.40$ and $P(P' \cap F) = 20/100 = 0.20$. Such probabilities of intersections are called *joint probabilities* because they are the probabilities of joint occurrences of two events. This is why such a probability table is called a *joint probability table*.

The probabilities along the bottom and right side of Table 9.3 are called *marginal probabilities* because of their marginal positions in a joint probability table. They are the probabilities of the events featured in the particular row or column, independent of the other variable. They are therefore unconditional, or simple, probabilities (see Section 9.2). Thus, the probability of the next potential customer being male is $P(M) = 0.70$ and the probability of the next potential customer being a car purchaser is $P(P) = 0.50$.

Table 9.3

Gender	Purchaser Purchases (P)	Does not purchase (P')	Marginal probability
Male (M)	0.40	0.30	0.70
Female (F)	0.10	0.20	0.30
Marginal probability	0.50	0.50	1.00

Note for this table that the simple probabilities are the sums of two joint probabilities. Thus, the probability that the next person who enters will be a purchaser is: $P(P) = P(P \cap M) + P(P \cap F)$. This result can be generalized to a rule called the *marginal probability formula*.

The marginal probability of an event B that can occur in k mutually exclusive and exhaustive ways A_i (for $i = 1, 2, \ldots, k$) is equal to the sum of the k joint probabilities

$$P(B) = P(A_1 \cap B) + P(A_2 \cap B) + \cdots + P(A_k \cap B)$$

or

$$P(B) = \sum_{i=1}^{k} P(A_i \cap B) \qquad (9.13)$$

which can be written [see equation (9.4)] as

$$P(B) = \sum_{i=1}^{k} P(A_i)P(B|A_i) \qquad (9.14)$$

From a joint probability table, it is also possible to determine conditional probabilities using equation (9.2). Thus, in our example

$$P(P|M) = \frac{P(P \cap M)}{P(M)} = \frac{0.40}{0.70} = 0.57$$

which indicates there is a 0.57 probability that the next potential customer will purchase a car given that he is male.

EXAMPLE 9.9 A computer company has three suppliers (A_1, A_2, A_3) of a component part for one of its personal computers. It gets 20% of these parts from A_1, 50% from A_2, and 30% from A_3. From past experience it is known that a percentage of each supplier's parts will be defective: 1% of A_1's, 0.5% of A_2's, and 0.9% of A_3's. The probability experiment is to select one of these parts at random and test to see if it is defective. Let A_1, A_2, A_3 represent the events of selecting a part from the given supplier, B represent the event of selecting a nondefective part, and B' the event of selecting a defective part. From this information develop a joint probability table that includes these joint probabilities: $P(A_1 \cap B)$, $P(A_2 \cap B)$, $P(A_3 \cap B)$, $P(A_1 \cap B')$, $P(A_2 \cap B')$, $P(A_3 \cap B')$; and these marginal probabilities: $P(A_1)$, $P(A_2)$, $P(A_3)$, $P(B)$, $P(B')$.

Solution

Again, we will treat relative frequency estimates as exact probabilities. Therefore

$$P(A_1) = 0.20, \qquad P(A_2) = 0.50, \qquad P(A_3) = 0.30$$

and

$$P(B'|A_1) = 0.01, \qquad P(B'|A_2) = 0.005, \qquad P(B'|A_3) = 0.009$$

From these conditional probabilities for defective parts, given the suppliers, we can calculate the conditional probabilities of nondefective parts, given the suppliers. Thus

$$P(B|A_1) = 1 - P(B'|A_1) = 1 - 0.010 = 0.990$$

$$P(B|A_2) = 1 - P(B'|A_2) = 1 - 0.005 = 0.995$$

$$P(B|A_3) = 1 - P(B'|A_3) = 1 - 0.009 = 0.991$$

We can now calculate the joint probabilities using equation (9.4). Thus

$$P(A_1 \cap B) = P(A_1)P(B|A_1) = (0.20)(0.990) = 0.1980$$

$$P(A_2 \cap B) = P(A_2)P(B|A_2) = (0.50)(0.995) = 0.4975$$

$$P(A_3 \cap B) = P(A_3)P(B|A_3) = (0.30)(0.991) = 0.2973$$

and

$$P(A_1 \cap B') = P(A_1)P(B'|A_1) = (0.20)(0.01) = 0.0020$$

$$P(A_2 \cap B') = P(A_2)P(B'|A_2) = (0.50)(0.005) = 0.0025$$

$$P(A_3 \cap B') = P(A_3)P(B'|A_3) = (0.30)(0.009) = 0.0027$$

Therefore, using equation (9.14),

$$P(B) = \sum_{i=1}^{k} P(A_i)P(B|A_i)$$

$$= P(A_1)P(B|A_1) + P(A_2)P(B|A_2) + P(A_3)P(B|A_3)$$

$$= 0.1980 + 0.4975 + 0.2973 = 0.9928$$

and

$$P(B') = \sum_{i=1}^{k} P(A_i)P(B'|A_i)$$

$$= P(A_1)P(B'|A_1) + P(A_2)P(B'|A_2) + P(A_3)P(B'|A_3)$$

$$= 0.0020 + 0.0025 + 0.0027 = 0.0072$$

We now have all of the required probabilities, and the completed joint probability table is shown in Table 9.4.

Table 9.4

Defects	Suppliers			Marginal probability
	(A_1)	(A_2)	(A_3)	
Nondefective (B)	0.1980	0.4975	0.2973	0.9928
Defective (B')	0.0020	0.0025	0.0027	0.0072
Marginal probability	0.20	0.50	0.30	1.00

9.9 BAYES' THEOREM

In Example 9.9, $P(B|A_1)$ represents the probability of a part being nondefective given that it comes from supplier A_1; and $P(B'|A_1)$ represents the probability of a part being defective given that it comes from A_1. In terms of cause-and-effect (see Section 1.19), we are asking: What is the probability of effect B (or B') given that the part comes from (was caused by) supplier A_1? Now, in this section, we will be determining these probabilities: $P(A_1|B)$ and $P(A_1|B')$, to answer the inverse of the causality question: Given that effect B (or B') has occurred, what is the probability that the part comes from (was caused by) supplier A_1? They are inverse questions, because while one asks forward in time from cause-to-effect, the other asks backward in time from effect-to-cause. To answer this new inverse question we will be deriving a formula called *Bayes' theorem*, which again assumes that event B (or B') can occur in k mutually exclusive and exhaustive ways A_i. Stated for events B and A_1, Bayes' theorem is

$$P(A_1|B) = \frac{P(A_1)P(B|A_1)}{\sum_{i=1}^{k} P(A_i)P(B|A_i)} \qquad (9.15)$$

As Bayes' theorem gives the result of this inverse of causality, it is also called *Bayes' theorem for the probability of causes*.

To derive Bayes' theorem for B and A_1 we begin with the general formula for conditional probabilities [equation (9.2)] for both A_1 given B and B given A_1.

$$P(A_1|B) = \frac{P(A_1 \cap B)}{P(B)}, \qquad \text{provided that } P(B) \neq 0$$

$$P(B|A_1) = \frac{P(A_1 \cap B)}{P(A_1)}, \qquad \text{provided that } P(A_1) \neq 0$$

Solving both equations for $P(A_1 \cap B)$,

$$P(A_1 \cap B) = P(B)P(A_1|B)$$

$$P(A_1 \cap B) = P(A_1)P(B|A_1)$$

Therefore

$$P(B)P(A_1|B) = P(A_1)P(B|A_1)$$

Dividing both sides by $P(B)$

$$P(A_1|B) = \frac{P(A_1)P(B|A_1)}{P(B)}$$

Finally, substituting the marginal probability formula [equation (9.14)] for $P(B)$ in the denominator we get Bayes' theorem

$$P(A_1|B) = \frac{P(A_1)P(B|A_1)}{\sum_{i=1}^{k} P(A_i)P(B|A_i)}$$

Clearly for these conditions the formula can be can be modified for B' or for any of the A_k events. Thus, for example, for A_2,

$$P(A_2|B) = \frac{P(A_2)P(B|A_2)}{\sum_{i=1}^{k} P(A_i)P(B|A_i)}$$

EXAMPLE 9.10 As a statistician working for the computer company in Example 9.9, you decide to use Bayes' theorem to determine the probability that a particular defective part comes from supplier A_1.

Solution

The question is: What is $P(A_1|B')$? Using Bayes' theorem [equation (9.15)] and the calculations from Example 9.9,

$$P(A_1|B') = \frac{P(A_1)P(B'|A_1)}{\sum\limits_{i=1}^{k} P(A_i)P(B'|A_i)}$$

$$P(A_1|B') = \frac{P(A_1)P(B'|A_1)}{P(A_1)P(B'|A_1) + P(A_2)P(B'|A_2) + P(A_3)P(B'|A_3)}$$

$$= \frac{0.0020}{0.0020 + 0.0025 + 0.0027} = \frac{0.0020}{0.0072} = 0.28$$

In the context of Bayes' theorem, the unconditional probability known before the experiment is called a *prior probability*. Thus in Example 9.9, we know in advance of selecting a part that the probability the selected part will be from supplier A_1 is $P(A_1) = 0.20$. After the experiment, if we have selected a defective part the probability of the selected part being from A_1 becomes $P(A_1|B') = 0.28$. This is called a *posterior probability;* a revised version of $P(A_1)$ after we learn the outcome of the experiment. Note the change in probability: in advance the probability of A_1 is 0.20 (i.e., supplier A_1 provides 20% of the parts), but after the experiment if we have selected a defective part the probability of the part being from A_1 becomes 0.28.

This revision of prior probabilities after the experiment has been done and new information is available is the primary use of Bayes' theorem. It is particularly important for subjective probabilities (see Section 8.7), where the prior probability of an event is assigned on the basis of personal degree of belief. Thus, a business executive can assign a subjective probability to the outcome of a financial decision based on available quantitative data, intuitions, judgments, beliefs, and can then use Bayes' theorem to revise this probability for future decisions as new evidence accumulates. In the field of business statistics the theorem is used as the basis for *Bayesian decision analysis*, which deals with the continual testing and revising of managerial probability assignments in financial decision situations.

9.10   TREE DIAGRAMS

A *tree diagram* is a tool for calculating the probability of a sequence of events. It is a visual representation of the multiplication rules, showing the formation of intersections and intersection probabilities (joint probabilities) of two or more events. In such a diagram, the probability of each event is shown as a line, called a *branch*, and the sequence of branches that form an intersection is called a *path*.

The tree diagram shown in Fig. 9-2(a) shows the intersection probabilities of dependent events (conditional probabilities). It could represent, for example, the experiment of drawing two cards, without replacement, from a deck of standard playing cards, with the following events $A_1 = \{$a king of hearts$\}$, $A_2 = \{$not a king of hearts$\}$, $B_1 = \{$a heart$\}$, and $B_2 = \{$not a heart$\}$. Consider the probability of drawing a king of hearts on the first draw and a card that is not a heart on the second draw. The unconditional probability of the first event $[P(A_1)]$ is shown above the first branch and the conditional probability of the second event $[P(B_1|A_1)]$ is shown above the second branch in the pathway. The outcome of both events occurring, which is the intersection $A_1 \cap B_1$, is listed along with the intersections of the other events in the column immediately to the right of the path. The formulas for the probabilities of the intersections [equation (9.4)] are listed in the second column to the right of the paths. Thus for events A_1 and B_1,

$$P(A_1 \cap B_1) = P(A_1)P(B_1|A_1)$$

The more typical version of a tree diagram, which we will use in subsequent problems, is shown in Fig. 9-2(b). It differs from the one shown in Fig. 9-2(a) in that it shows the probability values instead of the probability symbols above the branches and it gives the actual probability values for the intersections in the second column. The particular tree diagram shown in Fig. 9-2(b) represents two trials of the coin-

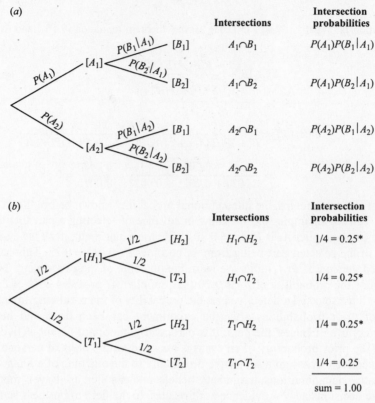

Fig. 9-2

tossing experiment. As the outcome of the first trial has no influence on the outcome of the second trial, the two events are independent and their intersection probabilities are calculated by the special multiplication rule [equation (9.8)]. Thus, the probability of getting a head on the first throw and a head on the second throw is

$$P(H_1 \cap H_2) = P(H_1)P(H_2)$$

$$= \left(\frac{1}{2}\right)\left(\frac{1}{2}\right) = \frac{1}{4}$$

As the listed intersections include all the mutually exclusive events in the sample space their probabilities sum to 1.00.

EXAMPLE 9.11   The experiment is to flip a coin twice. Use both a multiplication rule and a tree diagram to find the probability of getting a tail on both the first and second flips and to find the probability of not getting a tail on the first and second flips.

Solution

The probability of getting a tail on a single flip of a coin is $\frac{1}{2}$. The event of getting a tail on the first flip (T_1) and the event of getting a tail on the second flip (T_2) are independent events. Therefore, the special multiplication rule applies and

$$P(T_1 \cap T_2) = P(T_1)P(T_2) = \left(\frac{1}{2}\right)\left(\frac{1}{2}\right) = \frac{1}{4}$$

This is in agreement with what is shown for $P(T_1 \cap T_2)$ on the lowest path in Fig. 9-2(b).

The probability of not getting a tail on the first and the second flips can be found by using first the special multiplication rule to find all the intersection probabilities except $P(T_1 \cap T_2)$

$$P(H_1 \cap H_2) = P(H_1)P(H_2) = \left(\frac{1}{2}\right)\left(\frac{1}{2}\right) = \frac{1}{4}$$

$$P(H_1 \cap T_2) = P(H_1)P(T_2) = \left(\frac{1}{2}\right)\left(\frac{1}{2}\right) = \frac{1}{4}$$

$$P(T_1 \cap H_2) = P(T_1)P(H_2) = \left(\frac{1}{2}\right)\left(\frac{1}{2}\right) = \frac{1}{4}$$

and then using a variation of Property 4 in Section 8.6 to find the probability of this union

$$P[(H_1 \cap H_2) \cup (H_1 \cap T_2) \cup (T_1 \cap H_2)] = P(H_1 \cap H_2) + P(H_1 \cap T_2) + P(T_1 \cap H_2) = \frac{3}{4} = 0.75$$

Using the tree diagram shown in Fig. 9-2(b), the probability of not getting tails on both flips can be found by summing all the intersection probabilities except the probability of $T_1 \cap T_2$. The sum of these probabilities (each is marked with an asterisk in the table) is

$$P(H_1 \cap H_2) + P(H_1 \cap T_2) + P(T_1 \cap H_2) = \frac{1}{4} + \frac{1}{4} + \frac{1}{4} = \frac{3}{4} = 0.75$$

9.11 COUNTING RULES

From Property 7 in Section 8.6, we know that: If S contains N equally likely simple events (e_i), then $P(e_i) = \frac{1}{N}$ and $P(A) = N_A\left(\frac{1}{N}\right)$. The total number of events (N) in a sample space cannot, however, always be easily determined. *Counting rules* are mathematical formulas that describe how to count the total number of events in a set. They are of great use in calculating the probabilities of outcomes from a sequence of trials in an experiment.

There are three counting rules: the *multiplication principle, permutations*, and *combinations*. Any arrangement of the outcomes in a unique and defined order is a permutation of the outcomes. Any arrangement without regard to order is a combination of the outcomes. The fundamental tool for deriving the formulas for the number of permutations and the number of combinations is the multiplication principle.

9.12 COUNTING RULE: MULTIPLICATION PRINCIPLE

The *counting rule: multiplication principle* determines the total number of sample points when there are two or more trials of an experiment. In its simplest form, the multiplication principle states that: If an experiment has two consecutive trials, in which the first trial has n_1 possible outcomes and, after this has occurred, the second trial has n_2 possible outcomes, then the total number of sample points (i.e., the number of ways in which the combined trials can happen) is $n_1 \times n_2$. Extending the multiplication principle to cover more than two trials: If an experiment has k consecutive trials with n_1 possible outcomes for the first trial, n_2 for the second, ..., n_k for the kth, then the total number of sample points is given by the formula

$$\# \text{ sample points} = n_1 \times n_2 \times \cdots \times n_k \tag{9.16}$$

In words, the outcomes of a sequence of trials are multiplied to "count" the total number of sample points for the experiment. When a sample space is shown in a tree diagram, the multiplication principle counts the number of unique paths through the diagram.

EXAMPLE 9.12   How many three-letter words can be formed from the last four letters of the alphabet (W, X, Y, and Z), if each letter can be used more than once in a word? After you have completed your calculations, use a tree diagram to show that your calculations are correct.

Solution

Consider this experiment as drawing three times from a group of four letters. After a letter is drawn, it is returned to the group and thus is available again for the next draw. The first trial has four possible outcomes (W, X, Y, or Z), the second trial has the same four possible outcomes, as does the third trial. Thus, $k = 3$, $n_1 = 4$, $n_2 = 4$, $n_3 = 4$, and

$$\text{\# sample points} = n_1 \times n_2 \times n_3$$

$$= 4^3 = 64$$

The requested tree diagram is shown in Fig. 9-3, where each unique path through the diagram represents a unique three-letter word. In agreement with the calculations, there are 64 such paths.

9.13   COUNTING RULE: PERMUTATIONS

Any arrangement of objects in a given specific and unique order is a *permutation* of the objects. The *counting rule: permutations* is a special case of the counting rule: multiplication principle. It tells us the number of unique orderings that can come from selecting objects, one after the other, from a set of objects.

If we have a set of n distinct objects and choose, one after the other, r objects from the set ($r \leq n$), then for the first trial, we select from the entire set of n objects. For the second trial, the object selected in the first trial is no longer available (it has become the first object of the ordering) and so the number of objects from which to select is $n - 1$. For the third trial, the objects selected in the first and second trials are no longer available and so the number of objects from which to select is $n - 2$, and so on until the last object (on the rth trial) is selected from the remaining $n - r + 1$ objects. Thus, the number of possible orderings (permutations) of the n distinct objects taken r at a time is

$$_nP_r = n(n-1)\cdots(n-r+1) \tag{9.17}$$

This equation for calculating $_nP_r$ is the *counting rule: permutations formula*. (A second symbol, P_r^n is also commonly used to denote the formula.) It can be simplified through the use of factorials. Recall from Section 1.6 that n factorial (written $n!$) is calculated by

$$n! = n(n-1)\cdots(n-r+1)(n-r)(n-r-1)\cdots(3)(2)(1)$$

Therefore

$$n! = [n(n-1)\cdots(n-r+1)][(n-r)(n-r-1)\cdots(3)(2)(1)]$$

$$= {_nP_r} \times [(n-r)!]$$

Rearranging terms,

$$_nP_r = \frac{n!}{(n-r)!} \tag{9.18}$$

This simpler equation is another version of the counting rule: permutations formula.

EXAMPLE 9.13   Use equations (9.17) and (9.18) to determine how many unique three-letter words can be formed from the last four letters in the alphabet (W, X, Y, Z). The same letter cannot be used more than once in any given word (note how this differs from Example 9.12). Then use a tree diagram to show that your calculations are correct.

Solution

The question is: How many permutations are there of ($n = 4$) things taken ($r = 3$) at a time? Using

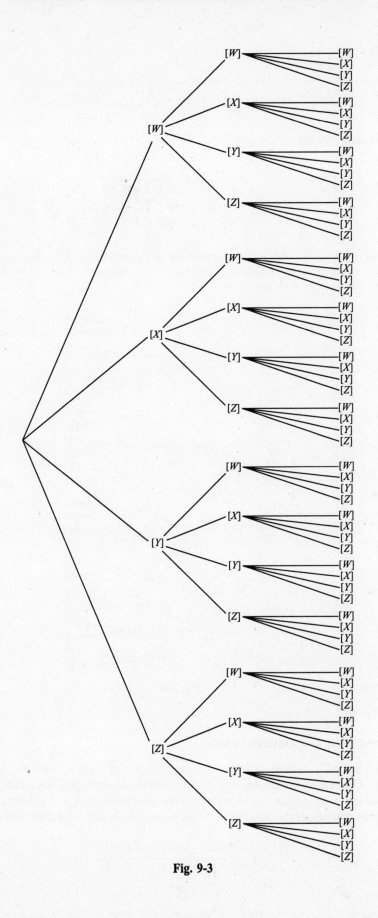

Fig. 9-3

equation (9.17),

$$_nP_r = n(n-1)\cdots(n-r+1)$$

$$_4P_3 = 4(4-1)(4-3+1) = 4 \times 3 \times 2 = 24$$

and using equation (9.18),

$$_nP_r = \frac{n!}{(n-r)!}$$

$$_4P_3 = \frac{4!}{(4-3)!} = \frac{4!}{1!} = \frac{4 \times 3 \times 2 \times 1}{1} = 24$$

The requested tree diagram (without probabilities) is shown in Fig. 9-4, where each unique path through the diagram represents a permutation of the four letters taken three at a time. In agreement with the calculations, there are 24 such paths.

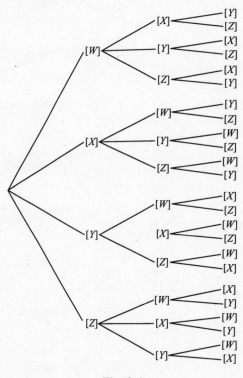

Fig. 9-4

9.14   COUNTING RULE: COMBINATIONS

If you have a set of n distinct objects and you select r of them (where $r \leq n$), and you are *not concerned about the order* in which the objects are selected or arranged, then each distinct group of r objects so-selected is a *combination*. The number of possible combinations of n distinct objects taken r at a time is given by the *counting rule: combinations formula*, which can be written either as

$$_nC_r = \frac{n(n-1)\cdots(n-r+1)}{r!} \tag{9.19}$$

or as

$$_nC_r = \frac{n!}{r!(n-r)!}$$
(9.20)

(The symbols $\begin{pmatrix} n \\ r \end{pmatrix}$ and C_r^n are also commonly used to denote these formulas.)

The counting rule: combinations can be visualized by converting the 24 unique three-letter words (which are paths, or permutations) shown in the tree diagram in Fig. 9-4 into the four rows of six words each shown in Fig. 9-5. Each row has only one unique combination of letters from which six permutations have been formed. From this figure, it can be seen that there are just four unique three-letter combinations (one in each row) of the four letters.

EXAMPLE 9.14   Using equation (9.20), determine how many unique three-letter combinations can be formed from the last four letters of the alphabet (W, X, Y, and Z).

WXY	WYX	XWY	XYW	YWX	YXW
WXZ	WZX	XWZ	XZW	ZWX	ZXW
WZY	WYZ	ZWY	ZYW	YWZ	YZW
XYZ	XZY	YXZ	YZX	ZXY	ZYX

Fig. 9-5

Solution

$$_nC_r = \frac{n!}{r!(n-r)!} = \frac{4!}{3!(4-3)!} = \frac{4 \times 3 \times 2 \times 1}{3 \times 2 \times 1(1)} = 4$$

This result confirms what is shown in Fig. 9-5.

The counting rule: combinations can be considered as a special case of the counting rule: permutations. Using the counting rule: permutations formula [equation (9.18)], we find that the number of permutations

of r objects from a set of r objects (i.e., $n = r$) is $\left[_rP_r = \frac{r!}{(r-r)!} = r! \right]$. Thus, for example, the number of three-letter words that can be formed from three letters is ($r! = 3! = 3 \times 2 = 6$). This result can be seen in Fig. 9-5, where each row contains all six permutations of three letters from a set of three letters. The first column contains the four combinations of the four letters taken three at a time. Together, all 24 words are the permutations of four letters (rather than three letters) taken three at a time. Thus, we can see that the total number of permutations ($_4P_3$) is equal to the number of rows ($_4C_3$) times the number of columns ($r!$).

$$_4P_3 = {_4C_3}(r!)$$
$$= \left[\frac{4!}{3!(4-3)!} \right](3!)$$
$$= \frac{4!}{1!} = 24$$

This illustrates what is true in general when r objects are selected from n distinct objects

$$_nP_r = {_nC_r} \times r!$$

and

$$_nC_r = \frac{_nP_r}{r!}$$

$$= \frac{n(n-1)\cdots(n-r+1)}{r!}$$

$$= \frac{n!}{r!(n-r)!}$$

Solved Problems

CONDITIONAL PROBABILITIES AND THE MULTIPLICATION RULES

9.1   Consider again the experiment described in Example 9.1 in which a vaccine is tested on a group of 160 volunteers. Eighty volunteers are vaccinated and the rest are not. After 12 months all 160 people are asked if they got a cold during the past year. The results are shown in Table 9.1. The experiment is to randomly select one of the 160 people. If for this experiment $S = \{$the 160 people$\}$, $A = \{$got a cold$\}$, $A' = \{$did not get a cold$\}$, $B = \{$vaccinated$\}$, and $B' = \{$not vaccinated$\}$, then find these probabilities: (a) $P(A|B')$,   (b) $P(B|A)$,   (c) $P(B'|A')$.

Solution

(a) $P(A|B') = \dfrac{P(A \cap B)}{P(B')}$

$$P(A \cap B') = N_{A \cap B'}\left(\frac{1}{N}\right) = 52\left(\frac{1}{160}\right) = \frac{52}{160} = \frac{13}{40}$$

$$P(B') = N_{B'}\left(\frac{1}{N}\right) = 80\left(\frac{1}{160}\right) = \frac{1}{2}$$

and thus

$$P(A|B') = \frac{P(A \cap B')}{P(B')} = \frac{13/40}{1/2} = \frac{13}{20} = 0.65$$

This states that the probability of selecting a person who got a cold from the nonvaccinated group is 0.65.

(b) $P(B|A) = \dfrac{P(A \cap B)}{P(A)}$

We know from Example 9.1 that

$$P(A \cap B) = \frac{3}{10}$$

Using equation (8.1),

$$P(A) = N_A\left(\frac{1}{N}\right) = 100\left(\frac{1}{160}\right) = \frac{5}{8}$$

and using equation (9.2),

$$P(B|A) = \frac{P(A \cap B)}{P(A)} = \frac{3/10}{5/8} = \frac{12}{25} = 0.48$$

This states that the probability of selecting a person who is vaccinated from the group that got colds is 0.48.

(c) $P(B'|A') = \dfrac{P(A' \cap B')}{P(A')}$

$P(A' \cap B') = N_{A' \cap B'}\left(\dfrac{1}{N}\right) = 28\left(\dfrac{1}{160}\right) = \dfrac{7}{40}$

$P(A') = N_{A'}\left(\dfrac{1}{N}\right) = 60\left(\dfrac{1}{160}\right) = \dfrac{3}{8}$

and thus

$$P(B'|A') = \frac{P(A' \cap B')}{P(A')} = \frac{7/40}{3/8} = \frac{7}{15} = 0.47$$

This states that the probability of selecting a person who is not vaccinated from the group that did not get a cold is 0.47.

9.2   If in Problem 9.1, two people are randomly selected, one after the other, from the 160, then what is the probability that: (a) both got colds,   (b) one got a cold and the other did not?

Solution

(a) If we let A_1 be the event that the first person got a cold and A_2 be the event that the second person got a cold, then using equation (9.4),

$$P(A_1 \cap A_2) = P(A_1)P(A_2|A_1)$$

From Table 9.1,

$$P(A_1) = N_{A_1}\left(\frac{1}{N}\right) = 100\left(\frac{1}{160}\right) = \frac{100}{160}$$

Having removed one person from the cold group,

$$P(A_2|A_1) = \frac{99}{159}$$

and thus

$$P(A_1 \cap A_2) = \frac{100}{160} \times \frac{99}{159} = \frac{9,900}{25,440} = 0.39$$

(b) If we let A_1 be the event that the first person got a cold and A'_2 be the event that the second person did not, then using equation (9.4),

$$P(A_1 \cap A'_2) = P(A_1)P(A'_2|A_1)$$

We know that

$$P(A_1) = P(A) = \frac{100}{160}$$

With 60 people still in the no-cold group and one person having been removed from the cold group

$$P(A'_2|A_1) = \frac{60}{159}$$

and

$$P(A_1 \cap A'_2) = P(A_1)P(A'_2|A_1)$$

$$= \frac{100}{160} \times \frac{60}{159} = \frac{6,000}{25,440} = 0.24$$

9.3   In the work force of a large factory, 70% of the employees are high-school graduates, 8% are supervisors, and 5% are both supervisors and high-school graduates. From this information, answer these questions.   (a) If the employee is a high-school graduate, what is the probability he is a supervisor?   (b) If the employee is not a high-school graduate, what is the probability he is a supervisor?

Solution

(a) $A \cap B = \{$supervisor (A) and high-school graduate $(B)\}$, $P(A \cap B) = 0.05$

   $B = \{$high-school graduate$\}$, $P(B) = 0.70$

      Using equation (9.2)

$$P(A|B) = \frac{P(A \cap B)}{P(B)} = \frac{0.05}{0.70} = 0.071$$

(b) $A \cap B' = \{$supervisor (A) and not high-school graduate $(B')\}$, $P(A \cap B') = 0.03$

   $B' = \{$not high-school graduate$\}$, $P(B') = 0.30$

$$P(A|B') = \frac{P(A \cap B')}{P(B')} = \frac{0.03}{0.30} = 0.10$$

9.4   A die is thrown two times. What is the probability that the total of both times is six, given that the first is twice as large as the second?

Solution

   Let event $A = \{$1st twice 2nd$\}$ and event $B = \{$a sum of six$\}$. Using equation (9.1),

$$P(A|B) = \frac{N_{A \cap B}}{N_B}$$

where

   $A \cap B$ can only occur with the sequence 4 and 2, so $N_{A \cap B} = 1$

   B can occur five ways (1 and 5, 5 and 1, 2 and 4, 4 and 2, 3 and 3), so $N_B = 5$

Thus

$$P(A|B) = \frac{1}{5} = 0.20$$

9.5   Two cards are drawn from a well-shuffled, standard deck of playing cards [four suits (diamonds, hearts, clubs, and spades) of 13 cards each (ace through king)]. If the first card is not replaced between selections, then what is the probability that: (a) both cards will be hearts,   (b) both will be queens,   (c) one will be a king and the other a queen?

Solution

(a)   If we let H_1 be the event that the first card is a heart and H_2 be the event that the second card is a heart, then using equation (9.4),

$$P(H_1 \cap H_2) = P(H_1)P(H_2|H_1)$$

and

$$P(H_1) = \frac{13}{52} = \frac{1}{4}$$

$$P(H_2|H_1) = \frac{12}{51}$$

Therefore

$$P(H_1 \cap H_2) = \frac{1}{4} \times \frac{12}{51} = \frac{12}{204} = 0.059$$

(b) If we let Q_1 be the event that the first card is a queen and Q_2 be the event that the second card is a queen, then using equation (9.4),

$$P(Q_1 \cap Q_2) = P(Q_1)P(Q_2|Q_1)$$

and

$$P(Q_1) = \frac{4}{52} = \frac{1}{13}$$

$$P(Q_2|Q_1) = \frac{3}{51}$$

Therefore

$$P(Q_1 \cap Q_2) = \frac{1}{13} \times \frac{3}{51} = \frac{3}{663} = 0.0045$$

(c) If we let K_1 be the event that the first card is a king and Q_2 be the event that the second card is a queen, then using equation (9.4),

$$P(K_1 \cap Q_2) = P(K_1)P(Q_2|K_1)$$

and

$$P(K_1) = \frac{4}{52} = \frac{1}{13}$$

$$P(Q_2|K_1) = \frac{4}{51}$$

Therefore

$$P(K_1 \cap Q_2) = \frac{1}{13} \times \frac{4}{51} = \frac{4}{663} = 0.0060$$

9.6 A manufacturer of automobile headlights has sent a shipment of 1,000 headlights to a customer, not knowing that three of the headlights are defective. The customer has a policy of testing a sample of three headlights from such a shipment, and if at least one of the three is defective he rejects the shipment. What is the probability he will reject this shipment?

Solution

The event we are interested in, {at least one in three is defective}, has as its complement {none of the three is defective}. If we use equation (9.5) (the generalization of the general multiplication rule) to solve for the probability of the complement, we can then solve for the probability of rejection by using Property 3 in Section 8.6.

For the complement to be true, all three sample headlights must be good (nondefective). Representing these events by N_1, N_2, N_3, we find that

$$P(N_1) = \frac{997}{1,000}$$

$$P(N_2|N_1) = \frac{996}{999}$$

$$P(N_3|N_1 \cap N_2) = \frac{995}{998}$$

and, using equation (9.5),

$$P(N_1 \cap N_2 \cap N_3) = P(N_1)P(N_2|N_1)P(N_3|N_1 \cap N_2)$$

$$= \frac{997}{1,000} \times \frac{996}{999} \times \frac{995}{998} = \frac{988,046,940}{997,002,000} = 0.991018$$

Therefore, using Property 3 from Section 8.6, the probability of rejection is

$$P(\text{at least one in three is defective}) = 1 - P(N_1 \cap N_2 \cap N_3)$$

$$= 1 - 0.991018 = 0.0090$$

9.7 In Chapter 8, we used the set theory interpretation of probability to show for the experiment of flipping a coin twice that

$$P(HH, HT, TH) = P(\text{at least one head}) = 0.75$$

Now, use equation (9.8) (the special multiplication rule) to show that this is true.

Solution

If we let H_1 and H_2 represent heads on the first and second trials, repectively, and T_1 and T_2 represent tails on the first and second trials, then

$$P(HH, HT, TH) = P[(H_1 \cap H_2) \cup (H_1 \cap T_2) \cup (T_1 \cap H_2)]$$

Property 4 in Section 8.6 states that: If events $A_1, A_2, \ldots, A_k$ in S are all mutually exclusive, then

$$P(A_1 \cup A_2 \cup \cdots \cup A_k) = P(A_1) + P(A_2) + \cdots + P(A_k)$$

Therefore,

$$P(HH, HT, TH) = P(H_1 \cap H_2) + P(H_1 \cap T_2) + P(T_1 \cap H_2)$$

As getting a head on the first trial has no effect on the outcome of the second trial, H_1 and H_2 are independent and therefore, using equation (9.8),

$$P(H_1 \cap H_2) = P(H_1)P(H_2) = \frac{1}{2} \times \frac{1}{2} = \frac{1}{4}$$

Similarly, as H_1 and T_2 are independent

$$P(H_1 \cap T_2) = P(H_1)P(T_2) = \frac{1}{2} \times \frac{1}{2} = \frac{1}{4}$$

And, as T_1 and H_2 and independent

$$P(T_1 \cap H_2) = P(T_1)P(H_2) = \frac{1}{2} \times \frac{1}{2} = \frac{1}{4}$$

Therefore

$$P(HH, HT, TH) = P(H_1 \cap H_2) + P(H_1 \cap T_2) + P(T_1 \cap H_2)$$

$$= \frac{1}{4} + \frac{1}{4} + \frac{1}{4} = \frac{3}{4} = 0.75$$

9.8 You know that $P(A) = 0.25$ and $P(A \cap B) = 0.20$. What is $P(B|A)$ if: (a) A and B are independent events, (b) A and B are dependent events?

Solution

(*a*) If *A* and *B* are independent events, then

$$P(B|A) = P(B)$$

and equation (9.8) is used

$$P(A \cap B) = P(A)P(B)$$

Therefore

$$P(B|A) = P(B) = \frac{P(A \cap B)}{P(A)} = \frac{0.20}{0.25} = 0.80$$

(*b*) If *A* and *B* are dependent events, then equation (9.4) is used

$$P(A \cap B) = P(A)P(B|A)$$

and

$$P(B|A) = \frac{P(A \cap B)}{P(A)} = \frac{0.20}{0.25} = 0.80$$

9.9 In the cold-vaccination study described in Example 9.1, 100 people got colds and 60 people did not get colds (see Table 9.1). Four consecutive random selections are made from the 160 people, with replacement after each selection. What is the probability that the first two had colds and the second two did not?

Solution

For determining the joint probability of *k* independent events, we use equation (9.9) (the generalization of the special multiplication rule). If we let C_1 and C_2 be the first two independent selections of people with colds and N_3 and N_4 be the next two independent selections of people without colds, then

$$P(C_1) = P(C_2) = \frac{100}{160} = \frac{5}{8}$$

and

$$P(N_3) = P(N_4) = \frac{60}{160} = \frac{3}{8}$$

and thus

$$P(C_1 \cap C_2 \cap N_3 \cap N_4) = P(C_1)P(C_2)P(N_3)P(N_4)$$

$$= \frac{5}{8} \times \frac{5}{8} \times \frac{3}{8} \times \frac{3}{8} = \frac{225}{4,096} = 0.055$$

9.10 You flip a coin seven times in a row and get seven heads. (*a*) What is the probability of this occurring? (*b*) What is the probability that if you go on to flip the coin an eighth time you will get a tail?

Solution

(*a*) If we let H_1, H_2, H_3, H_4, H_5, H_6, and H_7 be the seven independent heads, then

$$P(H_1) = P(H_2) = P(H_3) = P(H_4) = P(H_5) = P(H_6) = P(H_7) = \frac{1}{2}$$

and using equation (9.9)

$$P(H_1 \cap H_2 \cap H_3 \cap H_4 \cap H_5 \cap H_6 \cap H_7) = P(H_1)P(H_2)P(H_3)P(H_4)P(H_5)P(H_6)P(H_7)$$

$$= \left(\frac{1}{2}\right)^7 = \frac{1}{128} = 0.0078$$

(b) The probability of each individual flip is the same, and therefore the probability of getting a tail on the eighth flip is $\frac{1}{2}$. The error of thinking that because there were seven heads in a row, a tail is now more likely than a head is called the *gambler's fallacy*. It is the mistake of not accepting that all flips of a coin are independent events.

9.11 What is the probability of rolling a 2 at least once in three consecutive rolls of a die?

Solution

The event we are interested in, {at least one 2 in three rolls}, has as its complement {no 2s in three rolls}. We can use equation (9.9) to solve for the probability of the complement. We can then solve for the probability of {at least one 2 in three rolls} by using Property 3 in Section 8.6, which states: For event A and its complement A', $P(A) + P(A') = 1$.

For the complement to be true, there can be no 2s in any of the three rolls. Representing these independent rolls by N_1, N_2, and N_3, we find

$$P(N_1) = P(N_2) = P(N_3) = \frac{5}{6}$$

and using equation (9.9),

$$P(\text{no 2s in three rolls}) = P(N_1 \cap N_2 \cap N_3) = P(N_1)P(N_2)P(N_3)$$

$$= \left(\frac{5}{6}\right)^3 = \frac{125}{216} = 0.578704$$

Therefore, using Property 3,

$$P(\text{at least one 2 in three rolls}) = 1 - P(\text{no 2s in three rolls})$$

$$= 1 - 0.578704 = 0.42$$

ADDITION RULES

9.12 For the cold-vaccination study in Example 9.1 (see Table 9.1), if $S = \{$the 160 people$\}$, $A = \{$got a cold$\}$, $A' = \{$did not get a cold$\}$, $B = \{$was vaccinated$\}$, $B' = \{$was not vaccinated$\}$, then use equation (9.10) (the general addition rule) to determine: (a) $P(A \cup B)$, (b) $P(A' \cup B)$, (c) $P(A \cup B')$, (d) $P(A' \cup B')$, (e) $P(A \cup A')$.

Solution

(a) $A = \{$got a cold$\}$, $P(A) = \dfrac{100}{160}$

$B = \{$was vaccinated$\}$, $P(B) = \dfrac{80}{160}$

$A \cap B = \{$got a cold and was vaccinated$\}$, $P(A \cap B) = \dfrac{48}{160}$

Therefore, using equation (9.10),

$$P(A \cup B) = P(A) + P(B) - P(A \cap B)$$

$$= \frac{100}{160} + \frac{80}{160} - \frac{48}{160} = \frac{132}{160} = 0.82$$

(b) $A' = \{$did not get a cold$\}$, $P(A') = \dfrac{60}{160}$

   $B = \{$was vaccinated$\}$, $P(B) = \dfrac{80}{160}$

   $A' \cap B = \{$did not get a cold and was vaccinated$\}$, $P(A' \cap B) = \dfrac{32}{160}$

   Therefore

$$P(A' \cup B) = \frac{60}{160} + \frac{80}{160} - \frac{32}{160} = \frac{108}{160} = 0.68$$

(c) $A = \{$got a cold$\}$, $P(A) = \dfrac{100}{160}$

   $B' = \{$was not vaccinated$\}$, $P(B') = \dfrac{80}{160}$

   $A \cap B' = \{$got a cold and was not vaccinated$\}$, $P(A \cap B') = \dfrac{52}{160}$

   Therefore

$$P(A \cup B') = \frac{100}{160} + \frac{80}{160} - \frac{52}{160} = \frac{128}{160} = 0.80$$

(d) $A' = \{$did not get a cold$\}$, $P(A') = \dfrac{60}{160}$

   $B' = \{$was not vaccinated$\}$, $P(B') = \dfrac{80}{160}$

   $A' \cap B = \{$did not get a cold and was not vaccinated$\}$, $P(A' \cap B') = \dfrac{28}{160}$

   Therefore

$$P(A' \cup B') = \frac{60}{160} + \frac{80}{160} - \frac{28}{160} = \frac{112}{160} = 0.70$$

(e) $A = \{$got a cold$\}$, $P(A) = \dfrac{100}{160}$

   $A' = \{$did not get a cold$\}$, $P(A') = \dfrac{60}{160}$

   $A \cap A' = \{$got a cold and did not get a cold$\}$, $P(A \cap A') = 0$, as they are mutually exclusive
   Therefore

$$P(A \cup A') = \frac{100}{160} + \frac{60}{160} - 0 = \frac{160}{160} = 1.0$$

9.13   In the Venn diagram in Fig. 9-6, the numbers on the boundaries of the circles are the probabilities for the events represented by the circles, and the numbers within enclosed areas of the circles are the probabilities for events represented by those areas. Using this information and equation (9.10), find: (a) $P(A \cup B)$,   (b) $P(A' \cup B)$,   (c) $P(A \cup B')$,   (d) $P(A' \cup B')$.

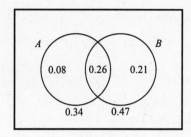

Fig. 9-6

Solution

(a) $P(A) = 0.34$, $P(B) = 0.47$, $P(A \cap B) = 0.26$
Therefore, using equation (9.10),

$$P(A \cup B) = P(A) + P(B) - P(A \cap B)$$
$$= 0.34 + 0.47 - 0.26 = 0.55$$

(b) $P(A') = 1 - 0.34 = 0.66$, $P(B) = 0.47$, $P(A' \cap B) = 0.21$
Therefore

$$P(A' \cup B) = 0.66 + 0.47 - 0.21 = 0.92$$

(c) $P(A) = 0.34$, $P(B') = 1 - 0.47 = 0.53$, $P(A \cap B') = 0.08$
Therefore

$$P(A \cup B') = 0.34 + 0.53 - 0.08 = 0.79$$

(d) $P(A') = 0.66$, $P(B') = 0.53$, $P(A' \cap B') = 1 - (0.08 + 0.26 + 0.21) = 0.45$
Therefore

$$P(A' \cup B') = 0.66 + 0.53 - 0.45 = 0.74$$

9.14 For the sample space shown in Fig. 9-6, are A and B independent events? Are A and B mutually exclusive events?

Solution

The special multiplication rule [equation (9.8)] can be used to test if two events are independent. If they are, then it must be true that

$$P(A \cap B) = P(A)P(B)$$

For this sample space $P(A) = 0.34$, $P(B) = 0.47$, and therefore $P(A)P(B) = (0.34)(0.47) = 0.1598$. But we can see from the diagram that

$$P(A \cap B) = 0.26$$

Therefore, $P(A \cap B) \neq P(A)P(B)$, and thus these events are dependent. A and B are not mutually exclusive because $P(A \cap B) \neq 0$.

9.15 In the Venn diagram in Fig. 9-7, the numbers in the circles are the probabilities for the events represented by the circles. Using this information and the appropriate addition rule, find: (a) $P(A \cup B)$, (b) $P(A' \cup B)$, (c) $P(A \cup B')$, (d) $P(A' \cup B')$.

Solution

(a) $P(A) = 0.39$, $P(B) = 0.27$
As these events are mutually exclusive we use equation (9.12)

$$P(A \cup B) = P(A) + P(B) = 0.39 + 0.27 = 0.66$$

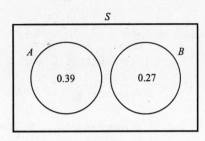

Fig. 9-7

(b) $P(A') = 1 - 0.39 = 0.61$, $P(B) = 0.27$, $P(A' \cap B) = 0.27$
As A' and B are not mutually exclusive we use equation (9.10)

$$P(A' \cup B) = P(A') + P(B) - P(A' \cap B) = 0.61 + 0.27 - 0.27 = 0.61$$

(c) $P(A) = 0.39$, $P(B') = 1 - 0.27 = 0.73$, $P(A \cap B') = 0.39$
Again as A and B' are not mutually exclusive we use equation (9.10)

$$P(A \cup B') = 0.39 + 0.73 - 0.39 = 0.73$$

(d) $P(A') = 0.61$, $P(B') = 0.73$, $P(A' \cap B') = 1 - (0.39 + 0.27) = 0.34$
Again as A' and B' are not mutually exclusive we use equation (9.10)

$$P(A' \cup B') = 0.61 + 0.73 - 0.34 = 1.0$$

9.16   For the sample space shown in Fig. 9-7, are A and B independent events?

Solution

Using equation (9.8) to test for independence,

$$P(A \cap B) = P(A)P(B)$$

For this sample space $P(A) = 0.39$, $P(B) = 0.27$, and therefore

$$P(A)P(B) = (0.39)(0.27) = 0.1053$$

But we can see from the diagram that

$$P(A \cap B) = 0$$

Therefore, $P(A \cap B) \neq P(A)P(B)$, and these events are dependent.

9.17   In the die-rolling experiment, if $S = \{1, 2, 3, 4, 5, 6\}$, $A = \{$even number$\}$, $B = \{$number $\geq 2\}$, $C = \{$number $\leq 4\}$, use equation (9.11) to determine $P(A \cup B \cup C)$.

Solution

We know from what is given that

$$A = \{\text{even number}\}, \qquad P(A) = \frac{3}{6}$$

$$B = \{\text{number} \geq 2\}, \qquad P(B) = \frac{5}{6}$$

$$C = \{\text{number} \leq 4\}, \qquad P(C) = \frac{4}{6}$$

$$A \cap B = \{\text{even number} \geq 2\}, \qquad P(A \cap B) = \frac{3}{6}$$

$$A \cap C = \{\text{even number} \leq 4\}, \qquad P(A \cap C) = \frac{2}{6}$$

$$B \cap C = \{2 \leq \text{number} \leq 4\}, \qquad P(B \cap C) = \frac{3}{6}$$

$$A \cap B \cap C = \{2 \leq \text{even number} \leq 4\}, \qquad P(A \cap B \cap C) = \frac{2}{6}$$

Therefore, using equation (9.11),

$$P(A \cup B \cup C) = P(A) + P(B) + P(C) - P(A \cap B) - P(A \cap C) - P(B \cap C) + P(A \cap B \cap C)$$

$$= \frac{3}{6} + \frac{5}{6} + \frac{4}{6} - \frac{3}{6} - \frac{2}{6} - \frac{3}{6} + \frac{2}{6}$$

$$= \frac{3 + 5 + 4 - 3 - 2 - 3 + 2}{6} = \frac{6}{6} = 1.0$$

JOINT PROBABILITY TABLES, MARGINAL PROBABILITIES, AND BAYES' THEOREM

9.18　An *urn* is an opaque vessel whose contents cannot be seen. Three such urns (A_1, A_2, A_3) each contain four balls. The balls are identical except for color: A_1 has three blue balls and one yellow, A_2 has two blue and two yellow, and A_3 has one blue and three yellow. The experiment is to randomly select one of the urns and then, without looking, select one ball from that urn. Let A_1, A_2, A_3 represent the events of selecting a ball from the given urn, B represent the event of selecting a blue ball, and B' the event of selecting a yellow ball. From this information develop a joint probability table.

Solution

We know from what is given that

$$P(A_1) = P(A_2) = P(A_3) = \frac{1}{3} = 0.3333$$

$$P(B|A_1) = \frac{3}{4} = 0.75$$

$$P(B|A_2) = \frac{2}{4} = 0.50$$

$$P(B|A_3) = \frac{1}{4} = 0.25$$

$$P(B'|A_1) = \frac{1}{4} = 0.25$$

$$P(B'|A_2) = \frac{2}{4} = 0.50$$

$$P(B'|A_3) = \frac{3}{4} = 0.75$$

We can now calculate the joint probabilities using equation (9.3). Thus

$$P(B \cap A_1) = P(A_1)P(B|A_1) = (0.3333)(0.75) = 0.249975$$

$$P(B \cap A_2) = P(A_2)P(B|A_2) = (0.3333)(0.50) = 0.166650$$

$$P(B \cap A_3) = P(A_3)P(B|A_3) = (0.3333)(0.25) = 0.083325$$

$$P(B' \cap A_1) = P(A_1)P(B'|A_1) = (0.3333)(0.25) = 0.083325$$

$$P(B' \cap A_2) = P(A_2)P(B'|A_2) = (0.3333)(0.50) = 0.166650$$

$$P(B' \cap A_3) = P(A_3)P(B'|A_3) = (0.3333)(0.75) = 0.249975$$

For the marginal probabilities, we know $P(A_1)$, $P(A_2)$, and $P(A_3)$ and we can now calculate $P(B)$ and $P(B')$ using equation (9.14) (the marginal probability formula). Thus

$$P(B) = \sum_{i=1}^{k} P(A_i)P(B|A_i)$$

$$= P(A_1)P(B|A_1) + P(A_2)P(B|A_2) + P(A_3)P(B|A_3)$$

$$= 0.249975 + 0.166650 + 0.083325 = 0.499950$$

$$P(B') = \sum_{i=1}^{k} P(A_i)P(B'|A_i)$$

$$= P(A_1)P(B'|A_1) + P(A_2)P(B'|A_2) + P(A_3)P(B'|A_3)$$

$$= 0.083325 + 0.166650 + 0.249975 = 0.499950$$

We now have all of the required probabilities, and the completed joint probability table is shown in Table 9.5.

Table 9.5

Balls	Urns			Marginal probability
	A_1	A_2	A_3	
B	0.249975	0.166650	0.083325	0.499950
B'	0.083325	0.166650	0.249975	0.499950
Marginal probability	0.3333	0.3333	0.3333	1.00

9.19   From the information in Problem 9.18, answer the following questions both directly from Table 9.5 and also by using equation (9.15) (Bayes' theorem). (*a*) Given that a blue ball was selected, what is the probability it came from urn A_1?   (*b*) Given that a yellow ball was selected, what is the probability it came from urn A_2?

Solution

(*a*)   The question is: What is $P(A_1|B)$?

We know from equation (9.2) that

$$P(A_1|B) = \frac{P(B \cap A_1)}{P(B)}$$

Therefore, from Table 9.5,

$$P(A_1|B) = \frac{0.249975}{0.499950} = 0.50$$

Using equation (9.15) and the information in Problem 9.18,

$$P(A_1|B) = \frac{P(A_1)P(B|A_1)}{\sum\limits_{i=1}^{k} P(A_i)P(B|A_i)}$$

$$= \frac{P(A_1)P(B|A_1)}{P(A_1)P(B|A_1) + P(A_2)P(B|A_2) + P(A_3)P(B|A_3)}$$

$$= \frac{0.249975}{0.249975 + 0.166650 + 0.083325} = \frac{0.249975}{0.499950} = 0.50$$

These results show that while the prior probability that urn A_1 will be selected was 0.3333, once it was known that a blue ball was selected this increased the probability (posterior) to 0.50 that urn A_1 had been selected.

(b)   The question is: What is $P(A_2|B')$?
      We know from equation (9.2) that

$$P(A_2|B') = \frac{P(B' \cap A_2)}{P(B')}$$

Therefore, from Table 9.5,

$$P(A_2|B') = \frac{0.166650}{0.499950} = 0.3333$$

Using equation (9.15) and the information in Problem 9.18,

$$P(A_2|B') = \frac{P(A_2)P(B'|A_2)}{P(A_1)P(B'|A_1) + P(A_2)P(B'|A_2) + P(A_3)P(B'|A_3)}$$

$$= \frac{0.166650}{0.083325 + 0.166650 + 0.249975} = \frac{0.166650}{0.499950} = 0.3333$$

These results show that the prior and posterior probabilities of selecting A_2 are both 0.3333.

9.20   A district sales manager for a textbook publishing company feels there is a 60% probability that a rival company will sell its chemistry textbook to the chemistry department of a large university. He also feels that if this happens, then there is an 80% probability that a community college in the same city as the university, which will choose a chemistry textbook after the university, will also adopt the rival's book. If the university does not adopt, then he feels there is still a 50% probability that the college will adopt the rival's book anyway. If U and U' represent the events of adoption and nonadoption by the university of the rival's book and C and C' represent these events for the college, then develop a joint probability table that includes these intersection probabilities: $P(C \cap U)$, $P(C \cap U')$, $P(C' \cap U)$, $P(C' \cap U')$; and these marginal probabilities: $P(U)$, $P(U')$, $P(C)$, $P(C')$.

Solution

We know from what is given that

$$P(U) = 0.60, \quad P(C|U) = 0.80, \quad P(C|U') = 0.50$$

Therefore we can calculate that

$$P(U') = 1 - 0.60 = 0.40$$

$$P(C'|U) = 1 - 0.80 = 0.20$$

$$P(C'|U') = 1 - 0.50 = 0.50$$

We can now calculate the intersection probabilities using equation (9.3)

$$P(C \cap U) = P(U)P(C|U) = (0.60)(0.80) = 0.48$$

$$P(C \cap U') = P(U')P(C|U') = (0.40)(0.50) = 0.20$$

$$P(C' \cap U) = P(U)P(C'|U) = (0.60)(0.20) = 0.12$$

$$P(C' \cap U') = P(U')P(C'|U') = (0.40)(0.50) = 0.20$$

For the marginal probabilities, we know $P(U)$ and $P(U')$ and we can now calculate $P(C)$ and $P(C')$ using equation (9.14)

$$P(C) = \sum_{i=1}^{k} P(U_i)P(C|U_i)$$

$$= P(U)P(C|U) + P(U')P(C|U')$$

$$= 0.48 + 0.20 = 0.68$$

$$P(C') = \sum_{i=1}^{k} P(U_i)P(C'|U_i)$$

$$= P(U)P(C'|U) + P(U')P(C'|U')$$

$$= 0.12 + 0.20 = 0.32$$

We now have all of the required probabilities, and the completed joint probability table is shown in Table 9.6.

Table 9.6

| College | University | | Marginal probability |
	U	U'	
C	0.48	0.20	0.68
C'	0.12	0.20	0.32
Marginal probability	0.60	0.40	1.00

9.21 From the information in Problem 9.20, answer the following questions both directly from Table 9.6 and by using equation (9.15). (*a*) Given that the college has adopted the textbook, what is the probability that the university has also adopted it? (*b*) Given that the college has not adopted the textbook, what is the probability that the university has also not adopted it?

Solution

(*a*) The question is: What is $P(U|C)$?
    We know from equation (9.2) that

$$P(U|C) = \frac{P(C \cap U)}{P(C)}$$

Therefore, from Table 9.6,

$$P(U|C) = \frac{0.48}{0.68} = 0.71$$

Using equation (9.15) and the information in Problem 9.20

$$P(U|C) = \frac{P(U)P(C|U)}{\sum\limits_{i=1}^{k} P(U_i)P(C|U_i)}$$

$$= \frac{P(U)P(C|U)}{P(U)P(C|U) + P(U')P(C|U')}$$

$$= \frac{0.48}{0.48 + 0.20} = \frac{0.48}{0.68} = 0.71$$

The results show that while the prior (subjective) probability that the university will adopt was 0.60, once it was known that the college had adopted the book this knowledge increased the probability (posterior) of the university adopting to 0.71.

(b)   This question is: What is $P(U'|C')$?
We know from equation (9.2) that

$$P(U'|C') = \frac{P(C' \cap U')}{P(C')}$$

So from Table 9.6,

$$P(U'|C') = \frac{0.20}{0.32} = 0.625, \text{ or } 0.62$$

Using equation (9.15) and the information in Problem 9.20,

$$P(U'C') = \frac{P(U')P(C'|U')}{P(U')P(C'|U') + P(U)P(C'|U)}$$

$$= \frac{0.20}{0.20 + 0.12} = \frac{0.20}{0.32} = 0.625, \text{ or } 0.62$$

These results show that while the prior (subjective) probability that the university will not adopt was 0.40, once it was known that the college had not adopted this increased the probability (posterior) that the university had not adopted to 0.62.

9.22   A new viral disease has infected approximately 25% of the pig population on farms in several southern states. There is a diagnostic test for the presence of the virus but it gives a positive result (virus present) only 84% of the time when the pig actually has the disease and a negative result (virus absent) only 80% of the time when the pig does not have the disease. The probability experiment is to take a pig from this population and test it for the presence of the virus. From this information, answer these questions.   (*a*) If the test result is positive, what is the probability the pig actually is infected by the virus?   (*b*) If the test result is negative, what is the probability the pig is actually not infected by the virus?   (*c*) What is the probability the test will give the correct diagnosis?

Solution

For answering these questions, let V and V' represent the events that the pig actually is or is not infected by the virus and R and R' represent the events of positive and negative results with the diagnostic test.

(*a*)   The question is: What is $P(V|R)$?
Using equation (9.15),

$$P(V|R) = \frac{P(V)P(R|V)}{P(V)P(R|V) + P(V')P(R|V')}$$

While the percentages given are relative frequency estimates, we will do all probability calculations as if they were exact percentages. Therefore, we know from what is given that

$$P(V) = 0.25, \quad P(R|V) = 0.84, \quad P(R'|V') = 0.80$$

Therefore we can calculate
$$P(V') = 1 - P(V) = 1 - 0.25 = 0.75$$
$$P(R|V') = 1 - P(R'|V') = 1 - 0.80 = 0.20$$

And thus
$$P(V|R) = \frac{(0.25)(0.84)}{(0.25)(0.84) + (0.75)(0.20)} = \frac{0.21}{0.21 + 0.15} = \frac{0.21}{0.36} = 0.58$$

Therefore, while it is true that if the pig has the disease there will be a positive test result 84% of the time, it is inversely true that if the test result is positive then there is only a 58% probability that the pig actually has the disease.

(b) The question is: What is $P(V'|R')$?
Using equation (9.15),
$$P(V'|R') = \frac{P(V')P(R'|V')}{P(V')P(R'|V') + P(V)P(R'|V)}$$

The only missing element is
$$P(R'|V) = 1 - P(R|V) = 1 - 0.84 = 0.16$$

Therefore
$$P(V'|R') = \frac{(0.75)(0.80)}{(0.75)(0.80) + (0.25)(0.16)} = \frac{0.60}{0.60 + 0.04} = \frac{0.60}{0.64} = 0.94$$

Thus, while it is true that if the pig does not have the disease there will be a negative result 80% of the time, it is inversely true that if the test result is negative there is a 94% probability the pig does not have the disease.

(c) The question is: What is $P[(V \cap R) \cup (V' \cap R')]$?
As $V \cap R$ and $V' \cap R'$ are mutually exclusive, we know from Property 4 in Section 8.6 that
$$P[(V \cap R) \cup (V' \cap R')] = P(V \cap R) + P(V' \cap R')$$

and from equation (9.4)
$$P(V \cap R) = P(V)P(R|V) = (0.25)(0.84) = 0.21$$
$$P(V' \cap R') = P(V')P(R'|V') = (0.75)(0.80) = 0.60$$

Therefore
$$P(\text{correct diagnosis}) = P[(V \cap R) \cup (V' \cap R')] = 0.21 + 0.60 = 0.81$$

Thus, with this test there is an 81% chance of a correct diagnosis.

9.23 An insurance company executive has developed an aptitude test for selling insurance. She knows that in the current sales force, 65% of the salespeople have good sales records and the remaining 35% have bad sales records. She gives her test to the entire sales force and finds that 73% of those with good records pass the test and 78% of those with bad records fail the test. The probability experiment is to select a salesperson at random and give them the test. From this information, answer these questions. (a) If someone passes the test, what is the probability they have a good sales record? (b) If someone fails the test, what is the probability they have a bad sales record? (c) What is the probability that performance on the test will correctly identify someone with either a good or a bad sales record?

Solution

For answering these questions, let T and T' represent the events of passing or failing the test and R and R' represent the events of having a good or bad sales record.

(a) The question is: What is $P(R|T)$?
Using equation (9.15),
$$P(R|T) = \frac{P(R)P(T|R)}{P(R)P(T|R) + P(R')P(T|R')}$$

From the information given, we know that

$$P(R) = 0.65, \quad (R') = 0.35, \quad P(T|R) = 0.73, \quad P(T'|R') = 0.78$$

Therefore we can calculate

$$P(T|R') = 1 - P(T'|R') = 1 - 0.78 = 0.22$$

Thus

$$P(R|T) = \frac{(0.65)(0.73)}{(0.65)(0.73) + (0.35)(0.22)} = \frac{0.4745}{0.4745 + 0.0770} = \frac{0.4745}{0.5515} = 0.86$$

Therefore, while there is a 65% probability that if a salesperson is picked at random from the sales force they will have a good sales record, if they have passed the test there is an 86% probability they will have a good sales record.

(b)   The question is: What is $P(R'|T')$?
      Using equation (9.15),

$$P(R'|T') = \frac{P(R')P(T'|R')}{P(R')P(T'|R') + P(R)P(T'|R)}$$

The only missing element is

$$P(T'|R) = 1 - P(T|R) = 1 - 0.73 = 0.27$$

Thus

$$P(R'|T') = \frac{(0.35)(0.78)}{(0.35)(0.78) + (0.65)(0.27)}$$

$$= \frac{0.2730}{0.2730 + 0.1755} = \frac{0.2730}{0.4485} = 0.61$$

Therefore, while there is a 35% probability that if a salesperson is selected at random they will have a bad sales record, if they have failed the test there is a 61% probability they will have a bad record.

(c)   The question is: What is $P[(R \cap T) \cup (R' \cap T')]$?
      As $R \cap T$ and $R' \cap T'$ are mutually exclusive, we know from Property 4 in Section 8.6 that

$$P[(R \cap T) \cup (R' \cap T')] = P(R \cap T) + P(R' \cap T')$$

and from equation (9.4)

$$P(R \cap T) = P(R)P(T|R) = (0.65)(0.73) = 0.4745$$
$$P(R' \cap T') = P(R')P(T'|R') = (0.35)(0.78) = 0.2730$$

Thus

$$P(\text{correctly identifying}) = P[(R \cap T) \cup (R' \cap T')]$$
$$= 0.4745 + 0.2730 = 0.75$$

Therefore, with this test there is a 75% chance of identifying someone with either a good or bad sales record.

TREE DIAGRAMS

9.24 Four balls in an urn are identical except for color; one is red (R), one white (W), one yellow (Y), and the fourth is blue (B). The experiment is to pick a ball from the urn and then, without replacing it, pick a second ball. Use a tree diagram to find P(at least one Y ball).

Solution

If we let Ra, Wa, Ya, and Ba represent possible outcomes for the first pick and Rb, Wb, Yb, and Bb represent possible outcomes for the second pick, then the tree diagram for this experiment is as shown in Fig. 9-8.

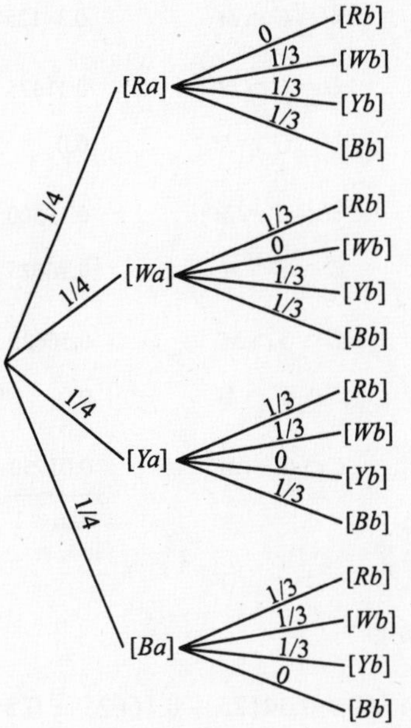

	Intersections	Intersection probabilities
$Ra \cap Rb$	$Ra \cap Rb$	0.0
$Ra \cap Wb$	$Ra \cap Wb$	1/12 = 0.08333
$Ra \cap Yb$	$Ra \cap Yb$	1/12 = 0.08333*
$Ra \cap Bb$	$Ra \cap Bb$	1/12 = 0.08333
$Wa \cap Rb$	$Wa \cap Rb$	1/12 = 0.08333
$Wa \cap Wb$	$Wa \cap Wb$	0.0
$Wa \cap Yb$	$Wa \cap Yb$	1/12 = 0.08333*
$Wa \cap Bb$	$Wa \cap Bb$	1/12 = 0.08333
$Ya \cap Rb$	$Ya \cap Rb$	1/12 = 0.08333*
$Ya \cap Wb$	$Ya \cap Wb$	1/12 = 0.08333*
$Ya \cap Yb$	$Ya \cap Yb$	0.0 *
$Ya \cap Bb$	$Ya \cap Bb$	1/12 = 0.08333*
$Ba \cap Rb$	$Ba \cap Rb$	1/12 = 0.08333
$Ba \cap Wb$	$Ba \cap Wb$	1/12 = 0.08333
$Ba \cap Yb$	$Ba \cap Yb$	1/12 = 0.08333*
$Ba \cap Bb$	$Ba \cap Bb$	0.0

sum = 1.00

Fig. 9-8

For P(at least one Y ball) the relevant probabilities are marked with an asterisk in the intersection probabilities column in Fig. 9-8. Therefore as these intersections are mutually exclusive we know from Property 4 in Section 8.6 that

$$P(\text{at least one } Y \text{ ball}) = P(Ra \cap Yb) + P(Wa \cap Yb) + P(Ya \cap Rb) + P(Ya \cap Wb)$$
$$+ P(Ya \cap Yb) + P(Ya \cap Bb) + P(Ba \cap Yb)$$

As $P(Ya \cap Yb) = 0.0$,

$$P(\text{at least one } Y \text{ ball}) = 6(0.08333) = 0.50$$

9.25   A friend in high school wants to go to medical school, but first she will either go to a local college where she has already been accepted, or to a prestigious university. She would prefer to go to the university but feels there is only a 65% probability she will be accepted. She further thinks that if she goes to the college there is a 95% probability she will graduate and then a 50% probability she will be accepted by a medical school. If instead she goes to the university, then she feels there is a 70% probability she will graduate followed by a 75% probability she will be accepted by a medical school. Of course she has to graduate from either the college or the university to be accepted by a medical school. Use a tree diagram to determine P(acceptance by a medical school).

Solution

If we let U and C represent her going to the university or the college, G and G' represent subsequent graduation or nongraduation, and M and M' represent her acceptance or nonacceptance by a medical school, then the tree diagram for this experiment is as shown in Fig. 9-9.

For $P(M)$ the relevant probabilities are marked with asterisks in the intersection probabilities column in Fig. 9-9. Therefore, as these intersections are mutually exclusive we know from Property 4 in Section 8.6 that

$$P(M) = P(U \cap G \cap M) + P(U \cap G' \cap M) + P(C \cap G \cap M) + P(C \cap G' \cap M)$$

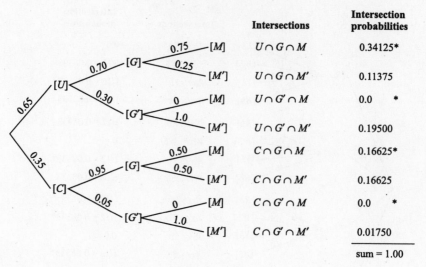

Fig. 9-9

and, as $P(U \cap G' \cap M) = P(C \cap G' \cap M) = 0$

$$P(M) = P(U \cap G \cap M) + P(C \cap G \cap M) = 0.34125 + 0.16625 = 0.51$$

9.26 A senator campaigning for reelection is trying to raise $100,000 for a last-minute series of television advertisements. He thinks there is a 55% probability he will be able to raise the money and that if he does there is then a 70% chance he will be reelected. He also feels that if he fails to raise the money then there will still be a 55% probability he will be reelected. Use a tree diagram to find P(he did not raise the money, given he was reelected).

Solution

If we let M and M' represent raising or not raising the money and R and R' represent being reelected or not reelected, then the tree diagram for this experiment is as shown in Fig. 9-10.

To find $P(M'|R)$ we use equation (9.15) and the values marked with asterisks in the intersection probabilities column in Fig. 9-10. Therefore

$$P(M'|R) = \frac{P(M')P(R|M')}{P(M')P(R|M') + P(M)P(R|M)}$$

$$= \frac{P(M' \cap R)}{P(M' \cap R) + P(M \cap R)}$$

$$= \frac{0.2475}{0.2475 + 0.3850} = \frac{0.2475}{0.6325} = 0.39$$

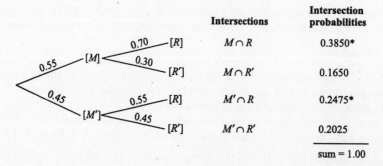

Fig. 9-10

COUNTING RULES

9.27   Use the *counting rule*: *multiplication principle* to determine the number of sample points in the sample spaces for these experiments: (a) flipping a coin eleven times,   (b) selecting a card from a standard 52-card deck five times in a row with replacement and reshuffling after each selection.

Solution

(a)   This experiment has $k = 11$ consecutive trials, with two possible outcomes for each trial. Using equation (9.16),

$$\text{\# sample points} = n_1 \times n_2 \times \cdots \times n_{11}$$
$$= 2^{11} = 2{,}048$$

(b)   This experiment has $k = 5$ consecutive trials, with 52 possible outcomes for each trial. Therefore

$$\text{\# sample points} = n_1 \times n_2 \times \cdots \times n_5$$
$$= 52^5 = 380{,}204{,}032$$

9.28   For the experiments in Problem 9.27, determine the following probabilities using both equation (9.9) and Property 7 from Section 8.6:   (a) P(getting a head on all 11 trials),   (b) P(getting a queen of hearts on all 5 trials).

Solution

(a)   If we let $H_1, H_2, \ldots, H_{11}$ represent the independent events of heads on each of the 11 flips, then from equation (9.9),

$$P(H_1 \cap H_2 \cap \cdots \cap H_{11}) = P(H_1)P(H_2)\cdots P(H_{11})$$

Therefore, as $P(H_1) = P(H_2) = \cdots = P(H_{11}) = \frac{1}{2}$

$$P(H_1 \cap H_2 \cap \cdots \cap H_{11}) = \left(\frac{1}{2}\right)^{11} = 0.00049$$

From Property 7, if $A = \{11 \text{ heads in a row}\}$, then there is only one way this can occur and $N_A = 1$. From Problem 9.27 (a) we know that $N = 2^{11} = 2{,}048$. Therefore

$$P(A) = N_A\left(\frac{1}{N}\right) = 1\left(\frac{1}{2{,}048}\right) = 0.00049$$

(b)   If we let $Q_1, Q_2, \ldots, Q_5$ represent the independent events of queen of hearts on each of the five card selections, then from equation (9.9)

$$P(Q_1 \cap Q_2 \cap \cdots \cap Q_5) = P(Q_1)P(Q_2)\cdots P(Q_5)$$

Therefore, as $P(Q_1) = P(Q_2) = \cdots = P(Q_5) = \frac{1}{52}$,

$$P(Q_1 \cap Q_2 \cap \cdots \cap Q_5) = \left(\frac{1}{52}\right)^5 = 0.0000000026$$

From Property 7, if $A = \{\text{queen of hearts on all 5 selections}\}$, then there is only one way this can occur and $N_A = 1$. From Problem 9.27(b) we know that $N = 52^5 = 380{,}204{,}032$. Therefore

$$P(A) = N_A\left(\frac{1}{N}\right) = 1\left(\frac{1}{380{,}204{,}032}\right) = 0.0000000026$$

9.29   A car manufacturer offers several options for a particular model of car: (a) two or four doors,   (b) one of six colors (red, yellow, blue, green, white, silver),   (c) AM or AM–FM radio,   (d) automatic or manual shifting. How many versions of this car are possible? If each of the versions is described on a separate card, all of the cards are put in a bowl, and then you blindly select one of the cards from the bowl, what is the probability it will read: a red car with two doors, manual shifting, and an AM–FM radio?

Solution

The experiment is picking a car that has a unique mixture of the four types of options. If we consider each option to be a trial of the experiment, then: $n_1 =$ door options $= 2$, $n_2 =$ color options $= 6$, $n_3 =$ radio options $= 2$, and $n_4 =$ shifting options $= 2$. Therefore, using equation (9.16),

$$\text{\# sample points} = \text{\# possible versions of the car} = n_1 \times n_2 \times n_3 \times n_4$$
$$= 2 \times 6 \times 2 \times 2 = 48$$

Using Property 7 from Section 8.6, if $A = \{$red car with two doors, manual shifting, and an AM–FM radio$\}$, then $N_A = 1$, $N = 48$, and

$$P(A) = N_A\left(\frac{1}{N}\right) = 1\left(\frac{1}{48}\right) = 0.021$$

9.30 Use both equations (9.17) and (9.18) to determine how many ways eight books can be arranged in a line along a shelf.

Solution

The question is: What is $_nP_n$?
Substituting n for r in equation (9.17) we get

$$_nP_n = n(n-1)\cdots(n-n+2)(n-n+1)$$
$$= n(n-1)\cdots(2)(1) = n!$$

and doing this in equation (9.18) we get

$$_nP_n = \frac{n!}{(n-n)!} = \frac{n!}{(0)!} = \frac{n!}{1} = n!$$

Therefore, in this problem where $n = 8$

$$_8P_8 = 8! = 8 \times 7 \times 6 \times 5 \times 4 \times 3 \times 2 \times 1 = 40,320$$

9.31 After shuffling a standard deck of 52 cards you deal three cards which you place on a table in a left-to-right sequence. What is the probability that the sequence is jack, queen, king of the same suit?

Solution

Using Property 7 from Section 8.6, if $A = \{$same-suit sequence of jack, queen, king$\}$, then as there are four suits, $N_A = 4$. And, using equation (9.18),

$$N = {}_nP_r = {}_{52}P_3 = \frac{n!}{(n-r)!} = \frac{52!}{(52-3)!} = \frac{52!}{49!} = 52 \times 51 \times 50 = 132,600$$

Therefore

$$P(A) = N_A\left(\frac{1}{N}\right) = 4\left(\frac{1}{132,600}\right) = 0.000030$$

9.32 You live in a state where the car license plates have three letters (without duplications on a plate) on the left followed by three numbers (again without duplications) on the right. The letters are randomly selected from the 26 letters of the alphabet, and the numbers are randomly selected from the ten integers 0 to 9. Assuming that all plates are available, and that you are assigned a plate randomly, what is the probability that you will get a plate that reads: ABC012?

Solution

Using Property 7 from Section 8.6, if $A = \{ABC012\}$, then $N_A = 1$. And, if we consider the selection of the three letters as the first trial of an experiment and the selection of the three numbers as the second trial, then using both equation (9.16) and equation (9.18),

$$N = {}_{26}P_3 \times {}_{10}P_3 = \left[\frac{26!}{(26-3)!}\right]\left[\frac{10!}{(10-3)!}\right]$$

$$= \left(\frac{26!}{23!}\right)\left(\frac{10!}{7!}\right) = (26 \times 25 \times 24)(10 \times 9 \times 8)$$

$$= (15,600)(720) = 11,232,000$$

Therefore

$$P(A) = N_A\left(\frac{1}{N}\right) = 1\left(\frac{1}{11,232,000}\right) = 0.000000089$$

9.33 In the house of representatives of your state legislature there are 90 Democrats and 70 Republicans. By a random process, a Majority Leader and an Assistant Majority Leader will be selected from the Democrats and a Minority Leader and Assistant Minority Leader will be selected from the Republicans. How many permutations are there of these four leadership positions?

Solution

Using equation (9.18), the permutations for the Democrats are

$$_nP_r = {}_{90}P_2 = \frac{n!}{(n-r)!} = \frac{90!}{(90-2)!} = 90 \times 89 = 8,010$$

and those for the Republicans are

$$_nP_r = {}_{70}P_2 = \frac{n!}{(n-r)!} = \frac{70!}{(70-2)!} = 70 \times 69 = 4,830$$

If we consider the selection of the two Democrats as the first trial of an experiment and the selection of the two Republicans as the second trial, then using equation (9.16),

$$(\text{leadership permutations}) = {}_{90}P_2 \times {}_{70}P_2 = (8,010)(4,830) = 38,688,300$$

9.34 If in the house of representatives described in Problem 9.33 there are six Democrats and four Republicans from the same city, what is the probability that representatives from this city will be selected to all four leadership positions?

Solution

Using equation (9.18), the leadership permutations for the Democrats from this city are

$$_nP_r = {}_6P_2 = \frac{n!}{(n-r)!} = \frac{6!}{(6-2)!} = 6 \times 5 = 30$$

and those for the city's Republicans are

$$_nP_r = {}_4P_2 = \frac{4!}{(4-2)!} = 4 \times 3 = 12$$

Then, calculating leadership permutations for the four positions as was done in Problem 9.33, but now restricted to the city's representatives,

$$(\text{leadership permutations}) = {}_6P_2 \times {}_4P_2 = (30)(12) = 360$$

Therefore, using Property 7 from Section 8.6, if $A = \{$all four leaders from the same city$\}$, $N_A = 360$, and from Problem 9.33, $N = 38,688,300$, then

$$P(A) = N_A\left(\frac{1}{N}\right) = 360\left(\frac{1}{38,688,300}\right) = 0.0000093$$

9.35 If all three-letter words that can be formed from the word *CLOVER* (with no letter duplications in a word) are each written on a separate card and all these cards are placed in a bowl, what is the probability of then selecting a card from the bowl that has a word on it that begins with a vowel?

Solution

Using equation (9.18), the total number of three-letter words from *CLOVER* is

$$_nP_r = {}_6P_3 = \frac{n!}{(n-r)!} = \frac{6!}{(6-3)!} = 6 \times 5 \times 4 = 120$$

As there are two vowels in *CLOVER* (*E, O*), the number of three-letter words that can be formed starting with a vowel is found by using equation (9.16)

$$2 \times {}_4P_2 = 2\left[\frac{4!}{(4-2)!}\right] = 2(4 \times 3) = 24$$

Therefore, using Property 7 from Section 8.6, if $A = \{$words starting with a vowel$\}$, $N_A = 24$, $N = 120$, then

$$P(A) = N_A\left(\frac{1}{N}\right) = 24\left(\frac{1}{120}\right) = 0.20$$

9.36 A manufacturer shows 40 bathing suits to a buyer. Use equation (9.20) to determine how many ways the buyer can choose five of the suits to sell in her store.

Solution

The question is: What is ${}_{40}C_5$? Using equation (9.20),

$$_{40}C_5 = \frac{n!}{r!(n-r)!} = \frac{40!}{5!(40-5)!} = \frac{40 \times 39 \times 38 \times 37 \times 36 \times 35!}{(5 \times 4 \times 3 \times 2 \times 1)(35!)} = 658,008$$

9.37 Your instructor in a college history course gives you a list of 20 possible essay questions from which he will randomly pick four for the final examination. Pressed for time, you prepare for only four of the questions. What is the probability these four will be on the examination?

Solution

Using equation (9.20), the number of possible four-question final examinations that can be taken from the list of 20 question is

$$_{20}C_4 = \frac{n!}{r!(n-r)!} = \frac{20!}{4!(20-4)!}$$
$$= \frac{20 \times 19 \times 18 \times 17 \times 16!}{(4 \times 3 \times 2 \times 1)(16!)} = 4,845$$

Therefore, using Property 7 from Section 8.6, if $A = \{$four questions you have prepared$\}$, $N_A = 1$, $N = 4,845$, then

$$P(A) = N_A\left(\frac{1}{N}\right) = 1\left(\frac{1}{4,845}\right) = 0.00021$$

You have a 0.00021 probability of selecting the four questions on the exam.

9.38 To play your state lottery, you select six numbers from 1 to 42. You win the grand prize if your six match the six winning numbers selected by the lottery. What is the probability that your ticket will win the grand prize?

Solution

   Using equation (9.20) to determine how many six-number combinations can be selected from the 42 numbers,

$$_{42}C_6 = \frac{n!}{r!(n-r)!} = \frac{42!}{6!(42-6)!}$$

$$= \frac{42 \times 41 \times 40 \times 39 \times 38 \times 37 \times 36!}{(6 \times 5 \times 4 \times 3 \times 2 \times 1)(36!)}$$

$$= 5{,}245{,}786$$

Therefore, using Property 7 from Section 8.6, if $A = \{$winning grand prize$\}$, $N_A = 1$, $N = 5{,}245{,}786$, then

$$P(A) = N_A\left(\frac{1}{N}\right) = 1\left(\frac{1}{5{,}245{,}786}\right) = 0.00000019$$

9.39 In the lottery in Problem 9.38, you win the second prize if you match five of the lottery's six winning numbers. What is the probability that your ticket will win the second prize?

Solution

   To win the second prize you must have selected five of the six winning numbers and one of the 36 losing numbers. If we consider selecting the five winners as the first trial of an experiment and selecting the one loser as the second trial, then using both equation (9.20) and equation (9.16), the number of ways you can pick five correct numbers is

$$_{6}C_5 \times {_{36}}C_1 = \left[\frac{6!}{5!(6-5)!}\right]\left[\frac{36!}{1!(36-1)!}\right]$$

$$= \left(\frac{6 \times 5!}{5!1!}\right)\left(\frac{36 \times 35!}{1!35!}\right) = 6 \times 36 = 216$$

Therefore, using Property 7 from Section 8.6, if $A = \{$winning 2nd prize$\}$, $N_A = 216$, and $N = 5{,}245{,}786$ (from Problem 9.38), then

$$P(A) = N_A\left(\frac{1}{N}\right) = 216\left(\frac{1}{5{,}245{,}786}\right) = 0.000041$$

9.40 What is the probability that if you are dealt a five-card poker hand, two of the cards will be diamonds, two will be hearts, and one will be a club?

Solution

   To get this hand you must have been given two of the 13 diamonds, two of the 13 hearts, and one of the 13 clubs. If we consider these as three trials of an experiment, then using both equation (9.20) and equation (9.16), the number of ways you can get this hand is

$$_{13}C_2 \times {_{13}}C_2 \times {_{13}}C_1 = \left[\frac{13!}{2!(13-2)!}\right]\left[\frac{13!}{2!(13-2)!}\right]\left[\frac{13!}{1!(13-1)!}\right]$$

$$= \left[\frac{13 \times 12 \times 11!}{(2 \times 1)(11!)}\right]\left[\frac{13 \times 12 \times 11!}{(2 \times 1)(11!)}\right]\left[\frac{13 \times 12!}{1!12!}\right]$$

$$= \left(\frac{13 \times 12}{2}\right)\left(\frac{13 \times 12}{2}\right)\left(\frac{13}{1}\right) = 78 \times 78 \times 13 = 79{,}092$$

The total number of five-card poker hands that are possible from a 52-card deck is

$$_{52}C_5 = \frac{52!}{5!(52-5)!} = \frac{52 \times 51 \times 50 \times 49 \times 48 \times 47!}{(5 \times 4 \times 3 \times 2 \times 1)(47!)} = 2{,}598{,}960$$

Therefore, using Property 7 from Section 8.6, if $A = \{$hand with 2 diamonds, 2 hearts, and 1 club$\}$, $N_A = 79{,}092$, $N = 2{,}598{,}960$, then

$$P(A) = N_A\left(\frac{1}{N}\right) = 79{,}092\left(\frac{1}{2{,}598{,}960}\right) = 0.030$$

9.41 You want to arrange eight books in a line along a shelf. How many unique combinations are there of these eight books?

Solution

The question is: What is $_nC_n$? In general, using equation (9.19),

$$_nC_n = \frac{n(n-1)\cdots(n-n+2)(n-n+1)}{n!}$$

$$= \frac{n(n-1)\cdots(2)(1)}{n!} = \frac{n!}{n!} = 1$$

Therefore, when $n = 8$

$$_8C_8 = 1$$

Supplementary Problems

CONDITIONAL PROBABILITIES AND THE MULTIPLICATION RULES

9.42 A single card is drawn from a well-shuffled, standard deck of playing cards. Given that a diamond card has been selected, what is the probability it is a *face card* (jack, queen, or king)?

Ans. $\frac{3}{13} = 0.23$

9.43 A single card is drawn from a well-shuffled, standard deck of playing cards. Given that a card numbered 2, 3, 4, or 5 has been selected, what is the probability it is a diamond?

Ans. $\frac{4}{16} = 0.25$

9.44 A single card is drawn from a well-shuffled, standard deck of playing cards. Given that a king has been selected, what is the probability it is a red card?

Ans. $\frac{2}{4} = 0.50$

9.45 Two cards are drawn, one after the other, from a well-shuffled, standard deck of playing cards. The first card is not returned to the deck after it has been drawn. Use equation (9.4) to determine the probability of selecting a diamond card in both draws.

Ans. $\left(\frac{13}{52}\right)\left(\frac{12}{51}\right) = 0.059$

9.46 An urn contains 60 marbles: 30 are white, 18 are red, and 12 are blue. Two marbles are removed, at random, from the urn. The first marble is not returned to the urn after it has been removed. What is the probability that both marbles are blue?

Ans. $\left(\dfrac{12}{60}\right)\left(\dfrac{11}{59}\right) = 0.037$

9.47 Determine the probability of not rolling a 4 on either of two consecutive repetitions of the die-rolling experiment.

Ans. $\left(\dfrac{5}{6}\right)^2 = 0.69$

9.48 In three repetitions of the card-selection experiment, if the cards are not replaced between selections, then what is the probability that they will all be queens?

Ans. $\left(\dfrac{4}{52}\right)\left(\dfrac{3}{51}\right)\left(\dfrac{2}{50}\right) = 0.00018$

9.49 In the coin-flipping experiment, what is the probability of rolling two heads in succession?

Ans. $\left(\dfrac{1}{2}\right)^2 = 0.25$

9.50 In six repetitions of the card-selection experiment, with replacement and reshuffling after each selection, what is the probability that all six cards will be red cards?

Ans. $\left(\dfrac{1}{2}\right)^6 = 0.016$

9.51 During the past year in a maternity ward of a hospital, 1,060 males were born and 1,000 females. Assuming these totals to be representative of all births, what is the probability that the next four babies born in the ward will be girls?

Ans. $\left(\dfrac{1,000}{2,060}\right)^4 = 0.056$

ADDITION RULES

9.52 A single card is drawn from a well-shuffled, standard deck of playing cards. What is the probability of drawing either a diamond or a red card?

Ans. $\dfrac{13}{52} + \dfrac{26}{52} - \dfrac{13}{52} = \dfrac{26}{52} = 0.50$

9.53 A single card is drawn from a well-shuffled, standard deck of playing cards. What is the probability of drawing either a diamond or a king?

Ans. $\dfrac{13}{52} + \dfrac{4}{52} - \dfrac{1}{52} = \dfrac{16}{52} = 0.31$

9.54 A single card is drawn from a well-shuffled, standard deck of playing cards. What is the probability of drawing either a 2, 3, or 4 and not a diamond?

Ans. $\dfrac{39}{52} + \dfrac{12}{52} - \dfrac{9}{52} = \dfrac{42}{52} = 0.81$

9.55   Determine $P(B|A)$ when: (a) A and B are mutually exclusive events,   (b) A and B are independent events.

Ans.   (a) 0,   (b) $P(B)$

9.56   Determine $P(A \cap B)$ when: (a) A and B are mutually exclusive events,   (b) A and B are independent events.

Ans.   (A) 0,   (b) $P(A)P(B)$

9.57   Determine $P(A \cup B)$ when: (a) A and B are mutually exclusive events,   (b) A and B are independent events.

Ans.   (a) $P(A) + P(B)$,   (b) $P(A) + P(B) - P(A \cap B) = P(A) + P(B) - [P(A)P(B)]$

9.58   One hundred mayors of U.S. cities are attending a conference on environmental issues. Fifty of the mayors are Democrats and 50 are Republicans. Sixty are men and 40 are women, and of the 60 men, 25 are Democrats. If one of the mayors is randomly chosen, then determine the probability that: (a) the mayor will be a male Democrat,   (b) the mayor will be either a male or a Democrat.

Ans.   (a) $\dfrac{25}{100} = 0.25$,   (b) $\dfrac{60}{100} + \dfrac{50}{100} - \dfrac{25}{100} = 0.85$

9.59   For the mayors in Problem 9.58, 15 of the 40 women are Republicans. If one mayor is randomly selected from all 100 mayors each day for two days (sampling with replacement), then determine the probability that: (a) a man will be chosen on day one and a woman on day two,   (b) on day two, either a male Democrat or a female Republican will be chosen.

Ans.   (a) $\dfrac{60}{100} \times \dfrac{40}{100} = 0.24$,   (b) $\dfrac{25}{100} + \dfrac{15}{100} = 0.40$

9.60   For the mayors in Problem 9.58, 15 of the 25 male Democrats are over 45 years of age. If one mayor is randomly selected from the 100, what is the probability of selecting a male Democrat who is over 45 years of age?

Ans. $\dfrac{60}{100} \times \dfrac{25}{60} \times \dfrac{15}{25} = 0.15$

9.61   For the mayors in Problem 9.58, 50 are over 45 years of age (30 males and 25 females). Of these 50, 25 are Democrats. If one mayor is randomly selected from the 100, what is the probability of selecting either a male or a Democrat or someone who is over 45 years of age?

Ans. $\dfrac{60}{100} + \dfrac{50}{100} + \dfrac{50}{100} - \dfrac{25}{100} - \dfrac{30}{100} - \dfrac{25}{100} + \dfrac{15}{100} = 0.95$

JOINT PROBABILITY TABLES, MARGINAL PROBABILITIES, AND BAYES' THEOREM

9.62   Events A_1, A_2, and A_3 are mutually exclusive and exhaustive, with probabilities $P(A_1) = 0.20$, $P(A_2) = 0.60$, and $P(A_3) = 0.20$. Given that $P(B|A_1) = 0.10$, $P(B|A_2) = 0.50$, and $P(B|A_3) = 0.40$, calculate $P(B)$.

Ans.   0.40

9.63   From the information provided in Problem 9.62, calculate: (a) $P(A_1|B)$,   (b) $P(A_2|B)$,   (c) $P(A_3|B)$.

Ans.   (a) 0.05,   (b) 0.75,   (c) 0.20

9.64   Events D_1, D_2, and D_3 are mutually exclusive and exhaustive causes of two effects, C and C'. The probabilities of the causes are $P(D_1) = 0.69$, $P(D_2) = 0.05$, and $P(D_3) = 0.26$. Given that $P(C|D_1) = 0.10$, $P(C|D_2) = 0.36$, and $P(C|D_3) = 0.54$, what is the probability of effect C'?

Ans.   0.77

9.65   For the information provided in Problem 9.64, what is the probability that effect C' is caused by D_1, D_2, or D_3? In other words, what are: (a) $P(D_1|C')$,   (b) $P(D_2|C')$,   (c) $P(D_3|C')$?

Ans.   (a) 0.81,   (b) 0.04,   (c) 0.16

9.66   At Easter, a father places colored eggs in baskets for his small daughter to find. He hides three baskets in three hiding places. The first basket contains two red eggs, the second basket contains a red egg and a blue egg, and the third basket contains two blue eggs. Given that the child finds a blue egg, what is the probability that it comes from the second basket?

Ans.   1/3

9.67   Colored marbles are placed in two urns, one black and the other white. The black urn contains 12 blue marbles and 6 red ones. The white urn contains 4 blue marbles and 8 red ones. An urn is selected at random and one marble is drawn from it. If the marble is blue, then what is the probability that the marble was drawn from the black urn?

Ans.   2/3

TREE DIAGRAMS

9.68   For the experiment of flipping a coin three times, use a tree diagram to find P(at least two tails). Let H_1, H_2, and H_3 represent heads on the first, second, and third flips, and T_1, T_2, and T_3 represent getting a tail on the first, second, and third flips.

Ans.   In the tree diagram, shown in Fig. 9-11, the asterisks represent all the ways of getting at least two tails. P(at least two tails) $= 0.50$

9.69   For the experiment of rolling a die twice, use a tree diagram to find P(total for both rolls of 7 or 11). Let $1a$, $2a$, $3a$, $4a$, $5a$, and $6a$ represent the possible outcomes for the first roll and $1b$, $2b$, $3b$, $4b$, $5b$, and $6b$ represent the outcomes for the second roll.

Ans.   In the tree diagram, shown in Fig. 9-12, the asterisks represent all the ways of getting a total for both rolls of 7 or 11. P(total for both rolls of 7 or 11) $= 0.22$

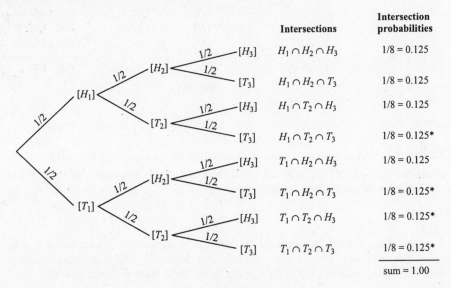

Fig. 9-11

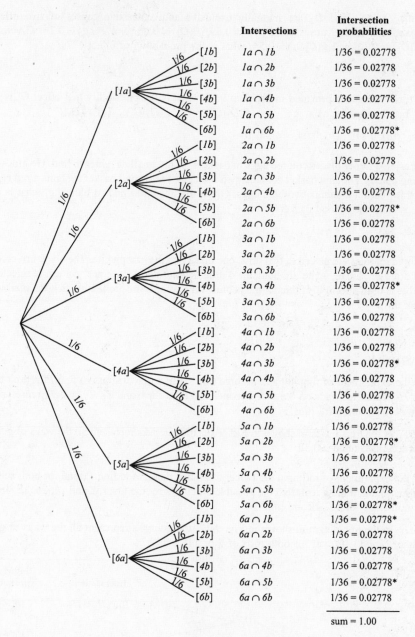

	Intersections	Intersection probabilities
[1a] $\to$ [1b]	$1a \cap 1b$	$1/36 = 0.02778$
[1a] $\to$ [2b]	$1a \cap 2b$	$1/36 = 0.02778$
[1a] $\to$ [3b]	$1a \cap 3b$	$1/36 = 0.02778$
[1a] $\to$ [4b]	$1a \cap 4b$	$1/36 = 0.02778$
[1a] $\to$ [5b]	$1a \cap 5b$	$1/36 = 0.02778$
[1a] $\to$ [6b]	$1a \cap 6b$	$1/36 = 0.02778*$
[2a] $\to$ [1b]	$2a \cap 1b$	$1/36 = 0.02778$
[2a] $\to$ [2b]	$2a \cap 2b$	$1/36 = 0.02778$
[2a] $\to$ [3b]	$2a \cap 3b$	$1/36 = 0.02778$
[2a] $\to$ [4b]	$2a \cap 4b$	$1/36 = 0.02778$
[2a] $\to$ [5b]	$2a \cap 5b$	$1/36 = 0.02778*$
[2a] $\to$ [6b]	$2a \cap 6b$	$1/36 = 0.02778$
[3a] $\to$ [1b]	$3a \cap 1b$	$1/36 = 0.02778$
[3a] $\to$ [2b]	$3a \cap 2b$	$1/36 = 0.02778$
[3a] $\to$ [3b]	$3a \cap 3b$	$1/36 = 0.02778$
[3a] $\to$ [4b]	$3a \cap 4b$	$1/36 = 0.02778*$
[3a] $\to$ [5b]	$3a \cap 5b$	$1/36 = 0.02778$
[3a] $\to$ [6b]	$3a \cap 6b$	$1/36 = 0.02778$
[4a] $\to$ [1b]	$4a \cap 1b$	$1/36 = 0.02778$
[4a] $\to$ [2b]	$4a \cap 2b$	$1/36 = 0.02778$
[4a] $\to$ [3b]	$4a \cap 3b$	$1/36 = 0.02778*$
[4a] $\to$ [4b]	$4a \cap 4b$	$1/36 = 0.02778$
[4a] $\to$ [5b]	$4a \cap 5b$	$1/36 = 0.02778$
[4a] $\to$ [6b]	$4a \cap 6b$	$1/36 = 0.02778$
[5a] $\to$ [1b]	$5a \cap 1b$	$1/36 = 0.02778$
[5a] $\to$ [2b]	$5a \cap 2b$	$1/36 = 0.02778*$
[5a] $\to$ [3b]	$5a \cap 3b$	$1/36 = 0.02778$
[5a] $\to$ [4b]	$5a \cap 4b$	$1/36 = 0.02778$
[5a] $\to$ [5b]	$5a \cap 5b$	$1/36 = 0.02778$
[5a] $\to$ [6b]	$5a \cap 6b$	$1/36 = 0.02778*$
[6a] $\to$ [1b]	$6a \cap 1b$	$1/36 = 0.02778*$
[6a] $\to$ [2b]	$6a \cap 2b$	$1/36 = 0.02778$
[6a] $\to$ [3b]	$6a \cap 3b$	$1/36 = 0.02778$
[6a] $\to$ [4b]	$6a \cap 4b$	$1/36 = 0.02778$
[6a] $\to$ [5b]	$6a \cap 5b$	$1/36 = 0.02778*$
[6a] $\to$ [6b]	$6a \cap 6b$	$1/36 = 0.02778$

sum = 1.00

Fig. 9-12

COUNTING RULES

9.70   A committee consisting of one man and one woman is to be chosen from a group of 8 men and 15 women. How many possible committees can be chosen?

   Ans.   120

9.71   A fraternity plans to send four students—a freshman, a sophomore, a junior, and a senior—to a national meeting. The volunteers for this group include four freshmen, four sophomores, eight juniors, and three seniors. How many different sets of four students, one from each class, are possible?

   Ans.   384

9.72 A man packs four slacks, six shirts, and three ties for a trip. How many different "outfits" of these three kinds of clothing can he form?

Ans. 72

9.73 A die is rolled nine times. What is the number of sample points for this experiment?

Ans. 10,077,696

9.74 For the experiment in Problem 9.73, determine the probability of getting a 5 on all 9 trials.

Ans. 0.000000099

9.75 An exam consists of ten multiple-choice questions, each of which has four choices and only one correct answer. What is the probability of selecting all ten correct answers by making random choices for each question?

Ans. 0.00000095

9.76 If each of the three-letter words represented by unique paths in Fig. 9-4 is written on a separate card and then all the cards are placed in a bowl, what is the probability that if you blindly selected one card from the bowl the word written on it would include the letter W?

Ans. 0.75

9.77 A basketball league has 18 teams. How many ways can the teams finish the season in a first, second, and third order?

Ans. 4,896

9.78 How many ways can nine graduate students be assigned as teaching assistants to nine courses (one graduate student per course)?

Ans. 362,880

9.79 How many ways can twelve 30-second commercials be arranged to be shown in a one-hour television program?

Ans. 479,001,600

9.80 How many ways can the first and second prize winners of a raffle be selected from 1,000 ticket holders?

Ans. 999,000

9.81 If a salesman must visit five cities, how many unique routes are there connecting the cities?

Ans. 120

9.82 How many ways can four distinct batteries be placed in the first through the fourth positions of a long flashlight?

Ans. 24

9.83 How many ways can 200 job candidates be rated: top candidate, second best, third best, fourth best?

Ans. 1,552,438,800

9.84 How many ways can your state income tax office randomly select three out of 100 tax forms to audit?

Ans. 161,700

9.85   A college plans to send five students to a national conference. Thirty students volunteer to go. How many unique combinations of five students can be chosen?

Ans.   142,506

9.86   A woman has 12 dresses and wants to choose four to take with her on a trip. Determine the number of combinations from which she can choose.

Ans.   495

9.87   If a salesperson must visit ten cities, how many unique combinations are there of these cities?

Ans.   1

Chapter 10

Random Variables, Probability Distributions, and Cumulative Distribution Functions

10.1 RANDOM VARIABLES

In this chapter we will introduce and examine three concepts from probability theory that are of fundamental importance in inferential statistics: *random variables, probability distributions*, and *cumulative distribution functions*. These are all functions that have a domain, a rule of association, and a range (see Section 1.17). We will begin with random variables.

The domain of a random variable (also called a *chance variable*) is the sample space that summarizes the outcomes of a randomly determined statistical experiment (see Section 8.1). The rule of association for a random variable assigns one and only one real number (see Section 1.17) to each sample point in the sample space. And the range of a random variable is the sample space of numbers defined by the rule of association.

EXAMPLE 10.1 The experiment is flipping a coin twice. If the random variable is the number of heads on the two flips, then what is: (*a*) its domain, (*b*) its rule of association, (*c*) its range?

Solution

(*a*) $S = \{HH, HT, TH, TT\}$

(*b*) Count the number of heads for each sample point.

(*c*) $S = \{0, 1, 2\}$

EXAMPLE 10.2 The experiment is rolling a die twice. If the random variable is the number of dots on the two rolls, then what is: (*a*) its domain, (*b*) its rule of association, (*c*) its range?

Solution

(*a*) The sample space for this experiment, shown as a tree diagram in Fig. 9-12, has 36 sample points

$$S = \{(1, 1), (1, 2), (1, 3), \ldots, (6, 4), (6, 5), (6, 6)\}$$

(*b*) Count the total number of dots for the two rolls.

(*c*) $S = \{2, 3, 4, 5, 6, 7, 8, 9, 10, 11, 12\}$

Strictly speaking, a random variable is this set theory definition (see Section 8.6): A real-number-valued function defined on the sample space of a randomly determined experiment. There is also, however, a "common usage" meaning for a random variable: If the outcomes of a statistical experiment are quantitative measurements (numerical values) and if these outcomes depend to some extent on chance, then the resulting measurement variable (see Section 1.16) is said to be a random variable. Thus, for example, if you take a random sample of children and measure the height of each child in inches, then this quantitative measurement variable is said in common usage to be a random variable.

The relationship between the two versions of a random variable is that they both deal with numerical measurement values that are the outcomes of statistical experiments. The common usage version deals with the set of all possible numerical measurements that can result from the experiment, and the set theory version deals with the measurement process itself, the function (or rule) that assigns a number to each outcome. In the remainder of the book, the version of a random variable that is being discussed will be apparent from the context.

As with the variables that were described earlier (see Sections 1.13 to 1.16), we typically denote random variables by capital letters from the end of the alphabet (Z, Y, X, etc.). Also, as before, we distinguish between the random variable and the real number value it can assume by denoting the value by the variable's lowercase letter (z, y, x, etc.). Thus, for example, the symbol $P(X=x)$ denotes the probability that the random variable X will assume the value of x. If, say, $X=1$, then $P(X=1)$ denotes the probability that the random variable X will assume the value of 1.

EXAMPLE 10.3   For each of the following, indicate whether it is or is not a random variable: (*a*) number of heart beats per minute,   (*b*) number of fives rolled in two rolls of a die that has fives on all of its faces,   (*c*) time in seconds between airplane takeoffs at an airport,   (*d*) amount of pesticide in a sample of lake water, measured in milligrams per liter,   (*e*) amount of money taken in each week by a department store,   (*f*) rating the size of objects on this three-valued scale: small, medium, large,   (*g*) number of people out of 100 surveyed who say yes in answer to a question,   (*h*) number of state lottery tickets you have to buy before you win the grand prize.

Solution

(*a*)   Random variable

(*b*)   Not a random variable because the outcome of the experiment is not determined by a random process; the outcome is always a pair of fives and thus the number 2.

(*c*)   Random variable

(*d*)   Random variable

(*e*)   Random variable

(*f*)   As is, this is not a random variable. If, however, one counted the number of objects that were rated "small" in a sample of 50, then that would be a random variable.

(*g*)   Random variable

(*h*)   Random variable

10.2   DISCRETE VERSUS CONTINUOUS RANDOM VARIABLES

A random variable may be discrete or continuous. The exact definitions of these two terms depend on whether we use the common usage or the set theory version of a random variable.

If we consider the common usage version of a random variable, then a random variable is a randomly determined quantitative measurement variable. Therefore, in common usage, a *discrete random variable* is the same as a randomly determined discrete quantitative measurement variable, and a *continuous random variable* is a randomly determined continuous quantitative measurement variable (see Section 2.8).

If we consider the set theory version of a random variable, then a random variable is a function defined on the sample space of a randomly determined experiment. In this version, a *discrete random variable* has a sample space that is *finite or countably infinite*, and a *continuous random variable* has a sample space that is *infinite and not countable*.

All the sample spaces discussed in Chapters 8 and 9 and so far in this chapter are finite; they have a finite number of sample points. The sample space $S = \{HH, HT, TH, TT\}$, for example, is finite with four sample points, and thus a random variable defined on this sample space (e.g., number of heads) is a discrete random variable. Similarly, the 36-point sample space for the experiment of rolling a die twice is finite, and thus the random variable of total number of dots for the two rolls is a discrete random variable.

In this chapter we deal for the first time with countably infinite sample spaces, which have an infinite number of sample points although the points can be listed and thus counted. An example of such a sample space would result from the experiment of counting the number of days from now until there is an earthquake where you live. It could happen today, tomorrow, or not in your lifetime, or not during the next million years. The random variable number of days to the earthquake can take on an infinite number of values yet, theoretically, the days can be listed and thus counted. In fact, the list is the set of all possible positive integers: 1, 2, 3, ..., ∞. Because this sample space is countably infinite, the random variable number of days to the earthquake is a discrete random variable.

Sample spaces that are infinite and not countable are introduced in this chapter and then are the principal topic of Chapter 12 in Volume 2. For such sample spaces, it is not possible to list and count the elements. Such a space, for example, would result from the experiment of measuring the height in inches of a random sample of children. This sample space would include all outcomes that are theoretically possible, which assumes complete accuracy of measurement (see Section 2.14) and sensitivity of measurement to an infinite number of decimal places. Such a set of outcomes is not listable and therefore not countable, and so the random variable height in inches is a continuous random variable. An infinite and uncountable sample space corresponds to all possible values in an uninterrupted interval along the real number line (see Section 1.20).

EXAMPLE 10.4 For each of the following, indicate whether the random variable is discrete or continuous: (a) number of heart beats per minute, (b) time in seconds between airplane takeoffs at an airport, (c) amount of pesticide in a sample of lakewater, measured in milligrams per liter, (d) amount of money taken in each week by a department store, (e) number of people out of 100 surveyed who say yes in answer to a question, (f) number of state lottery tickets you have to buy before you win the grand prize.

Solution

(a) Discrete random variable

(b) Continuous random variable

(c) Continuous random variable

(d) Discrete random variable

(e) Discrete random variable

(f) Discrete random variable that can take on a countably infinite number of values

10.3 DISCRETE PROBABILITY DISTRIBUTIONS

In Section 8.6 we defined a probability function as any mathematical function that both assigns real numbers called probabilities to events in a sample space and satisfies the three axioms of probability theory. The function has a domain, which is all the events in the sample space, and it has a range, which is all the probabilities assigned to these events. When the sample space has been defined by a discrete random variable, the domain of the probability function consists of all values that the random variable can assume $(X=x)$, and the range of the probability function consists of the probabilities assigned to these values $[P(X=x)]$. In this case, where the sample space is defined by a discrete random variable, the probability function may be called either a *discrete probability distribution* or a *probability mass function*. The two terms are synonymous. The term mass function refers to the fact that the probability is "massed" at discrete values of the random variable. Both terms are denoted by the symbols $P(X=x)=f(x)$.

A discrete probability distribution is defined by a mathematical formula, and all probabilities in the distribution are calculated with that formula. However, the distribution may be presented in four ways: As the formula itself or as a list, table, or graph of the probabilities calculated from the formula. Examples of all four presentations for the experiment of rolling a die once are shown in Table 10.1 and Fig. 10-1 where the discrete random variable that defines the sample space is the number of dots on the final upward face of the die. The presentation as a *probability list* is shown in Fig. 10-1(a) where all values of the variable

$(X = x)$ are given, along with the probability assigned to each value $[P(X = x) = f(x)]$. The same information is presented as a *probability table* in Table 10.1. Note the symbol at the base of the table $\sum_x f(x)$. This means sum the values of $f(x)$ for all values of x that the random variable can take on. This symbol has the same meaning as $\sum f(x)$ and they are used interchangeably throughout the book.

The graphs used to present discrete probability distributions are essentially the same as those used in Chapter 5 to present relative frequency distributions. Thus, if the random variable is a discrete ratio-level measurement variable, as here, we can use a probability version of a bar chart, histogram, or rod graph (see Problems 5.9 and 5.10). A *probability histogram* for the experiment of rolling a die once is shown in Fig.

Table 10.1

Number of dots x	Probability $f(x)$
1	1/6
2	1/6
3	1/6
4	1/6
5	1/6
6	1/6
	$\sum_x f(x) = 1.00$

(a) For the possible values of the random variable $x = 1, 2, 3, 4, 5, 6$

$P(X = 1) = f(1) = 1/6;\ P(X = 2) = f(2) = 1/6;\ P(X = 3) = f(3) = 1/6;$
$P(X = 4) = f(4) = 1/6;\ P(X = 5) = f(5) = 1/6;\ P(X = 6) = f(6) = 1/6$

(b)

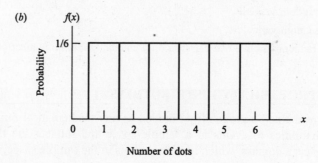

(c)

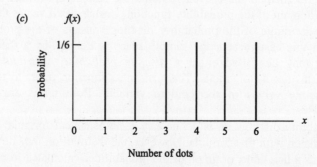

(d) $P(X = x) = f(x) = 1/6$, for $x = 1, 2, 3, 4, 5, 6$

Fig. 10-1

10-1(b), where the bars represent values of the random variable ($X = x$). Each bar has a height that is equal to $f(x)$ and, on the X axis, a base of one that is centered on the value. A *probability rod graph* of the same distribution is shown in Fig. 10-1(c), where now each value ($X = x$) is represented by a vertical line of height $f(x)$. It is in the rod graph form that one sees the "massing" of the probabilities above discrete values along the X axis, which is why the discrete probability distribution is also called a probability mass function. Finally, the presentation of the discrete probability distribution as a formula is shown in Fig. 10-1(d).

Because a discrete probability distribution is a probability function, it obeys the rules and properties of probability theory (see Section 8.6). This means, among other things, that $[P(X = x) = f(x)]$ is always greater than or equal to zero $[f(x) \geq 0]$, and that the sum of the probabilities for any given discrete probability distribution is always equal to one $\left[\sum_x f(x) = P(S) = 1.00; \text{ see Table 10.1} \right]$.

EXAMPLE 10.5   For the experiment of flipping a coin three times, present the discrete probability distribution for the random variable number of heads as both a probability table and a probability histogram.

Solution

To solve this problem, we first convert the probabilities for the simple events in the sample space (shown as a tree diagram in Fig. 9-11) to probabilities for simple events in the new sample space defined by the random variable $S = \{0, 1, 2, 3\}$. This means we must find $f(0), f(1), f(2), f(3)$.

As all simple events (paths) shown in the tree diagram are mutually exclusive events, we can use Property 4 from set theory (Section 8.6) to find the new probability values. Thus

$$f(0) = P(T_1 \cap T_2 \cap T_3) = 0.125$$

$$f(1) = P(H_1 \cap T_2 \cap T_3) + P(T_1 \cap H_2 \cap T_3) + P(T_1 \cap T_2 \cap H_3) = 3(0.125) = 0.375$$

$$f(2) = P(H_1 \cap H_2 \cap T_3) + P(H_1 \cap T_2 \cap H_3) + P(T_1 \cap H_2 \cap H_3) = 3(0.125) = 0.375$$

$$f(3) = P(H_1 \cap H_2 \cap H_3) = 0.125$$

This discrete probability distribution is presented as a probability table in Table 10.2 and as a probability histogram in Fig. 10-2.

Table 10.2

Number of heads x	Probability $f(x)$
0	0.125
1	0.375
2	0.375
3	0.125
$\sum$	1.00

10.4   CONTINUOUS PROBABILITY DISTRIBUTIONS

A *continuous probability distribution* (or *probability density function*) assigns probabilities to events in the sample space of a continuous random variable, which can take on an infinite and not countable number of specific values. Chapter 12 in Volume 2 is devoted entirely to such distributions, but we discuss them briefly in this chapter as part of a general introduction to probability distributions.

We begin our presentation of continuous probability distributions by comparing them with discrete probability distributions. The graph of a discrete probability distribution is shown in Fig. 10-3(a), together

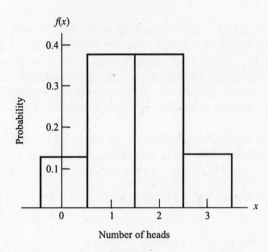

Fig. 10-2

with a summary of its important properties (from Sections 10.1, 10.2, and 10.3). Each discrete probability in the distribution, denoted by $f(x)$, is based on a discrete random variable X that takes on a specific value x [Properties (1), (2), and (3)]. The domain of the function consists of all sample points ($X=x$) in the sample space defined by X, and the range consists of the probabilities assigned to these sample points [$P(X=x)=f(x)$] [Property (4)]. The discrete probability distribution can be presented as a list, table, graph, or formula [Property (5)]. In the histogram version of the graph [see Fig. 10-3(a)], the height of

(a)

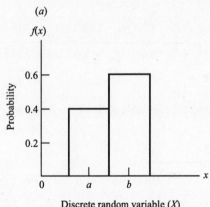

Discrete random variable (X)

(b)

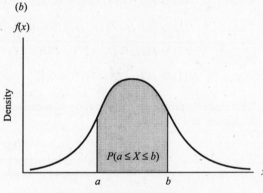

Continuous random variable (X)

Properties

(1) Discrete random variable denoted by X

(2) Specific values of random variable denoted by x

(3) Discrete probability distributions (also called probability mass functions) denoted by $f(x)$

(4) $P(X = x) = f(x)$

(5) Can be presented as list, table, graph, or formula

(6) $P(X = x) = f(x) \geq 0$

(7) $\sum_x f(x) = P(S) = 1.00$

Properties

(1) Continuous random variable denoted by X

(2) Specific values of random variable denoted by x

(3) Continuous probability distributions (also called probability density functions) denoted by $f(x)$

(4) $P(X = x) = 0$

(5) Can be presented as graph or formula

(6) $P(a \leq X \leq b) = \int_a^b f(x)\,dx \geq 0$ where $f(x \geq 0)$

(7) $P(-\infty < X < \infty = \int_{-\infty}^{\infty} f(x)\,dx = P(S) = 1.00$

Fig. 10-3

each bar represents a probability $[P(X=x)=f(x)]$, and therefore we know from Axiom I of set theory (Section 8.6) that a bar can never extend below the X axis; that is, $P(X=x)=f(x) \geq 0$ [Property (6)].

Finally, it is always true for a discrete probability distribution that $\sum_x f(x) = P(S) = 1.00$ [Property (7)]; that is, it is certain for any trial of the experiment that one of the events in S must occur (see Section 8.6, Axiom II). Further, for such a histogram version of a discrete probability distribution, if the distance between consecutive scale x values on the X axis is always 1.0 (e.g., if $a=2$ and $b=3$), then the area of each bar above an x value is equal to $f(x)$, and the total area under the histogram is equal to $\sum f(x) = 1.0$.

The graph of a continuous probability distribution is shown in Fig. 10-3(b), together with a summary of its important properties. Such a continuous probability distribution, denoted also by $f(x)$, is based on a continuous random variable X that can take on an infinite and not countable number of specific values x [Properties (1), (2), and (3)]. This continuous random variable X defines a sample space that has an infinite and uncountable number of sample points. Because of this, there is zero probability that the random variable will assume any one specific value x $[P(X=x)=0;$ Property (4)]. While the continuous probability distribution cannot assign probability values to each sample point, what it does do is assign a real number called a *probability density* to each sample point. Thus, the distribution is a function whose domain consists of all the sample points in the sample space defined by X, and whose range consists of all $f(x)$ values assigned to the sample points. These $f(x)$ values represent theoretical measurements of the "density" or "concentration" of probability for each x value. Because there is an infinite and uncountable number of sample points, and thus a corresponding number of probability densities, continuous probability distributions cannot be presented as lists or tables, but only as graphs of continuous curves [see Fig. 10-3(b)] or as the mathematical formulas that generate the curve [Property (5)].

If the height of the curve above the X axis in Fig. 10-3(b) represents probability density $[f(x)]$ and not probability $[P(X=x)]$, then where is probability in the graph of the continuous probability distribution? It is in the *area under the curve*. When a probability is calculated from a continuous probability distribution, it is the probability that X will take on some value in the interval between a and b $[P(a \leq X \leq b)]$, and this probability is the area above the interval from a to b that is bounded by the curve, the X axis, and the vertical lines above a and b [the shaded area in Fig. 10-3(b)].

To calculate the probability that X will take on some value in the interval between a and b (the area above it under the curve) requires techniques from *integral calculus*. These techniques are not required for this book, but if you are familiar with calculus you will recognize in Property (6) that $P(a \leq X \leq b) = \int_a^b f(x)\, d(x)$ states symbolically that to find the probability that X will take on some value in the interval from a to b, $f(x)$ must be integrated from a to b. For a crude idea of what this means, for those who have not had calculus, consider the vertical density lines above a and b in Figure 10-3(b). These lines have height but no width because the base point representing each x value is infinitesimally small, and thus the lines have no area. However, if the infinite number of such lines above the points in the interval from a to b are added together (integrated) then this process of summation does produce an area and this area is a probability $[P(a \leq X \leq b)]$. Property (6) also states that Axiom I from set theory (Section 8.6) holds true for this probability $[P(a \leq X \leq b) \geq 0]$ and that for this to be true it must also be true that $f(x) \geq 0$.

Whether or not the vertical density lines above a and b in Fig. 10-3(b) are included in the interval from a to b has no effect on the area over that interval, since these lines have no area. Stated more formally: Whether or not the endpoints of an interval are included in the calculations does not affect the probability that X will take on a value in that interval. Thus for a continuous probability distribution

$$P(a < X \leq b) = P(a \leq X \leq b) = P(a < X < b) = P(a \leq X < b) \qquad (10.1)$$

Finally, Property (7) states, again in the symbolic language of integral calculus, that the total area (total probability) under a continuous probability distribution over the interval extending from minus infinity to plus infinity ($-\infty$ to ∞) is always equal to $P(S) = 1.00$. [This assumes that $f(x)=0$ for values of x in the interval from $-\infty$ to ∞ that are not in S.]

Note in Fig. 10-3(*b*) that no scale is given on the vertical axis (the *Y* axis) that would allow direct reading of $f(x)$. This is typical, as only areas under the probability density function are of importance for statistical procedures.

While techniques from integral calculus are required to calculate probabilities for continuous probability distributions, a knowledge of these techniques is not required for general statistics because summary tables of these relationships are available for all important probability distributions. Several of these tables are provided in the Appendix to this book.

10.5 THE RELATIONSHIP BETWEEN DISCRETE PROBABILITY DISTRIBUTIONS AND DESCRIPTIVE DISTRIBUTIONS

By the common usage definition (see Section 10.1), a random variable is the set of possible numerical outcomes of a randomly determined statistical experiment; it is the quantitative measurement variable that is being used in the experiment. Therefore, if such an experiment is performed, then the numerical data that result represent observed values of the random variable. Such data can be an entire measurement population or they can be measurement samples from such a population (see Sections 3.1 and 3.3). Either way, we saw in earlier chapters how such data can be organized into descriptive distributions (Chapter 4), presented in graphs (Chapter 5), and described by statistical measures of central tendency (Chapter 6) and dispersion (Chapter 7).

From the first introduction of the concepts of population and sample (see Sections 3.1 through 3.5) we indicated that because populations are rarely available, samples are taken and used to make statistical inferences about their populations. Such samples of real-world data allow us to *estimate* the characteristics of populations that would otherwise be unavailable because the populations are too large or too separated to be measured, or because they are hypothetical. Sample information also allows us to go to statistical theory to find an appropriate theoretical, mathematical description or *model* of how the unavailable population is distributed. This, in fact, is what probability distributions really are: theoretical models of population distributions; specifically of population relative frequency distributions. Discrete probability distributions, therefore, are theoretical models of population relative frequency distributions of discrete random variables.

EXAMPLE 10.6   You have formed a new species of rat by means of genetic engineering and now have the entire 500-member physical-population of these rats in your laboratory. Noting that they have one, two, or three black spots on their white coats, you measure all 500 on the discrete quantitative measurement variable (discrete random variable) number of spots. The resulting measurement population is presented as both a frequency and a relative frequency rod graph in Fig. 10-4(*a*). If you select a rat at random from the population, what are the probabilities it will have one spot, two spots, or three spots?

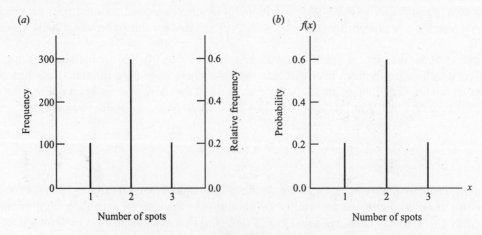

Fig. 10-4

Solution

The sample space defined by the random variable is $S = \{1, 2, 3\}$, and the associated probabilities [calculated with the classical probability function (see Section 8.6) from the frequencies in Fig. 10-4(a)] are: $P(1) = 100/500 = 0.2$, $P(2) = 300/500 = 0.6$, and $P(3) = 100/500 = 0.2$. This discrete probability distribution is presented as a probability rod graph in Fig. 10-4(b). If you now compare the rod graphs in Fig. 10-4(a) and Fig. 10-4(b), it is clear that the population relative frequency distribution is identical to the discrete probability distribution. The point is, for this *real* and *finite* measurement-population, the relative frequency of a measurement in the population is also the probability that this measurement will be randomly selected from the population. The probability that a randomly selected rat will have one spot is 0.2, the probability it will have two spots is 0.6, and the probability it will have three spots is 0.2.

The measurement population described in Example 10.6 was real, finite, and relatively small. This is not, however, the case with most statistical analyses. More typically, the measurement population cannot be measured completely or it is hypothetical. An example of a hypothetical measurement population is dot counts for every possible roll of a die. This population will never be available for analysis, but what is available are: (1) An *empirical sample* (observed rolls of the die) from this population, and (2) a mathematical model for its relative frequency distribution: The discrete probability distribution for the random variable number of dots (see Fig. 10-1).

If in a statistical analysis the most appropriate probability distribution has been selected as the model for a population relative frequency distribution, then sample relative frequency distributions from this population should become increasingly similar to the probability distribution as sample size increases. We have already seen this happen in the die-rolling experiment where, in Fig. 8-1, the sample relative frequency estimates got closer and closer to the theoretical probabilities as sample size increased. For the same 240 die rolls illustrated in Fig. 8-1, Fig. 10-5 shows, for all six dot numbers, a comparison between sample relative frequencies and the theoretical discrete probability distribution for this experiment.

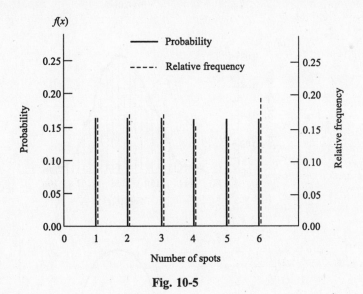

Fig. 10-5

10.6   THE RELATIONSHIP BETWEEN CONTINUOUS PROBABILITY DISTRIBUTIONS AND DESCRIPTIVE DISTRIBUTIONS

In Section 10.5 we indicated that discrete probability distributions are theoretical, mathematical models of population relative frequency distributions of discrete random variables. Similarly, continuous probability distributions are theoretical, mathematical models of population relative frequency distributions of continuous random variables.

To understand this new interpretation of continuous probability distributions, let us return to the genetically engineered rats of Example 10.6. It is now several years later and there are many thousands of these rats. We will consider them to be a sample from an infinitely large hypothetical population of all such rats in the present and future. From these rats we take a random physical-sample of 100 adult males and measure them on the continuous quantitative measurement variable (continuous random variable)—weight in grams. While weight is a continuous variable, it is only measured to the nearest gram. The resulting data are shown as a grouped relative frequency distribution (see Section 4.5) in Fig. 10-6(*a*). The data has been grouped into nine classes, each having a class width of 3 g. The resulting histogram and related polygon (see Sections 5.4 and 5.6) are symmetrical and unimodal.

After returning the 100-male sample to the population, we now take a new random sample of 500 adult males and this time weigh them to the nearest 0.1 g. The resulting data, grouped into 28 equal-width classes (1 g each), are shown as a relative frequency histogram and its related polygon in Fig. 10-6(*b*). Again the graphs are symmetrical and unimodal.

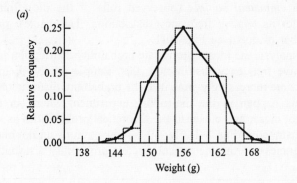

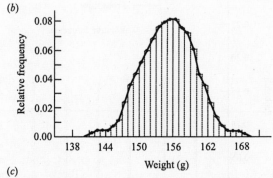

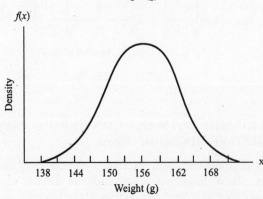

Fig. 10-6

If this process of sampling were continued indefinitely, and each time the sample size n was increased, the measurement taken to more decimal places, and the class width made progressively smaller, then as sample size approached population size N the relative frequency polygon would become a smooth curve called a relative frequency curve, as shown in Fig. 10-6(c). The ultimate limit to this process is when $n = N$, at which point the sample distribution has become the population distribution. While such a population distribution is almost never available, statistical theory provides theoretical models for it in the form of continuous probability distributions. This is why continuous probability distributions are also called the *limiting form* of a relative frequency distribution; their theoretical curves are the curves that sample relative frequency polygons should approach as a limit as n approaches N, as measurement is taken to more decimal places, and as class width is progressively decreased.

A model probability distribution is selected to achieve the closest possible approximation of the population relative frequency distribution. This selection is done from information about the statistical experiment that generated the data and from analysis of the sample characteristics. If the selected probability distribution is a good approximation, then it can be used to estimate population relative frequencies. This is possible because the probability available from a probability distribution—the probability that the random variable X will take on some value in the interval from a to b—is also an estimate of the population relative frequency in that interval. Thus, such probabilities also give the *expected relative frequency* in future samples of values of X in the interval from a to b.

The relative frequency polygons in Fig. 10-6(a) and (b) are essentially symmetrical and bell-shaped, with tails that approach the X axis. These traits characterize the most important and useful continuous probability distribution: the *normal distribution* (or *normal probability distribution*, or *normal probability density function*). From the version of a normal probability distribution shown in Fig. 10-6(c), you can see that it would be a good approximation for the descriptive (empirical) distributions shown in Fig. 10-6(a) and (b). Because many real-world continuous random variables generate relative frequency distributions that can be fit by normal distributions, this distribution is the main topic of a chapter (Chapter 12) and then becomes of great importance throughout the remainder of the book.

10.7 CUMULATIVE DISTRIBUTION FUNCTION OF A DISCRETE RANDOM VARIABLE

In Section 10.5 we indicated that discrete probability distributions (or probability mass functions) are theoretical, mathematical models of population relative frequency distributions of discrete random variables. Similarly, the *cumulative distribution function* (also called a *distribution function* or *cumulative probability distribution*) of a discrete random variable is the theoretical, mathematical model of the population "or less" cumulative distribution for that variable (see Section 4.9 and Problem 5.26). Thus, as the population "or less" cumulative relative frequency distribution of a random variable X gives the relative frequency in the population of values equal to or less than $X = x$, the cumulative distribution function gives the probability that the random variable X will take on a value that is equal to or less than $X = x$. This function, denoted by $F(x)$, is defined for all real numbers $(-\infty < x < \infty)$ by

$$F(x) = P(X \le x) \tag{10.2}$$

If X is a discrete random variable and we want to know $P(X \le a)$ for any real number a, then this probability can be calculated with this formula

$$F(a) = \sum_{x \le a} f(x) \tag{10.3}$$

where the symbol $\sum_{x \le a} f(x)$ means: Take the sum of $f(x)$ (the discrete probability distribution) for all values of x less than or equal to a.

EXAMPLE 10.7   For the experiment of rolling a die with the random variable number of dots, convert the discrete probability distribution in Table 10.1 into a cumulative distribution function. Summarize the function with the standard symbolic notation and then graph the function.

Solution

The cumulative distribution function for this experiment and random variable can be summarized as shown in either Fig. 10-7(a) or Fig. 10-7(b), with the abbreviated version in Fig. 10-7(b) the more typical version.

These summaries utilize the two formulas described above: $F(x) = P(X \leq x)$, for all real numbers in the interval $-\infty < x < \infty$, and $F(a) = \sum_{x \leq a} f(x)$ for any real number a. Together, the two formulas indicate that the probability $F(x)$ of the random variable X taking on any value x equal to or less than any real number a is the sum of all values of $f(x)$ for $x \leq a$. In the summary in Fig. 10-7(a), the summation of $f(x)$ is shown on the left for any value of x in the interval shown to the right. Thus, for the top line, as the random variable cannot take on any values that are less than or equal to any value in the interval $-\infty < x < 1$, $F(x)$ for all values of x in that interval will always be zero. For the second line from the top, as X can now take on one value ($x = 1$) that is less than or equal to any value in the interval $1 \leq x < 2$, $F(x)$ for any value in that interval is $f(1)$, which from Table 10.1 is 1/6. Then, for the third line from the top, as X can now take on two values ($x = 1$, and $x = 2$) that are less than or equal to any values in the interval $2 \leq x < 3$, $F(x)$ is now equal to $f(1) + f(2) = 2/6$ for all x values in that interval. This process of cumulation continues to the bottom line, which states that

(a)
$$F(x) = \begin{cases} 0 & -\infty < x < 1 \\ f(1) = 1/6 & 1 \leq x < 2 \\ f(1) + f(2) = 2/6 & 2 \leq x < 3 \\ f(1) + f(2) + f(3) = 3/6 & 3 \leq x < 4 \\ f(1) + f(2) + f(3) + f(4) = 4/6 & 4 \leq x < 5 \\ f(1) + f(2) + f(3) + f(4) + f(5) = 5/6 & 5 \leq x < 6 \\ f(1) + f(2) + f(3) + f(4) + f(5) + f(6) = 6/6 & 6 \leq x < \infty \end{cases}$$

(b)
$$F(x) = \begin{cases} 0 & x < 1 \\ 1/6 & 1 \leq x < 2 \\ 2/6 & 2 \leq x < 3 \\ 3/6 & 3 \leq x < 4 \\ 4/6 & 4 \leq x < 5 \\ 5/6 & 5 \leq x < 6 \\ 6/6 & 6 \leq x \end{cases}$$

(c)

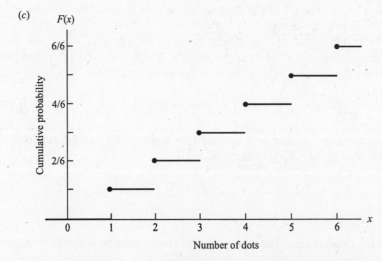

Fig. 10-7

$F(x) = f(1) + f(2) + f(3) + f(4) + f(5) + f(6) = 6/6$ for all x values that are in the interval $6 \leq x < \infty$. The version of this summary in Fig. 10-7(b) simply gives the calculated $F(x)$ values on the left for all x values in the interval on the right.

The graph of this cumulative distribution function is shown in Fig. 10-7(c). It is very similar to the "or less" cumulative graphs for discrete variables in Chapter 5 (see Fig. 5-32 and 5-33). However, while in that type of graph the dot above random-variable values indicated "or less" cumulative frequency, relative frequency, or percentage, now the dot above a random-variable value indicates $F(x)$ at that value: The probability that X can take on the value of x or less. The straight line extending to the right from the dot indicates that $F(x)$ remains constant up to the next random-variable value. Thus, for example, the dot above $x = 3$ indicates that $F(3) = 3/6$, that the probability of X taking on a value equal to or less than 3 is 3/6; and the height of the line above $x = 3.5$ indicates that $F(3.5)$ also equals 3/6, that the probability of X taking on a value equal to or less than 3.5 is 3/6.

The graph in Fig. 10-7(c) shows why the cumulative distribution function for a discrete random variable is called a *step function*: its values change in discrete steps at each value of the random variable. The size of the step at each value of the random variable ($X = x$) is equal to $P(X = x) = f(x)$. Thus, the step between $x = 2$ and $x = 3$ is equal to $P(x = 3) = f(3) = 1/6$.

You can see two other properties of the cumulative distribution function from Fig. 10-7(c): (1) $F(x)$ is always greater than or equal to zero [$F(x) \geq 0$], and (2) $F(x)$ always increases as $X = x$ increases, arriving at $F(x) = 1$ for the largest possible value that the random variable can take on.

General rules that are true for any discrete-variable cumulative distribution function $F(x)$ are, given any two real numbers a and b with $a < b$,

$$P(a < X \leq b) = F(b) - F(a) \tag{10.4}$$
$$P(a \leq X \leq b) = F(b) - F(a) + f(a) \tag{10.5}$$
$$P(a < X < b) = F(b) - F(a) - f(b) \tag{10.6}$$
$$P(a \leq X < b) = F(b) - F(a) + f(a) - f(b) \tag{10.7}$$

where $F(x)$ is a value of the cumulative distribution function and $f(x)$ is a value of the discrete probability distribution. Applications of these rules are provided in Problems 10.7 and 10.8.

10.8   CUMULATIVE DISTRIBUTION FUNCTION OF A CONTINUOUS RANDOM VARIABLE

The *cumulative distribution function of a continuous random variable* is defined in the same way as the cumulative distribution function of a discrete random variable (see Section 10.7). Thus, for all real numbers $(-\infty < x < \infty)$, $F(x) = P(X \leq x)$. Where the cumulative distribution function of a continuous random variable differs from the cumulative distribution function of a discrete random variable is in its calculation and presentation. For discrete random variables, we find $P(X \leq a)$ for any real number a by calculating $F(a) = \sum_{x \leq a} f(x)$. For continuous random variables, by contrast, we find $P(X \leq a)$ by calculating

$$F(a) = \int_{-\infty}^{a} f(x)\,dx \tag{10.8}$$

This formula uses the techniques from integral calculus that we introduced in Section 10.4. It indicates that to find $F(a)$, $f(x)$ must be integrated from $-\infty$ to a. We have illustrated what this means in Fig. 10-8 where $F(a)$, the shaded area that extends above the X axis from $-\infty$ to a, is the probability that the continuous random variable X will take on a value that is equal to or less than a. To understand what this means, recall from Section 10.4 that for a continuous distribution, the probability is calculated only for intervals: The probability that X will take on some value in the interval from a to b. Here we are determining the probability that X will take on a value in the interval from $-\infty$ to a, and to do this we are summing (integrating) the infinite number of vertical density lines between $-\infty$ and a.

The graph of a cumulative distribution function for a continuous random variable is the smooth-curve theoretical version of the population "or less" relative frequency ogive. (An ogive is a graphical

(a)

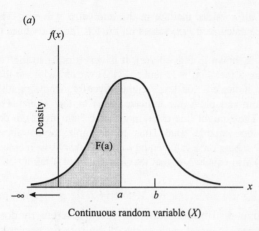

Continuous random variable (X)

(b)

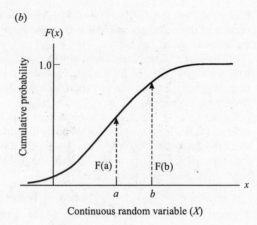

Continuous random variable (X)

Fig. 10-8

representation of a continuous cumulative distribution, as described in Section 5.11.) For the continuous probability distribution shown in Fig. 10-8(a), the smooth-curve ogive is shown in Fig. 10-8(b), where now the shaded area under the density curve in Fig. 10-8(a) over the interval from $-\infty$ to a has become the height of $F(x)$ above the X axis at a. Note that because the total area (probability) under the continuous probability distribution is 1.0, $F(x)$ rises continuously to the right to a maximum cumulative probability of 1.0.

It is always true for continuous random variables that $P(X=x)=0$ and, therefore, that $P(a < X \leq b) = P(a \leq X \leq b) = P(a < X < b) = P(a \leq X < b)$. As a consequence, the four equations for calculating the cumulative distribution function $F(x)$ of a discrete random variable [equations (10.4) through (10.7)] become $F(b) - F(a)$ for the cumulative distribution of a continuous random variable. Stated formally, for any continuous-variable cumulative distribution function $F(x)$, given for any two real numbers a and b that $a < b$, then

$$P(a < X \leq b) = P(a \leq X \leq b) = P(a < X < b) = P(a \leq X < b) = F(b) - F(a) \qquad (10.9)$$

This relationship for continuous random variables, together with the rules for discrete random variables, were used to construct the probability tables presented in the Appendix.

10.9   THE EXPECTED VALUE OF A DISCRETE RANDOM VARIABLE

A probability distribution has a *mean* that is also known as the *expected value* (or *mathematical expectation*, or *expectation*). To understand this concept, recall that the probability distribution of a random variable is the theoretical, mathematical model of the population relative frequency distribution of that variable (see Sections 10.5 and 10.6). Because relative frequency distributions can be described by statistical measures, probability distributions can also be described by comparable statistical measures. Both types of distributions have means as measures of central tendency, and both types have variances and standard deviations as measures of dispersion.

The expected value (or mean) of the probability distribution of a discrete random variable is defined as follows:

If X is a discrete random variable that can take on the values $x_1, x_2, \ldots, x_k$ with the respective probabilities $f(x_1), f(x_2), \ldots, f(x_k)$, then the expected value of X, denoted by $E(X)$, is

$$E(X) = \mu = \sum_{i=1}^{k} x_i f(x_i) = \sum_{x} x f(x) \qquad (10.10)$$

EXAMPLE 10.8   For the rat-selection experiment in Example 10.6, with the discrete random variable number of spots, determine the means for both the frequency/relative-frequency distribution in Fig. 10-4(a) and the discrete probability distribution in Fig. 10-4(b).

Solution

Applying the nongrouped frequency-distribution formula for arithmetic means [equation (6.11)] to the population data in Fig. 10-4(a),

$$\mu = \frac{\sum_{i=1}^{k} f_i x_i}{N} = \frac{(100 \times 1) + (300 \times 2) + (100 \times 3)}{500} = \frac{100 + 600 + 300}{500} = 2.0$$

This arithmetic-mean formula can also be written to make use of the relative frequency scale in Fig. 10-4(a)

$$\mu = \sum_{i=1}^{k} \times \left[(x_i) \times \left(\frac{f_i}{N} \right) \right] = \sum \left[(\text{value of } x_i) \times (\text{relative frequency of } x_i \text{ in the population}) \right] \qquad (10.11)$$

And thus

$$\mu = (1 \times 0.2) + (2 \times 0.6) + (3 \times 0.2) = 2.0$$

Applying the formula for E(X) to the data in Fig. 10-4(b),

$$E(X) = \sum_x x f(x) = (1 \times 0.2) + (2 \times 0.6) + (3 \times 0.2) = 0.2 + 1.2 + 0.6 = 2.0$$

You will probably have noted, in this example, that the E(X) calculation is identical to the second calculation of μ. This is because, as we stated in Example 10.6, for a real and finite measurement population such as this, the relative frequency of a measurement in the population is also the probability that this measurement will be randomly selected from the population.

The expected value E(X) is considered to be a mean because it indicates that in many repetitions of this experiment, you can "expect" this value to be the average of the results. Because E(X) is the mean of the mathematical model of the population distribution it is also given the symbol μ, or the symbols μ_x or μ_y if it is necessary to specify it is for random variable X or Y.

The expected value of a discrete random variable is considered to be a *weighted mean* of that variable, because the formula for the expected value is a reduced version of such a population weighted-mean formula. E(X) is the weighted mean of all possible values that X can take on with each value weighted by its probability. In Section 6.9, a weighted mean for a population was defined by equation (6.18)

$$\mu_w = \frac{\sum_{i=1}^{k} w_i x_i}{\sum_{i=1}^{k} w_i}$$

Thus for E(X)

$$\mu_w = \mu_x = \mu = E(X) = \frac{\sum_{i=1}^{k} x_i f(x_i)}{\sum_{i=1}^{k} f(x_i)} \qquad (10.12)$$

However, with expected values we are always dealing with complete probability distributions, so the sum of the probabilities in the denominator of the formula will always be equal to one. Therefore, this formula always reduces to equation (10.10).

The *expected value of a function of a discrete random variable* can also be calculated. If X is a discrete random variable that can take on the values $x_1, x_2, \ldots, x_k$ with the respective probabilities $f(x_1)$, $f(x_2), \ldots, f(x_k)$, and $g(X)$ is a function of X, then the expected value of $g(X)$, or $E[g(X)]$ is

$$E[g(X)] = \sum_{i=1}^{k} g(x_i) f(x_i) = \sum_x g(x) f(x) \tag{10.13}$$

In words, to get the expected value of the function of a random variable, take the sum of the products of the function-value at x_i times the probability of $X = x_i$.

EXAMPLE 10.9 What are the expected values of the following functions of the discrete random variable X: (a) X^2; (b) $a + bX$, where a and b are constants; (c) $(X - a)^2$, where a is a constant?

Solution

(a) $E(X^2) = \sum\limits_{i=1}^{k} x_i^2 f(x_i) = \sum\limits_x x^2 f(x)$

(b) $E(a + bX) = \sum\limits_{i=1}^{k} (a + bx_i) f(x_i) = \sum\limits_x (a + bx) f(x)$

This equation can be simplified by using summation-notation manipulations (see Section 1.22 and Problems 1.41 to 1.46)

$$E(a + bX) = \sum_x (a + bx) f(x) = \sum_x a f(x) + \sum_x bx f(x) = a \sum_x f(x) + b \sum_x x f(x)$$

As $\sum\limits_x f(x) = 1$ (see Section 10.3) and $\sum\limits_x x f(x) = E(X)$ [equation (10.10)]

$$E(a + bX) = a + bE(X)$$

(c) $E[(X - a)^2] = \sum\limits_{i=1}^{k} (x_i - a)^2 f(x_i) = \sum\limits_x (x - a)^2 f(x)$

Again using summation notation manipulations,

$$E[(X - a)^2] = \sum_x (x - a)^2 f(x) = \sum_x (x^2 - 2ax + a^2) f(x)$$

$$= \sum_x x^2 f(x) - \sum_x 2ax f(x) + \sum_x a^2 f(x)$$

$$= \sum_x x^2 f(x) - 2a \sum_x x f(x) + a^2 \sum_x f(x)$$

As $\sum\limits_x x^2 f(x) = E(X^2)$, $\sum\limits_x x f(x) = E(X)$, and $\sum\limits_x f(x) = 1$,

$$E[(X - a)^2] = E(X^2) - 2aE(X) + a^2$$

10.10 EXPECTED VALUE OF A CONTINUOUS RANDOM VARIABLE

The expected value $E(X)$ of the probability distribution of a continuous random variable is comparable to the $E(X)$ of a discrete random variable. The only difference is that the discrete expected value is defined with summation notation whereas the continuous expected value is defined with integral calculus.

If X is a continuous random variable with density function $f(x)$, then the expected value of X is defined by

$$E(X) = \mu_x = \mu = \int_{-\infty}^{\infty} x f(x) \, dx \tag{10.14}$$

Note that, as with $E(X)$ for a discrete variable, $E(X)$ for a continuous variable is considered to be the mean ($\mu_x = \mu$) of both the continuous variable X and the continuous probability distribution $f(x)$.

It should be emphasized again that knowledge of calculus is not required for this book. You will not be asked to use the formulas involving calculus in this or other chapters, but the formulas will provide precalculated results that we will make use of.

10.11 THE VARIANCE AND STANDARD DEVIATION OF A DISCRETE RANDOM VARIABLE

An important measure of dispersion is the *variance of a discrete random variable*. For a discrete random variable X, the variance of X (and of its probability distribution) is the *expected value of the squared deviation of X from its mean*.

$$\text{Var}(X) = \sigma_x^2 = \sigma^2 = E[(X - E\langle X\rangle)^2] = E[(X - \mu)^2] \tag{10.15}$$

Now, we know from Section 10.9 that $E[g(X)] = \sum_x g(x)f(x)$, so we can define the variance as follows:

If X is a discrete random variable that can take on the values $x_1, x_2, \ldots, x_k$ with respective probabilities $f(x_1), f(x_2), \ldots, f(x_k)$, then the variance of X is

$$E\big[(X - \mu)^2\big] = \sigma^2 = \sum_{i=1}^{k}(x_i - \mu)^2 f(x_i) = \sum_x (x - \mu)^2 f(x) \tag{10.16}$$

In Section 7.9 we indicated that the standard deviation of a population of measurements is the positive square root of the variance of those measurements [equation (7.20)]

$$\sigma = \sqrt{\sigma^2}$$

This is also true for the *standard deviation of the discrete random variable X* (or its probability distribution)

$$\sigma = \sqrt{\sigma^2} = \sqrt{\sum_x (x - \mu)^2 f(x)} \tag{10.17}$$

EXAMPLE 10.10 For the rat-selection experiment in Example 10.6, in which the discrete random variable is the number of spots, determine the variance for both the frequency/relative frequency distribution in Fig. 10-4(*a*) and the discrete probability distribution in Fig. 10-4(*b*).

Solution

We calculate the variance for the frequency/relative frequency distribution with the nongrouped frequency-distribution formula for variances. In Section 7.12, we developed this frequency distribution formula [equation (7.29)] for the standard deviation

$$\sigma = \sqrt{\frac{\sum_{i=1}^{k} f_i(x_i - \mu)^2}{N}}$$

which for the variance becomes

$$\sigma^2 = \frac{\sum_{i=1}^{k} f_i(x_i - \mu)^2}{N} \tag{10.18}$$

Applying this equation to the population data in Fig. 10-4(a), and using $\mu = \dfrac{\sum_{i=1}^{k} f_i x_i}{N} = 2.0$ from Example 10.8, we get

$$\sigma^2 = \frac{\sum_{i=1}^{k} f_i(x_i - \mu)^2}{N} = \frac{100(1-2.0)^2 + 300(2-2.0)^2 + 100(3-2.0)^2}{500}$$

$$= \frac{100 + 0 + 100}{500} = 0.4$$

This variance formula can also be written to make use of the relative frequency scale in Fig. 10-4(a)

$$\sigma^2 = \sum_{i=1}^{k} \left[(x_i - \mu)^2 \times \left(\frac{f_i}{N} \right) \right] \tag{10.19}$$

$$= \sum \left[(\text{deviation of } x_i \text{ from } \mu)^2 \times \left(\begin{array}{c} \text{relative frequency of this} \\ \text{deviation in the population} \end{array} \right) \right]$$

$$= [(1-2.0)^2 \times (0.2)] + [(2-2.0)^2 \times (0.6)] + [(3-2.0)^2 \times (0.2)] = 0.4$$

Applying equation (10.16) to the data in Fig. 10-4(b) and now using $E(X) = \mu = 2.0$ from Example 10.8,

$$\sigma^2 = \sum_x (x - \mu)^2 f(x)$$

$$= [(1-2.0)^2 \times (0.2)] + [(2-2.0)^2 \times (0.6)] + [(3-2.0)^2(0.2)]$$

$$= 0.4$$

Here for variances, as was true for means, the relative frequency formula is identical to the probability formula. Again it is because for such a real and finite population, relative frequency equals probability.

10.12 COMPUTATIONAL FORMULAS FOR THE VARIANCE AND STANDARD DEVIATION OF A DISCRETE RANDOM VARIABLE

In Chapters 6 and 7 we made the distinction between definitional and computational formulas for statistical measures. Thus from Section 7.12 we know that the definitional frequency-distribution formula for the population variance [equation (10.18)]

$$\sigma^2 = \frac{\sum_{i=1}^{k} f_i(x_i - \mu)^2}{N}$$

can be modfied to form a computational formula [equation (7.31) squared]

$$\sigma^2 = \frac{\sum_{i=1}^{k} f_i x_i^2}{N} - \mu^2 \tag{10.20}$$

and the relative frequency version of this computational formula is

$$\sigma^2 = \sum_{i=1}^{k} \left[x_i^2 \left(\frac{f_i}{N} \right) \right] - \mu^2 \tag{10.21}$$

There are comparable formulas for probability distributions. Thus, the definitional formula for the variance of a discrete probability distribution can be used to derive a computational formula. The derivation begins with equation (10.16)

$$\sigma^2 = \sum_x (x - \mu)^2 f(x)$$

$$= \sum_x (x^2 - 2x\mu + \mu^2) f(x)$$

$$= \sum_x x^2 f(x) - \sum_x 2x\mu f(x) + \sum_x \mu^2 f(x)$$

$$= \sum_x x^2 f(x) - 2\mu \sum_x x f(x) + \mu^2 \sum_x f(x)$$

Knowing that $\sum_x f(x) = 1$ and that $\sum_x x f(x) = \mu$,

$$\sigma^2 = \sum_x x^2 f(x) - 2\mu^2 + \mu^2$$

$$= \sum_x x^2 f(x) - \mu^2$$

Knowing that $E(X) = \mu$ and $E(X^2) = \sum_x x^2 f(x)$, we get this computational formula

$$\sigma^2 = \sum_x x^2 f(x) - \mu^2 = E(X^2) - [E(X)]^2 \tag{10.22}$$

Using this computational formula on the random variable described in Example 10.10, with $E(X) = \mu = 2.0$

$$\sigma^2 = \sum_x x^2 f(x) - \mu^2 = [(1 \times 0.2) + (4 \times 0.6) + (9 \times 0.2)] - (2.0)^2 = 0.4$$

This is the same value we got in Example 10.10 using the definitional formula.

The computational formula for the standard deviation, then, is

$$\sigma = \sqrt{\sum_x x^2 f(x) - \mu^2} = \sqrt{E(X^2) - [E(X)]^2} \tag{10.23}$$

EXAMPLE 10.11   For the experiment of rolling a die with the discrete random variable number of dots (see Table 10.1), determine the variance of that variable using both the definitional and the computational formulas described above.

Solution

First we need to calculate the expected value of the random variable

$$E(X) = \sum_x x f(x) = \left(1 \times \frac{1}{6}\right) + \left(2 \times \frac{1}{6}\right) + \left(3 \times \frac{1}{6}\right) + \left(4 \times \frac{1}{6}\right) + \left(5 \times \frac{1}{6}\right) + \left(6 \times \frac{1}{6}\right)$$

$$= (0.166667) + (0.333333) + (0.5) + (0.666667) + (0.833333) + (1.0)$$

$$= 3.5$$

Thus, $E(X) = \mu = 3.5$. We then apply the definitional formula [equation (10.16)] to the probability distribution for this variable shown in Table 10.1.

$$\sigma^2 = \sum_x (x - \mu)^2 f(x)$$

$$= \left[(1 - 3.5)^2 \times \frac{1}{6}\right] + \left[(2 - 3.5)^2 \times \frac{1}{6}\right] + \left[(3 - 3.5)^2 \times \frac{1}{6}\right]$$

$$+ \left[(4 - 3.5)^2 \times \frac{1}{6}\right] + \left[(5 - 3.5)^2 \times \frac{1}{6}\right] + \left[(6 - 3.5) \times \frac{1}{6}\right]$$

$$= \frac{6.25}{6} + \frac{2.25}{6} + \frac{0.25}{6} + \frac{0.25}{6} + \frac{2.25}{6} + \frac{6.25}{6}$$

$$= \frac{17.50}{6} = 2.92$$

Applying the computational formula [equation (10.22)] to the same data

$$\sigma^2 = \sum_x x^2 f(x) - \mu^2$$

$$= \left[\left(1 \times \frac{1}{6}\right) + \left(4 \times \frac{1}{6}\right) + \left(9 \times \frac{1}{6}\right) + \left(16 \times \frac{1}{6}\right) + \left(25 \times \frac{1}{6}\right) + \left(36 \times \frac{1}{6}\right)\right] - (3.5)^2$$

$$= \frac{91}{6} - 12.25 = 2.92$$

10.13   THE VARIANCE AND STANDARD DEVIATION OF A CONTINUOUS RANDOM VARIABLE

As you would expect from earlier sections of this chapter, a continuous random variable has a variance and a standard deviation that are comparable to σ^2 and σ for a discrete random variable. However, while the discrete values are defined with summation notation

$$\text{Var}(X) = \sigma_x^2 = \sigma^2 = E[(X - \mu)^2] = \sum_x (x - \mu)^2 f(x)$$

and

$$\sigma = \sqrt{\sigma^2}$$

the continuous values are defined with integral calculus:

If X is a continuous random variable with density function $f(x)$, then the variance of X is defined by

$$\text{Var}(X) = \sigma_x^2 = \sigma^2 = E[(X - \mu)^2] = \int_{-\infty}^{\infty} (x - \mu)^2 f(x)\, dx \qquad (10.24)$$

and the standard deviation is

$$\sigma = \sqrt{\sigma^2}$$

10.14   CHEBYSHEV'S THEOREM AND THE EMPIRICAL RULE

Chebyshev's theorem describes the relation between the standard deviation of a distribution and the concentration of values about the mean of the distribution (see Section 7.15). A version of the theorem that applies to probability distributions is:

For any number $k \geq 1$, the probability that a random variable X with mean μ and standard deviation σ will take on a value in the interval $\mu \pm k\sigma$ is *at least* $1 - \dfrac{1}{k^2}$.

EXAMPLE 10.12 The probability distribution in Table 10.3 is for the experiment of rolling a die twice with the discrete random variable total number of dots for the two rolls. (See Problem 10.2 for how this table was determined.) What is the probability that the random variable will take on a value in the interval $\mu \pm 2\sigma$?

Table 10.3

Total of dots x	Probability $f(x)$
2	0.02778
3	0.05556
4	0.08334
5	0.11112
6	0.13890
7	0.16668
8	0.13890
9	0.11112
10	0.08334
11	0.05556
12	0.02778
$\sum$	1.00

Solution

A solution using Chebyshev's theorem ($k = 2$), is that the probability is at least

$$1 - \frac{1}{k^2} = 1 - \frac{1}{4} = \frac{3}{4} = 0.75$$

The exact solution to this problem requires a summation of the probabilities for all possible values of X in the interval $\mu \pm 2\sigma$. The mean and standard deviation for the distribution of this variable, shown in Table 10.4, are $\mu = 7.00$ and $\sigma = 2.41$. (See Problem 10.15 for how this table and the calculations were determined.) Therefore

$$\mu \pm 2\sigma \text{ is } 7.00 \pm 2(2.41), \text{ or } 7.00 \pm 4.82, \text{ or } 2.18 \text{ to } 11.82$$

As X can take on the values 3 through 11 in this interval, we can say from the probability distribution in Tables 10.3 and 10.4 that the probability of X taking on any one of these values is

$$\sum_{x=3}^{11} f(x) = 0.94452, \text{ or } 0.94$$

The empirical rule also describes the relation between the standard deviation and the concentration of values about the mean of a distribution. A version of this rule that applies to normal probability distributions, which are continuous distributions, is:

For a random variable X with mean μ and standard deviation σ that has a probability distribution that is approximately normally distributed, there is ≈ 0.68 probability that X will take on a value in the interval $\mu \pm \sigma$, ≈ 0.95 probability that X will take on a value in the interval $\mu \pm 2\sigma$, and ≈ 1.00 probability that X will take on a value in the interval $\mu \pm 3\sigma$.

Table 10.4

Total of dots x	x^2	Probability f(x)	xf(x)	$x^2f(x)$
2	4	0.02778	0.05556	0.11112
3	9	0.05556	0.16668	0.50004
4	16	0.08334	0.33336	1.33344
5	25	0.11112	0.55560	2.77800
6	36	0.13890	0.83340	5.00040
7	49	0.16668	1.16676	8.16732
8	64	0.13890	1.11120	8.88960
9	81	0.11112	1.00008	9.00072
10	100	0.08334	0.83340	8.33400
11	121	0.05556	0.61116	6.72276
12	144	0.02778	0.33336	4.00032
$\sum$		1.00	7.00056	54.83772

$$\sigma^2 = \sum x^2 f(x) - \mu^2 = 54.83772 - (7.00056)^2 = 54.83772 - 49.00784 = 5.82988, \text{ or } 5.83$$

$$\sigma = \sqrt{\sigma^2} = \sqrt{5.82988} = 2.41451, \text{ or } 2.41$$

EXAMPLE 10.13 If we treat the discrete random variable number-of-dots in the experiment of rolling a die twice (Example 10.12) "as if it were continuous" (see Problems 5.9 and 5.26) and assume that its probability distribution is approximately normally distributed (unimodal, roughly mound-shaped, essentially symmetrical), then this version of the empirical rule should apply to this variable. How well does this rule apply for the interval $\mu \pm 2\sigma$?

Solution

It can be seen from the results in Example 10.12 that the exact probability is 0.94, the probability predicted by the empirical rule is ≈ 0.95, and Chebyshev's theorem only says that the probability is at least 0.75.

Solved Problems

RANDOM VARIABLES AND THEIR PROBABILITY DISTRIBUTIONS

10.1 For each of the following, indicate whether it is a random variable and, if so, whether it is discrete or continuous: (a) determining whether airplanes arrive on time, (b) classifying birds by their species, (c) age of female applicants to a medical school, (d) number of measureable earthquakes in California in a six-month period, (e) number of five-card poker hands you must be dealt before you get a hand with four queens, (f) wind velocity in miles per hour, (g) number of correct answers achieved on a twenty-question examination, (h) weight in grams of each of 400 melons.

Solution

(a) As is, this is not a random variable. If, however, one counted the number of planes that arrive "on time" in a sample of 200, then that would be a discrete random variable.

(b) As is, this is not a random variable. If, however, one counted the number of sparrows in a sample of 80 birds, then that would be a discrete random variable.

(c) Continuous random variable

(d) Discrete random variable

(e) Discrete random variable that can take on a countably infinite number of values

(f) Continuous random variable

(g) Discrete random variable

(h) Continuous random variable

10.2 For the experiment of rolling a die twice, present the discrete probability distribution for the random variable total number of dots for the two rolls as both a probability table and a probability rod graph.

Solution

To solve this problem, we first convert the probabilities for the simple events in the sample space (shown as a tree diagram in Fig. 9-12) to probabilities for the simple events in the new sample space defined by the random variable $S = \{2, 3, 4, 5, 6, 7, 8, 9, 10, 11, 12\}$. This means we must find $f(2), f(3), \ldots, f(11), f(12)$.

As all simple events (paths) shown in the tree diagram in Fig. 9-12 are mutually exclusive events, we can use Property 4 from set theory (Section 8.6) to find the new probability values. Thus

$$f(2) = P(1a \cap 1b) = 0.02778$$

$$f(3) = P(1a \cap 2b) + P(2a \cap 1b) = 2(0.02778) = 0.05556$$

$$f(4) = P(1a \cap 3b) + P(2a \cap 2b) + P(3a \cap 1b) = 3(0.02778) = 0.08334$$

$$f(5) = P(1a \cap 4b) + P(2a \cap 3b) + P(3a \cap 2b) + P(4a \cap 1b) = 4(0.02778) = 0.11112$$

$$f(6) = P(1a \cap 5b) + P(2a \cap 4b) + P(3a \cap 3b) + P(4a \cap 2b) + P(5a \cap 1b) = 5(0.02778) = 0.13890$$

$$f(7) = P(1a \cap 6b) + P(2a \cap 5b) + P(3a \cap 4b) + P(4a \cap 3b) + P(5a \cap 2b) + P(6a \cap 1b) = 6(0.02778)$$

$$= 0.16668$$

$$f(8) = P(2a \cap 6b) + P(3a \cap 5b) + P(4a \cap 4b) + P(5a \cap 3b) + P(6a \cap 2b) = 5(0.02778) = 0.13890$$

$$f(9) = P(3a \cap 6b) + P(4a \cap 5b) + P(5a \cap 4b) + P(6a \cap 3b) = 4(0.02778) = 0.11112$$

$$f(10) = P(4a \cap 6b) + P(5a \cap 5b) + P(6a \cap 4b) = 3(0.02778) = 0.08334$$

$$f(11) = P(5a \cap 6b) + P(6a \cap 5b) = 2(0.02778) = 0.05556$$

$$f(12) = P(6a \cap 6b) = 0.02778$$

This discrete probability distribution was presented as a table in Table 10.3 and is shown as a rod graph in Fig. 10-9.

10.3 Two participants were randomly selected from the 160 who took part in the cold-vaccination study of Example 9.1 and Table 9.1. There was no replacement between selections. Construct a probability table that presents the discrete probability distribution for the random variable number of people selected who got a cold during the year.

Solution

If we let C_1 and C_2 represent first and second selections of people who got colds and N_1 and N_2 represent first and second selections of people who did not get colds, then the completed tree diagram for the sample space of this experiment is shown in Fig. 10-10(a) and (b).

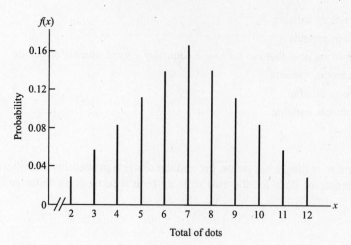

Fig. 10-9

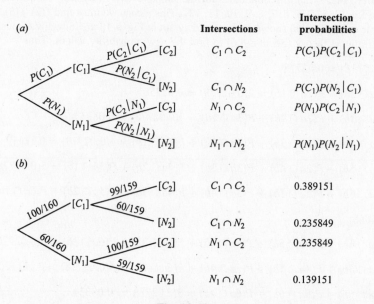

Fig. 10-10

We must now convert the probabilities for this sample space to probabilities for the new sample space defined by the random variable $S = \{0, 1, 2\}$. This means we must find $f(0), f(1)$, and $f(2)$, and to do this we can again use Property 4 from set theory (Section 8.6).

Thus

$$f(0) = P(N_1 \cap N_2) = 0.139151$$

$$f(1) = P(C_1 \cap N_2) + P(N_1 \cap C_2) = 2(0.235849) = 0.471698$$

$$f(2) = P(C_1 \cap C_2) = 0.389151$$

This discrete probability distribution is presented in Table 10.5.

10.4 Explain why the following lists are not discrete probability distributions: (a) $P(X = \text{small}) = f(\text{small}) = 0.5$, $P(X = \text{large}) = f(\text{large}) = 0.5$; (b) $P(X = 0) = f(0) = 0.2$, $P(X = 1) = f(1) = 0.6$, $P(X = 2) = f(2) = 0.3$; (c) $P(X = 3) = f(3) = -0.2$, $P(X = 4) = f(4) = 0.8$.

Table 10.5

Number got colds x	Probability $f(x)$
0	0.139151
1	0.471698
2	0.389151
$\sum$	1.00

Solution

(a) In Section 10.1 it is stated that the values of a random variable must be real numbers, not "small" or "large", so this is not a random variable. As discrete probability distributions are based on random variables (see Section 10.3), this is therefore not a discrete probability distribution.

(b) In Section 10.3 it is stated that for every discrete probability distribution $\sum_x f(x) = P(S) = 1.00$. Here $\sum_x f(x) = 1.1$, so this is not a discrete probability distribution.

(c) In Section 10.3 we said that it must be true for all possible values x of a random variable X that $f(x) \geq 0$. Here $f(3) = -0.2$, so this is not a discrete probability distribution.

CUMULATIVE DISTRIBUTION FUNCTIONS

10.5 For the experiment of rolling a die twice with the random variable total number of dots for the two rolls, convert the discrete probability distribution in Table 10.3 into a cumulative distribution function. Summarize the function with the abbreviated standard summary [see Fig. 10-7(b)] and then graph the function.

Solution

The requested summary and graph are shown in Figs. 10-11(a) and 10-11(b) respectively.

10.6 For the cold-vaccination study with the random variable number of people who got a cold, convert the discrete probability distribution in Table 10.5 into a cumulative distribution function. Summarize the function with the abbreviated standard summary [see Fig. 10-7(b)] and then graph the function.

Solution

The requested summary and graph are shown in Figs. 10-12(a) and 10-12(b) respectively.

10.7 From the graph of the cumulative distribution function in Fig. 10-13, determine the following: (a) $F(3)$, (b) $f(3)$, (c) $P(X > 3)$, (d) $F(3) - F(2)$, (e) $F(3.8)$, (f) $f(3.8)$.

Solution

(a) $F(3) = 0.4$

(b) From Section 10.7 we know that the probability of 3, or $f(3)$, is the size of the step $F(x)$ takes at 3. Therefore, $f(3) = 0.2$.

(c) $P(X) > 3$, the probability that X will take on a value greater than 3, is equal to $\sum_{x>3} f(x) = f(4) + f(5) + f(6)$, which is the same as $1 - F(3)$. Therefore

$$P(X > 3) = 1 - F(3) = 1 - 0.4 = 0.6$$

(a)

$$F(x) = \begin{cases} 0 & x < 2 \\ 0.02778 & 2 \le x < 3 \\ 0.08334 & 3 \le x < 4 \\ 0.16668 & 4 \le x < 5 \\ 0.27780 & 5 \le x < 6 \\ 0.41670 & 6 \le x < 7 \\ 0.58338 & 7 \le x < 8 \\ 0.72228 & 8 \le x < 9 \\ 0.83340 & 9 \le x < 10 \\ 0.91674 & 10 \le x < 11 \\ 0.97230 & 11 \le x < 12 \\ 1.00008 & 12 \le x \end{cases}$$

(b)

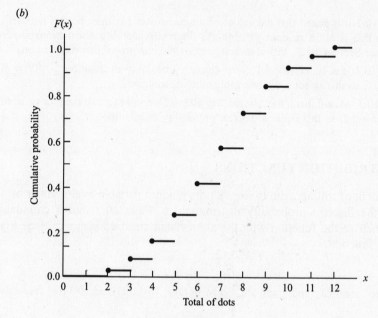

Fig. 10-11

(a)

$$F(x) = \begin{cases} 0 & x < 0 \\ 0.139151 & 0 \le x < 1 \\ 0.610849 & 1 \le x < 2 \\ 1.000000 & 2 \le x \end{cases}$$

(b)

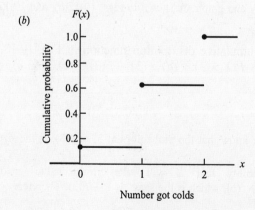

Fig. 10-12

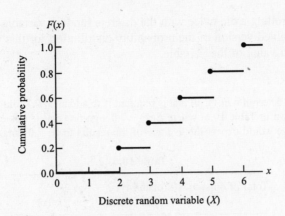

Fig. 10-13

(d) $F(3) - F(2) = f(3) = 0.2$

(e) $F(3.8) = 0.4$

(f) $f(3.8) = 0$

10.8 From the graph of the cumulative distribution function in Fig. 10-13, determine the following:
(a) $P(3 < X \leq 5)$, (b) $P(3 \leq X \leq 5)$, (c) $P(3 < X < 5)$, (d) $P(3 \leq X < 5)$.

Solution

The discrete random variable X in Fig. 10-13 can take on five values: $x = 2, 3, 4, 5, 6$. These five events are mutually exclusive so we can use Property (4) from set theory (Section 8.6) to determine unions of the events

$$P(A_1 \cup A_2 \cup \cdots \cup A_k) = P(A_1) + P(A_2) + \cdots + P(A_k)$$

(a) $P(3 < X \leq 5) = P(4 \cup 5) = f(4) + f(5) = F(5) - F(3) = 0.8 - 0.4 = 0.4$

(b) $P(3 \leq X \leq 5) = P(3 \cup 4 \cup 5) = f(3) + f(4) + f(5) = F(5) - F(3) + f(3) = 0.8 - 0.4 + 0.2 = 0.6$

(c) $P(3 < X < 5) = P(4) = f(4) = F(5) - F(3) - f(5) = 0.8 - 0.4 - 0.2 = 0.2$

(d) $P(3 \leq X < 5) = P(3 \cup 4) = f(3) + f(4) = F(5) - F(3) + f(3) - f(5) = 0.8 - 0.4 + 0.2 - 0.2 = 0.4$

EXPECTED VALUE OF A RANDOM VARIABLE

10.9 For the experiment of rolling a die with the discrete random variable number of dots (Section 10.3), determine the expected value of that variable.

Solution

Applying equation (10.10) for the expected value to the discrete probability distribution in Table 10.1

$$E(X) = \sum_x x f(x) = \left(1 \times \frac{1}{6}\right) + \left(2 \times \frac{1}{6}\right) + \left(3 \times \frac{1}{6}\right) + \left(4 \times \frac{1}{6}\right) + \left(5 \times \frac{1}{6}\right) + \left(6 \times \frac{1}{6}\right)$$

$$= (0.166667) + (0.333333) + (0.5) + (0.666667) + (0.833333) + (1.0)$$

$$= 3.5$$

This means that if the experiment were repeated a large number of times we would expect the average of the results to be 3.5 dots.

10.10 For the experiment of rolling a die twice with the discrete random variable total number of dots for the two rolls, use the tabled version of the probability distribution for this variable in Table 10.3 to determine the expected value of the variable.

Solution

To use Table 10.3 to determine $E(X)$, all that is required is to add an $xf(x)$ column. The resulting table and $E(X)$ calculation are shown in Table 10.6, where $E(X) = 7.0$ signifies that if this experiment were repeated a large number of times, we would expect the average of the results to be 7.0 dots.

Table 10.6

Total of dots x	Probability $f(x)$	$xf(x)$
2	0.02778	0.05556
3	0.05556	0.16668
4	0.08334	0.33336
5	0.11112	0.55560
6	0.13890	0.83340
7	0.16668	1.16676
8	0.13890	1.11120
9	0.11112	1.00008
10	0.08334	0.83340
11	0.05556	0.61116
12	0.02778	0.33336
$\sum$	1.00	7.00056, or 7.0

10.11 In Section 8.8 we indicated that for a given bet if the odds that an event will occur are the same as the betting odds, then it is a fair bet. Now we can define a fair bet in terms of expected values:

If X is a discrete random variable that represents possible outcomes of an experiment as winnings and losses on a bet $x_1, x_2, \ldots, x_k$, where the respective probabilities are $f(x_1), f(x_2), \ldots, f(x_k)$, then then the bet is considered to be a fair bet if $E(X) = \sum_x xf(x) = 0$.

In essence, this means that if the experiment/bet is repeated many times it is a fair bet if in the long run you can expect to come out even—neither winning nor losing money. From this expected-value definition, which of the following would be considered to be a fair bet: (a) on the flip of a coin, if heads you win \$3, if tails you lose \$2; (b) on the single roll of a die, if an even number you win \$3, if an odd number you lose \$3; (c) on drawing a single card from a deck, if a red card you win \$1, if a black card you lose \$2?

Solution

(a) $E(X) = \sum_x xf(x) = \left(\$3 \times \frac{1}{2}\right) + \left(-\$2 \times \frac{1}{2}\right) = \$1.5 - \$1 = \$0.5$

This is not a fair bet because over the long run you can expect to win, on average, \$0.5 per flip.

(b) $E(X) = \sum_x xf(x) = \left(-\$3 \times \frac{1}{6}\right) + \left(\$3 \times \frac{1}{6}\right) + \left(-\$3 \times \frac{1}{6}\right) + \left(\$3 \times \frac{1}{6}\right) + \left(-\$3 \times \frac{1}{6}\right) + \left(\$3 \times \frac{1}{6}\right)$

$= 3\left(-\frac{\$3}{6}\right) + 3\left(\frac{\$3}{6}\right) = -\$1.5 + \$1.5 = \$0$

This is a fair bet; in the long run you can expect to come out even.

(c) $E(X) = \sum_x xf(x) = \left(\$1 \times \dfrac{26}{52}\right) + \left(-\$2 \times \dfrac{26}{52}\right) = \$0.5 - \$1.0 = -\$0.5$

This is not a fair bet; in the long run you can expect to lose, on average, \$0.5 per draw.

10.12 A charity organization is holding a lottery for the following prizes: one \$500 prize, five \$100 prizes, and fifty \$50 prizes. They plan to sell 5,000 tickets for the lottery, and you are asked to set a ticket price that is three times as large as the fair price. What price do you recommend?

Solution

First we determine the probabilities for each prize: $f(\$500) = 1/5,000 = 0.0002$; $f(\$100) = 5/5,000 = 0.001$; $f(\$50) = 50/5,000 = 0.01$. The expected value per ticket is

$$E(X) = \sum_x xf(x) = (\$500 \times 0.0002) + (\$100 \times 0.001) + (\$50 \times 0.01)$$

$$= \$0.10 + \$0.10 + \$0.50 = \$0.70$$

Therefore, the fair price per ticket would be \$0.70, but to give the charity the profit they seek you would recommend that each ticket be \$2.10.

10.13 Table 10.7 is a 4-year (208-week) summary, for an electronics store, of how many of their most popular computer model they sold per week. For inventory purposes, the store manager wants to know: How many of these computers can the store expect to sell in the next 6 months (26 weeks)?

Table 10.7

Computers sold per week x_i	Number of weeks (out of 208) f_i
1	6
2	33
3	50
4	70
5	25
6	17
7	7
$\sum$	208

Solution

First we find the expected value for the discrete variable number of computers sold per week. The necessary table with its $f(x)$ and $xf(x)$ columns and the resulting calculation of $E(X)$ is shown in Table 10.8. Therefore, in the next 6 months the store can expect to sell $26 \times [E(X) = 3.740385] = 97.250010$, or 97 computers.

THE VARIANCE AND STANDARD DEVIATION OF A RANDOM VARIABLE

10.14 For the experiment of flipping a coin three times with the discrete random variable number of heads, use equations (10.22) and (10.23) to determine the variance and the standard deviation of this variable.

Table 10.8

Computers sold per week x	Probability $f(x)$	$xf(x)$
1	0.028846	0.028846
2	0.158654	0.317308
3	0.240385	0.721155
4	0.336538	1.346152
5	0.120192	0.600960
6	0.081731	0.490386
7	0.033654	0.235578
$\sum$	1.00	3.740385

Solution

We must first determine the expected value for the experiment by applying equation (10.10) for $E(X)$ to the probability distribution in Table 10.2

$$E(X) = \sum_x xf(x) = (0 \times 0.125) + (1 \times 0.375) + (2 \times 0.375) + (3 \times 0.125)$$

$$= (0) + (0.375) + (0.750) + (0.375) = 1.5$$

Thus, $E(X) = \mu = 1.5$. We then apply equation (10.22) to the probability distribution in Table 10.2

$$\sigma^2 = \sum_x x^2 f(x) - \mu^2$$

$$= [(0 \times 0.125) + (1 \times 0.375) + (4 \times 0.375) + (9 \times 0.125)] - (1.5)^2$$

$$= 0.75$$

Therefore, using equation (10.23) the standard deviation is

$$\sigma = \sqrt{\sigma^2} = \sqrt{0.75} = 0.87$$

10.15 For the experiment of rolling a die twice with the discrete random variable total number of dots for the two rolls, determine the variance and standard deviation of the variable by applying equations (10.22) and (10.23) to the probability distribution in Table 10.3.

Solution

To use Table 10.3 to determine σ^2 and σ, all that is required is to add x^2 and $x^2 f(x)$ columns. The resulting table and the σ^2 and σ calculations are shown in Table 10.4.

10.16 For the computer-sales experiment with the discrete random variable computers sold per week, determine the variance and standard deviation of this variable by applying equations (10.22) and (10.23) to the probability distribution shown in Table 10.8.

Solution

The necessary table with its x^2 and $x^2 f(x)$ columns and the resulting calculations of σ^2 and σ is shown in Table 10.9.

Table 10.9

Computers sold per week x	x^2	Probability f((x)	xf(x)	$x^2 f(x)$
1	1	0.028846	0.028846	0.028846
2	4	0.158654	0.317308	0.634616
3	9	0.240385	0.721155	2.163465
4	16	0.336538	1.346152	5.384608
5	25	0.120192	0.600960	3.004800
6	36	0.081731	0.490386	2.942316
7	49	0.033654	0.235578	1.649046
$\sum$		1.00	3.740385	15.807697

$$\sigma^2 = \sum x^2 f(x) - \mu^2 = 15.807697 - (3.740385)^2 = 15.807697 - 13.990480 = 1.817217, \text{ or } 1.82$$

$$\sigma = \sqrt{\sigma^2} = \sqrt{1.817217} = 1.348042, \text{ or } 1.35$$

CHEBYSHEV'S THEOREM AND THE EMPIRICAL RULE

10.17 For the computer-sales experiment with the discrete random variable computers sold per week, what is the probability that the random variable will take on a value in the interval $\mu \pm 2\sigma$?

Solution

Using the version of Chebyshev's theorem from Section 10.14, the probability is at least

$$1 - \frac{1}{k^2} = 1 - \frac{1}{2^2} = 0.75$$

Using the exact technique from Example 10.12, knowing that $\mu = 3.74$ and $\sigma = 1.35$ (see Table 10.9),

$$\mu \pm 2\sigma \text{ is } 3.74 \pm 2(1.35), \text{ or } 3.74 \pm 2.70, \text{ or } 1.04 \text{ to } 6.44$$

As X can take on the values 2 through 6 in this interval, we can say from the probability distribution in Tables 10.8 and 10.9 that the probability of X taking on any one of these values is

$$\sum_{x=2}^{6} f(x) = 0.937500, \text{ or } 0.94$$

If we treat this discrete variable as if it were continuous and consider its probability distribution to be approximately normally distributed, then we can apply the probability version of the empirical rule (see Section 10.14). Thus, the probability that X will take on a value in the interval $\mu \pm 2\sigma$ is ≈ 0.95.

Supplementary Problems

RANDOM VARIABLES AND PROBABILITY DISTRIBUTIONS

10.18 For each of the following random variables, indicate whether it is discrete or continuous:　(a) number of cars passing an intersection per hour,　(b) the time each car spends at a stop sign,　(c) weight of sugar (in grams) put on a bowl of cereal,　(d) hours of sunlight each day,　(e) number of babies born each year in a hospital,　(f) heights of plants (in inches) in a meadow.

Ans. (a) Discrete,　(b) continuous,　(c) continuous,　(d) continuous,　(e) discrete,　(f) continuous

10.19 At a company dinner, fifty bills are placed in a hat: three bills for $1, ten bills for $5, 23 bills for $10, 13 bills for $20, and one bill for $50. The bills are thoroughly mixed and then an employee chosen at random draws one bill at random from the hat. What are the probability values for the different kinds of bills in the hat?

Ans. $f(\$1) = 0.06, f(\$5) = 0.20, f(\$10) = 0.46, f(\$20) = 0.26, f(\$50) = 0.02.$

10.20 A sociologist wants to know how many people live in each house in a particular subdivision of 815 houses. He finds that 93 houses have one resident each, 160 houses have two residents each, 320 houses have three residents each, 110 houses have four residents each, 82 houses have five residents each, and 50 houses have six residents each. If one house is chosen at random, what are the probability values for the number of residents per house?

Ans. $f(1) = 0.114, f(2) = 0.196, f(3) = 0.393, f(4) = 0.135, f(5) = 0.101, f(6) = 0.061$

10.21 An ornithologist wants to know the clutch sizes (number of eggs per nest) of song sparrows on an island. Of the 150 song sparrow nests on the island, she finds 4 nests with two eggs, 36 nests with three eggs, 66 nests with four eggs, 40 nests with five eggs, and 4 nests with six eggs. If one nest is chosen at random, what are the probabilities for number of eggs in the nest?

Ans. $f(2) = 0.027, f(3) = 0.240, f(4) = 0.440, f(5) = 0.267, f(6) = 0.027$

CUMULATIVE DISTRIBUTION FUNCTION OF A RANDOM VARIABLE

10.22 For the experiment of flipping a coin three times with the random variable number of heads, convert the discrete probability distribution in Example 10.5 (Table 10.2) into a cumulative distribution function. Summarize the function with the abbreviated standard summary [see Fig. 10-7(b)] and then graph the function.

Ans. The requested summary and graph are shown in Fig. 10-14(a) and (b) respectively.

10.23 From the summary of the cumulative distribution function in Fig. 10-15, determine the following: (a) $F(4)$, (b) $f(4)$, (c) $P(X \geq 3)$, (d) $F(5) - F(4)$, (e) $F(4.4)$, (f) $f(4.4)$.

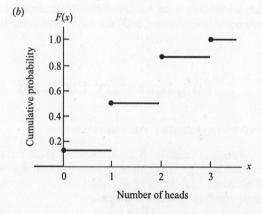

Fig. 10-14

$$F(x) = \begin{cases} 0 & x < 2 \\ 0.10 & 2 \le x < 3 \\ 0.25 & 3 \le x < 4 \\ 0.45 & 4 \le x < 5 \\ 0.70 & 5 \le x < 6 \\ 1.00 & 6 \le x \end{cases}$$

Fig. 10-15

Ans. (a) $F(4) = 0.45$, (b) $f(4) = 0.20$, (c) $P(X \ge 3) = \sum_{x \ge 3} f(x) = f(3) + f(4) + f(5) + f(6) = 1 - F(2)$; therefore, $P(X \ge 3) = 1 - F(2) = 1 - 0.10 = 0.90$, (d) $F(5) - F(4) = f(5) = 0.25$, (e) $F(4.4) = 0.45$, (f) $f(4.4) = 0$

10.24 For the bill-from-hat experiment described in Problem 10.19, convert the probabilities $[f(x)]$ into the cumulative distribution function $[F(x)]$ by cumulating from smallest to largest denomination.

Ans. $F(\$1) = 0.06$, $F(\$5) = 0.26$, $F(\$10) = 0.72$, $F(\$20) = 0.98$, $F(\$50) = 1.00$

10.25 For the number-of-residents experiment described in Problem 10.20, convert the probabilities $[f(x)]$ into the cumulative distribution function $[F(x)]$ by cumulating from smallest to largest number.

Ans. $F(1) = 0.114$, $F(2) = 0.310$, $F(3) = 0.703$, $F(4) = 0.838$, $F(5) = 0.939$, $F(6) = 1.000$

10.26 For the number-of-eggs experiment described in Problem 10.21, convert the probabilities $[f(x)]$ into the cumulative distribution function $[F(x)]$ by cumulating from smallest to largest number.

Ans. $F(2) = 0.027$, $F(3) = 0.267$, $F(4) = 0.707$, $F(5) = 0.974$, $F(6) = 1.001$, or 1

EXPECTED VALUE OF A RANDOM VARIABLE

10.27 For the experiment of flipping a coin three times with the discrete random variable number of heads, if the experiment is repeated 500 times, what is the total number of heads you would expect to have flipped?

Ans. 750

10.28 What would be a fair price to charge for the following bet: on the single roll of a die, if a one or a six you win $6, if a two or a five you win $3, if a three or a four you get nothing? (See Problem 10.11 for a demonstration of how this problem is solved.)

Ans. $3

10.29 An insurance agent sells a 35-year-old woman a $10,000 life insurance policy for an annual premium of $130. If the agent's company knows from past records that of the 35-year-old women, 3 out of every 1,000 will die between the ages of 35 and 36, what does the company expect to gain in the first year of this policy?

Ans. $100.00

10.30 For the bill-from-hat experiment described in Problem 10.19, if many such draws are done with the bill replaced after each draw, then what amount of money, on average, can the company expect to pay on each draw?

Ans. $11.86

10.31 For the number-of-residents experiment described in Problem 10.20, if the experiment is repeated many times, each time choosing from all 815 houses, then how many people, on average, can one expect to live in a house?

Ans. 3.10

10.32 For the number-of-eggs experiment described in Problem 10.21, if the experiment is repeated many times, each time choosing from all 150 nests, then how many eggs, on average, can one expect to find in a nest?

Ans. 4.03

THE VARIANCE AND STANDARD DEVIATION OF A RANDOM VARIABLE

10.33 For the repeated experiment in Problem 10.30, what are the variance and standard deviation of amount of money per draw?

Ans. $\sigma^2 = 64.400$, $\sigma = 8.02$

10.34 For the repeated experiment in Problem 10.31, what are the variance and the standard deviation of number of residents per house?

Ans. $\sigma^2 = 1.706$, $\sigma = 1.31$

10.35 For the repeated experiment in Problem 10.32, what are the variance and the standard deviation of number of eggs per nest?

Ans. $\sigma^2 = 0.714$, $\sigma = 0.85$

CHEBYSHEV'S THEOREM AND THE EMPIRICAL RULE

10.36 For the repeated experiment in Problems 10.31 and 10.34, the probability that the random variable (number of residents per house) will take on a value in the interval $\mu \pm 2\sigma$ is at least 0.75 when Chebyshev's theorem is used and, if the distribution is a normal distribution, approximately 0.95 when the empirical rule is applied. What is the exact solution?

Ans. $P = 0.939$

10.37 For the repeated experiment in Problems 10.32 and 10.35, the probability that the random variable (number of eggs per nest) will take on a value in the interval $\mu \pm 2\sigma$ is at least 0.75 when Chebyshev's theorem is used and, if the distribution is a normal distribution, approximately 0.95 when the empirical rule is applied. What is the exact solution?

Ans. $P = 0.947$

Appendix

Table A.1 Random Numbers

	1	2	3	4	5	6	7	8	9	10	11	12	13	14	15	16	17	18	19	20	21	22	23	24	25
1	10	09	73	25	33	76	52	01	35	86	34	67	35	48	76	80	95	90	91	17	39	29	27	49	45
2	37	54	20	48	05	64	89	47	42	96	24	80	52	40	37	20	63	61	04	02	00	82	29	16	65
3	08	42	26	89	53	19	64	50	93	03	23	20	90	25	60	15	95	33	47	64	35	08	03	36	06
4	99	01	90	25	29	09	37	67	07	15	38	31	13	11	65	88	67	67	43	97	04	43	62	76	59
5	12	80	79	99	70	80	15	73	61	47	64	03	23	66	53	98	95	11	68	77	12	17	17	68	33
6	66	06	57	47	17	34	07	27	68	50	36	69	73	61	70	65	81	33	98	85	11	19	92	91	70
7	31	06	01	08	05	45	57	18	24	06	35	30	34	26	14	86	79	90	74	39	23	40	30	97	32
8	85	26	97	76	02	02	05	16	56	92	68	66	57	48	18	73	05	38	52	47	18	62	38	85	79
9	63	57	33	21	35	05	32	54	70	48	90	55	35	75	48	28	46	82	87	09	83	49	12	56	24
10	73	79	64	57	53	03	52	96	47	78	35	80	83	42	82	60	93	52	03	44	35	27	38	84	35
11	98	52	01	77	67	14	90	56	86	07	22	10	94	05	58	60	97	09	34	33	50	50	07	39	98
12	11	80	50	54	31	39	80	82	77	32	50	72	56	82	48	29	40	52	42	01	52	77	56	78	51
13	83	45	29	96	34	06	28	89	80	83	13	74	67	00	78	18	47	54	06	10	68	71	17	78	17
14	88	68	54	02	00	86	50	75	84	01	36	76	66	79	51	90	36	47	64	93	29	60	91	10	62
15	99	59	46	73	48	87	51	76	49	69	91	82	60	89	28	93	78	56	13	68	23	47	83	41	13
16	65	48	11	76	74	17	46	85	09	50	58	04	77	69	74	73	03	95	71	86	40	21	81	65	44
17	80	12	43	56	35	17	72	70	80	15	45	31	82	23	74	21	11	57	82	53	14	38	55	37	63
18	74	35	09	98	17	77	40	27	72	14	43	23	60	02	10	45	52	16	42	37	96	28	60	26	55
19	69	91	62	68	03	66	25	22	91	48	36	93	68	72	03	76	62	11	39	90	94	40	05	64	18
20	09	89	32	05	05	14	22	56	85	14	46	42	75	67	88	96	29	77	88	22	54	38	21	45	98
21	91	49	91	45	23	68	47	92	76	86	46	16	28	35	54	94	75	08	99	23	37	08	92	00	48
22	80	33	69	45	98	26	94	03	68	58	70	29	73	41	35	53	14	03	33	40	42	05	08	23	41
23	44	10	48	19	49	85	15	74	79	54	32	97	92	65	75	57	60	04	08	81	22	22	20	64	13
24	12	55	07	37	42	11	10	00	20	40	12	86	07	46	97	96	64	48	94	39	28	70	72	58	15
25	63	60	64	93	29	16	50	53	44	84	40	21	95	25	63	43	65	17	70	82	07	20	73	17	90
26	61	19	69	04	46	26	45	74	77	74	51	92	43	37	29	65	39	45	95	93	42	58	26	05	27
27	15	47	44	52	66	95	27	07	99	53	59	36	78	38	48	82	39	61	01	18	33	21	15	94	66
28	94	55	72	85	73	67	89	75	43	87	54	62	24	44	31	91	19	04	25	92	92	92	74	59	73
29	42	48	11	62	13	97	34	40	87	21	16	86	84	87	67	03	07	11	20	59	25	70	14	66	70
30	23	52	37	83	17	73	20	88	98	37	68	93	59	14	16	26	25	22	96	63	05	52	28	25	62
31	04	49	35	24	94	75	24	63	38	24	45	86	25	10	25	61	96	27	93	35	65	33	71	24	72
32	00	54	99	76	54	64	05	18	81	59	96	11	96	38	96	54	69	28	23	91	23	28	72	95	29
33	35	96	31	53	07	26	89	80	93	54	33	35	13	54	62	77	97	45	00	24	90	10	33	93	33
34	59	80	80	83	91	45	42	72	68	42	83	60	94	97	00	13	02	12	48	92	78	56	52	01	06
35	46	05	88	52	36	01	39	09	22	86	77	28	14	40	77	93	91	08	36	47	70	61	74	29	41
36	32	17	90	05	97	87	37	92	52	41	05	56	70	70	07	86	74	31	71	57	85	39	41	18	38
37	69	23	46	14	06	20	11	74	52	04	15	95	66	00	00	18	74	39	24	23	97	11	89	63	38
38	19	56	54	14	30	01	75	87	53	79	40	41	92	15	85	66	67	43	68	06	84	96	28	52	07
39	45	15	51	49	38	19	47	60	72	46	43	66	79	45	43	59	04	79	00	33	20	82	66	95	41
40	94	86	43	19	94	36	16	81	08	51	34	88	88	15	53	01	54	03	54	56	05	01	45	11	76

(Continued)

Table A.1 Random Numbers (*Continued*)

	26	27	28	29	30	31	32	33	34	35	36	37	38	39	40	41	42	43	44	45	46	47	48	49	50
1	98	08	62	48	26	45	24	02	84	04	44	99	90	88	96	39	09	47	34	07	35	44	13	18	80
2	33	18	51	62	32	41	94	15	09	49	89	43	54	85	81	88	69	54	19	94	37	54	87	30	43
3	80	95	10	04	06	96	38	27	07	74	20	15	12	33	87	25	01	62	52	98	94	62	46	11	71
4	79	75	24	91	40	71	96	12	82	96	69	86	10	25	91	74	85	22	05	39	00	38	75	95	79
5	18	63	33	25	37	98	14	50	65	71	31	01	02	46	74	05	45	56	14	27	77	93	89	19	36
6	74	02	94	39	02	77	55	73	22	70	97	79	01	71	19	52	52	75	80	21	80	81	45	17	48
7	54	17	84	56	11	80	99	33	71	43	05	33	51	29	69	56	12	71	92	55	36	04	09	03	24
8	11	66	44	98	83	52	07	98	48	27	59	38	17	15	39	09	97	33	34	40	88	46	12	33	56
9	48	32	47	79	28	31	24	96	47	10	02	29	53	68	70	32	30	75	75	46	15	02	00	99	94
10	69	07	49	41	38	87	63	79	19	76	35	58	40	44	01	10	51	82	16	15	01	84	87	69	38
11	09	18	82	00	97	32	82	53	95	27	04	22	08	63	04	83	38	98	73	74	64	27	85	80	44
12	90	04	58	54	97	51	98	15	06	54	94	93	88	19	97	91	87	07	61	50	68	47	66	46	59
13	73	18	95	02	07	47	67	72	52	69	62	29	06	44	64	27	12	46	70	18	41	36	18	27	60
14	75	76	87	64	90	20	97	18	17	49	90	42	91	22	72	95	37	50	58	71	93	82	34	31	78
15	54	01	64	40	56	66	28	13	10	03	00	68	22	73	98	20	71	45	32	95	07	70	61	78	13
16	08	35	86	99	10	78	54	24	27	85	13	66	15	88	73	04	61	89	75	53	31	22	30	84	20
17	28	30	60	32	64	81	33	31	05	91	40	51	00	78	93	32	60	46	04	75	94	11	90	18	40
18	53	84	08	62	33	81	59	41	36	28	51	21	59	02	90	28	46	66	87	95	77	76	22	07	91
19	91	75	75	37	41	61	61	36	22	69	50	26	39	02	12	55	78	17	65	14	83	48	34	70	55
20	89	41	59	26	94	00	39	75	83	91	12	60	71	76	46	48	94	97	23	06	94	54	13	74	08
21	77	51	30	38	20	86	83	42	99	01	68	41	48	27	74	51	90	81	39	80	72	89	35	55	07
22	19	50	23	71	74	69	97	92	02	88	55	21	02	97	73	74	28	77	52	51	65	34	46	74	15
23	21	81	85	93	13	93	27	88	17	57	05	68	67	31	56	07	08	28	50	46	31	85	33	84	52
24	51	47	46	64	99	68	10	72	36	21	94	04	99	13	45	42	83	60	91	91	08	00	74	54	49
25	99	55	96	83	31	62	53	52	41	70	69	77	71	28	30	74	81	97	81	42	43	86	07	28	34
26	33	71	34	80	07	93	58	47	28	69	51	92	66	47	21	58	30	32	98	22	93	17	49	39	72
27	85	27	48	68	93	11	30	32	92	70	28	83	43	41	37	73	51	59	04	00	71	14	84	36	43
28	84	13	38	96	40	44	03	55	21	66	73	85	27	00	91	61	22	26	05	61	62	32	71	84	23
29	56	73	21	62	34	17	39	59	61	31	10	12	39	16	22	85	49	65	75	60	81	60	41	88	80
30	65	13	85	68	06	87	64	88	52	61	34	31	36	58	61	45	87	52	10	69	85	64	44	72	77
31	38	00	10	21	76	81	71	91	17	11	71	60	29	29	37	74	21	96	40	49	65	58	44	96	98
32	37	40	29	63	97	01	30	47	75	86	56	27	11	00	86	47	32	46	26	05	40	03	03	74	38
33	97	12	54	03	48	87	08	33	14	17	21	81	53	92	50	75	23	76	20	47	15	50	12	95	78
34	21	82	64	11	34	47	14	33	40	72	64	63	88	59	02	49	13	90	64	41	03	85	65	45	52
35	73	13	54	27	42	95	71	90	90	35	85	79	47	42	96	08	78	98	81	56	64	69	11	92	02
36	07	63	87	79	29	03	06	11	80	72	96	20	74	41	56	23	82	19	95	38	04	71	36	69	94
37	60	52	88	34	41	07	95	41	98	14	59	17	52	06	95	05	53	35	21	39	61	21	20	64	55
38	83	59	63	56	55	06	95	89	29	83	05	12	80	97	19	77	43	35	37	83	92	30	15	04	98
39	10	85	06	27	46	99	59	91	05	07	13	49	90	63	19	53	07	57	18	39	06	41	01	93	62
40	39	82	09	89	52	43	62	26	31	47	64	42	18	08	14	43	80	00	93	51	31	02	47	31	67

(*Continued*)

Table A.1 Random Numbers (*Continued*)

	51	52	53	54	55	56	57	58	59	60	61	62	63	64	65	66	67	68	69	70	71	72	73	74	75
1	59	58	00	64	78	75	56	97	88	00	88	83	55	44	86	23	76	80	61	56	04	11	10	84	08
2	38	50	80	73	41	23	79	34	87	63	90	82	29	70	22	17	71	90	42	07	95	95	44	99	53
3	30	69	27	06	68	94	68	81	61	27	56	19	68	00	91	82	06	76	34	00	05	46	26	92	00
4	65	44	39	56	59	18	28	82	74	37	49	63	22	40	41	08	33	76	56	76	96	29	99	08	36
5	27	26	75	02	64	13	19	27	22	94	07	47	74	46	06	17	98	54	89	11	97	34	13	03	58
6	91	30	70	69	91	19	07	22	42	10	36	69	95	37	28	28	82	53	57	93	28	97	66	62	52
7	68	43	49	46	88	84	47	31	36	22	62	12	69	84	08	12	84	38	25	90	09	81	59	31	46
8	48	90	81	58	77	54	74	52	45	91	35	70	00	47	54	83	82	45	26	92	54	13	05	51	60
9	06	91	34	51	97	42	67	27	86	01	11	88	30	95	28	63	01	19	89	01	14	97	44	03	44
10	10	45	51	60	19	14	21	03	37	12	91	34	23	78	21	88	32	58	08	51	43	66	77	08	83
11	12	88	39	73	43	65	02	76	11	84	04	28	50	13	92	17	97	41	50	77	90	71	22	67	69
12	21	77	83	09	76	38	80	73	69	61	31	64	94	20	96	63	28	10	20	23	08	81	64	74	49
13	19	52	35	95	15	65	12	25	96	59	86	28	36	82	58	69	57	21	37	98	16	43	59	15	29
14	67	24	55	26	70	35	58	31	65	63	79	24	68	66	86	76	46	33	42	22	26	65	59	08	02
15	60	58	44	73	77	07	50	03	79	92	45	13	42	65	29	26	76	08	36	37	41	32	64	43	44
16	53	85	34	13	77	36	06	69	48	50	58	83	87	38	59	49	36	47	33	31	96	24	04	36	42
17	24	63	73	87	36	74	38	48	93	42	52	62	30	79	92	12	36	91	86	01	03	74	28	38	73
18	83	08	01	24	51	38	99	22	28	15	07	75	95	17	77	97	37	72	75	85	51	97	23	78	67
19	16	44	42	43	34	36	15	19	90	73	27	49	37	09	39	85	13	03	25	52	54	84	65	47	59
20	60	79	01	81	57	57	17	86	57	62	11	16	17	85	76	45	81	95	29	79	65	13	00	48	60
21	03	99	11	04	61	93	71	61	68	94	66	08	32	46	53	84	60	95	82	32	88	61	81	91	61
22	38	55	59	55	54	32	88	65	97	80	08	35	56	08	60	29	73	54	77	62	71	29	92	38	53
23	17	54	67	37	04	92	05	24	62	15	55	12	12	92	81	59	07	60	79	36	27	95	45	89	09
24	32	64	35	28	61	95	81	90	68	31	00	91	19	89	36	76	35	59	37	79	80	86	30	05	14
25	69	57	26	87	77	39	51	03	59	05	14	06	04	06	19	29	54	96	96	16	33	56	46	07	80
26	24	12	26	65	91	27	69	90	64	94	14	84	54	66	72	61	95	87	71	00	90	89	97	57	54
27	61	19	63	02	31	92	96	26	17	73	41	83	95	53	82	17	26	77	09	43	78	03	87	02	67
28	30	53	22	17	04	10	27	41	22	02	39	68	52	33	09	10	06	16	88	29	55	98	66	64	85
29	03	78	89	75	99	75	86	72	07	17	74	41	65	31	66	35	20	83	33	74	87	53	90	88	23
30	48	22	86	33	79	85	78	34	76	19	53	15	26	74	33	35	66	35	29	72	16	81	86	03	11
31	60	36	59	46	53	35	07	53	39	49	42	61	42	92	97	01	91	82	83	16	98	95	37	32	31
32	83	79	94	24	02	56	62	33	44	42	34	99	44	13	74	70	07	11	47	36	09	95	81	80	65
33	32	96	00	74	05	36	40	98	32	32	99	38	54	16	00	11	13	30	75	86	15	91	70	62	53
34	19	32	25	38	45	57	62	05	26	06	66	49	76	86	46	78	13	86	65	59	19	64	09	94	13
35	11	22	09	47	47	07	39	93	74	08	48	50	92	39	29	27	48	24	54	76	85	24	43	51	59
36	31	75	15	72	60	68	98	00	53	39	15	47	04	83	55	88	65	12	25	96	03	15	21	92	21
37	88	49	29	93	82	14	45	40	45	04	20	09	49	89	77	74	84	39	34	13	22	10	97	85	08
38	30	93	44	77	44	07	48	18	38	28	73	78	80	65	33	28	59	72	04	05	94	20	52	03	80
39	22	88	84	88	93	27	49	99	87	48	60	53	04	51	28	74	02	28	46	17	82	03	71	02	68
40	78	21	21	69	93	35	90	29	13	86	44	37	21	54	86	65	74	11	40	14	87	48	13	72	20

Reprinted from Hubert M. Blalock. *Social Statistics* (2d ed), McGraw-Hill, New York, 1979, pp. 598–601.

Original source: The RAND Corporation, *A Million Random Digits*, Free Press, Glencoe, Ill., 1955, pp. 1–3, with the kind permission of the publisher. (Numbered guidelines on *X* and *Y* axes added.)

Table A.2 Satistics Class Data

This table summarizes data from the statistics class introduced in Example 3.5. In two parts (females and males) it gives: student initials (col. 1); assigned numbers (col. 2); score out of 100 possible points on the second lecture exam (col. 3); height in inches to nearest 1/4 inch (col. 4); weight in pounds (col. 5); household income to nearest \$100 (col. 6); hair color (black, blonde, brown, red) (col. 7); letter grade on term paper (*A, B, C, D, F*) (col. 8); and, whether selected (*) for the simple random sample (SRS) (col. 9), the proportional stratified random sample (PSRS) (col. 10), the systematic random sample (SYRS) (col. 11), or the single-stage cluster random sample (SCRS) (col. 12).

Initials (1)	Number (2)	2d Exam (3)	Height (4)	Weight (5)	Household income (6)	Hair color (7)	Grade (8)	SRS (9)	PSRS (10)	SYRS (11)	SCRS (12)
FEMALES											
LB	03	83	67.75	127	31,500	brown	B	–	–	*	–
AA	09	88	60.25	109	25,600	blonde	B	–	–	–	*
AE	17	57	63.75	117	76,500	black	F	*	*	–	–
MJ	27	78	65.25	123	20,200	blonde	B	*	*	*	*
NO	28	97	62.00	105	37,800	blonde	A	–	–	–	*
LT	30	82	63.50	119	15,400	black	B	–	–	–	*
JD	33	91	65.25	129	71,800	blonde	A	–	–	–	–
DD	34	90	65.50	123	31,700	blonde	B	–	–	–	–
EF	44	80	65.25	124	30,100	brown	B	*	–	–	–
AC	49	64	64.75	121	34,700	brown	C	–	*	–	–
JH	51	87	67.00	134	40,500	blonde	A	–	*	*	–
AH	53	79	64.25	115	36,900	black	C	*	–	–	–
GY	56	91	69.25	136	20,400	red	A	–	–	–	–
MZ	57	79	66.25	131	30,400	brown	C	–	–	–	–
BJ	58	65	63.00	111	28,500	blonde	D	–	–	–	–
TM	64	94	64.75	121	46,100	blonde	A	*	–	–	–
MALES											
CA	01	90	69.25	180	21,200	brown	B	–	*	–	*
FE	02	94	65.25	138	145,000	brown	A	–	–	–	*
HE	04	59	69.00	152	29,300	brown	D	–	–	–	–
LW	05	91	73.00	172	26,600	blonde	B	–	–	–	–
OA	06	84	69.25	163	20,900	blonde	B	–	*	–	–
PS	07	96	70.25	170	26,200	blonde	A	*	*	*	–
OF	08	84	67.00	158	33,700	blonde	B	–	–	–	–
HC	10	79	71.75	190	54,200	brown	C	–	–	–	*
EB	11	84	66.25	148	28,600	brown	B	–	–	*	–
MA	12	90	66.25	157	29,200	black	B	–	–	–	–
ME	13	72	70.25	156	58,400	brown	C	*	–	–	–
HK	14	93	68.00	164	21,700	black	A	–	–	–	–
AD	15	69	72.00	175	27,700	black	C	–	–	*	–
RE	16	87	69.00	160	24,200	blonde	B	–	–	–	–
FA	18	93	72.50	172	42,500	blonde	A	–	–	–	–
CE	19	90	65.25	152	28,100	brown	B	–	–	*	*
BP	20	74	72.25	184	22,200	red	C	–	–	–	*
EO	21	88	67.25	142	24,100	blonde	B	–	–	–	*

(*Continued*)

Table A.2   Statistics Class Data (*Continued*)

Initials (1)	Number (2)	2d Exam (3)	Height (4)	Weight (5)	Household income (6)	Hair color (7)	Grade (8)	SRS (9)	PSRS (10)	SYRS (11)	SCRS (12)
RA	22	86	67.00	124	49,000	brown	*B*	*	–	–	*
DA	23	71	68.25	152	25,100	brown	*C*	–	–	*	–
GK	24	80	71.00	169	31,700	black	*B*	–	–	–	–
JA	25	94	68.00	150	39,200	blonde	*A*	–	–	–	–
GB	26	98	69.25	147	35,600	blonde	*A*	*	–	–	–
JW	29	81	67.00	140	15,700	brown	*B*	–	*	–	*
HO	31	59	73.00	191	66,900	blonde	*D*	–	–	*	–
WA	32	86	67.50	138	14,100	brown	*B*	*	–	–	–
NA	35	90	69.00	159	33,300	red	*B*	–	–	*	–
SM	36	67	67.50	160	28,300	blonde	*C*	–	–	–	–
MQ	37	85	66.25	131	30,700	brown	*B*	*	–	–	*
JT	38	64	68.50	147	25,600	blonde	*D*	*	–	–	*
TS	39	69	71.25	186	32,400	brown	*C*	*	–	*	–
MU	40	83	69.00	152	12,700	blonde	*C*	–	–	–	–
GM	41	68	64.25	143	103,600	brown	*C*	–	*	–	–
BC	42	99	70.75	173	17,300	red	*A*	–	*	–	–
CI	43	95	68.25	151	37,200	brown	*A*	–	*	*	–
JL	45	64	70.00	160	43,700	brown	*D*	*	–	–	*
JQ	46	92	68.75	149	88,000	black	*A*	–	*	–	*
FV	47	87	69.50	144	31,600	black	*B*	–	*	*	–
DW	48	78	70.25	158	24,300	brown	*B*	–	*	–	–
DM	50	55	64.75	119	18,000	blonde	*F*	–	–	–	–
BF	52	88	74.00	194	23,900	red	*B*	–	*	–	–
NP	54	76	70.75	176	49,000	black	*C*	–	–	–	–
GT	55	90	70.00	166	35,000	brown	*A*	–	*	*	–
LR	59	49	69.00	165	58,600	brown	*F*	*	–	*	–
CR	60	80	68.75	161	52,100	black	*B*	–	–	–	–
PB	61	91	69.00	170	25,400	black	*A*	–	–	–	–
EJ	62	90	70.25	151	36,100	brown	*A*	–	–	–	–
AT	63	84	65.25	124	42,300	black	*B*	*	–	*	–

Index

A priori probabilities, 228
Abscissa, 13
Absolute constants, 8
Absolute dispersion, 199
Absolute values, 4, 18, 30, 183
Absolute zeros, 35, 36
Accuracy:
  in physical sciences, 41, 46–47, 50
  in statistics, 40–41
  of a measurement, 40–41
Addition rules, 284–288, 303–304
Algebraic expressions, 7, 22, 32
Approximate arithmetic mean, calculation from grouped
    frequency distributions, 144–145, 159–160, 179
Approximate mean deviation, 185–186
Approximate measurements, 38–39, 43–44, 49
Approximate standard deviation, calculation from grouped
    frequency distributions, 192–193, 210–211
Arbitrary constants, 8
Arithmetic mean, 52–53, 140–141, 157–158, 178–179
  as measure of average value, 143
  calculation from nongrouped frequency distributions,
    143–144
  calculation with coded data, 146, 161, 179
  deviations from, 142
  rounding-off guidelines, 141
Arithmetic numbers, 7
Arrays, 72, 82, 95–96
  quantile-locating formula for, 150–151
Average deviation, 183
Average value, 139
Axioms, 59, 234–237
Axis of symmetry, 14

Bar charts, 103, 111–114, 132–133
Bar graphs, 102, 198
Base, 5
Baseline data, 56
Bayes' theorem, 269–271, 288–294, 304–305
Bayesian decision analysis, 271
Bernoulli's theorem, 229
Betting odds, 238
Between-class widths, 76
Bias, 40, 63
Biased sampling designs, 62
Bimodal distribution, 176
Bimodal frequency histogram, 176
Biostatistics, 1
Box plots, 201–202, 221–222, 225
Branch, 271

Cartesian coordinates, 13, 24, 32
Categorical data, 38
Cause and effect, 11, 54

Celsius (or centigrade) scale for temperature, 35
Censuses, 57
Center of gravity of distribution, 143, 183
Central tendency, 139
  graphs, 198, 215–218, 225
Certain event, 235
Chebyshev's inequality, 195
Chebyshev's theorem, 195–196, 213–214, 224, 328–330,
    339, 342
Check count, 109
Class, 74
Class boundaries, 75
Class interval, 74
Class limits, 74
Class widths, 76
Classical probability function, 234
Classificatory data, 38
Cluster random sampling, 62
Clusters, 62
Coding formula, 146
Coefficient of dispersion, 199
Coefficient of variation, 199, 219, 225
Combinations, 276–278
Common logarithm, 21
Complement of an event, 232
Complete quadratic equation, 8
Component-parts frequency bar chart, 113
Component-parts frequency histogram, 134
Composite event, 231
Compound event, 231–233
Conditional probability, 258–261, 268, 278–284,
    302–303
  general formula, 260, 270
Confidence interval, 54
Confounding variable, 56
Consecutive-parts frequency bar chart, 113
Consecutive-parts relative frequency pictograph, 107
Consecutive values, 156
Constant, 8
Contingency table, 267
Continuous data, 38
Continuous measurement variables, 36
Continuous probability distribution, 313–319
Continuous random variable, 310–311, 321–322, 328
Continuous ratio-level measurement, 37
Continuum without gaps, 36–37
Control groups, 56
Controlled experiments, 56–57
Conversion factors, 42, 47–48, 50
Coordinates, 13
Counting rules, 273–278, 297–302, 306–308
  combinations, 276–278
  multiplication principle, 273–274, 297
  permutations, 274–276